Solid Waste Treatment Technologies

Sustainable waste management is a major step towards the attainment of Sustainable Development Goals. This book covers all technical, managerial, and legislative aspects of waste management at a global scale, providing a detailed description about different types of wastes, their characteristics, legal perspectives, and sustainable practices for their management. It explains developments in waste treatment technologies (classified based on waste type) and understanding the fundamentals of circular economy in waste management, supported by various case studies.

Features:

- Discusses fundamentals of solid waste management for sustainable waste management practices
- Describes technological aspects of waste management covering various physicochemical, biochemical, and thermochemical processes
- Summarizes regulatory framework for waste management at the global level
- Highlights the scope for circular economy in managing solid wastes
- Includes dedicated chapters on case studies imperative for capacity building in waste management

This book is aimed at researchers, graduate students, and professionals in environmental engineering, and waste management.

Environmental Nexus in Waste Management

About the Series:
The book series on "Environmental Nexus for Waste Management" addresses novel approaches and techniques linked to sustainable development paths with an interdisciplinary focus on waste resources monitoring, assessment, management, water/wastewater reclamation, and recycling as well as discussing solutions for more sustainable use of natural resources. It is a broad compendium of waste management related to research and development and discussions on the most up-to-date tackling strategies. The series of books constitutes an information source and facilitator for the transfer of knowledge, focusing on practical solutions, and better understanding towards achieving sustainable development.

Series Editors: *Vineet Kumar, Sunil Kumar*

Solid Waste Treatment Technologies
Challenges and Perspectives
Edited by Pratibha Gautam, Vineet Kumar and Sunil Kumar

For more information on the series, please visit our website link: https://www.routledge.com/Environmental-Nexus-in-Waste-Management/book-series/CRCENWM

Solid Waste Treatment Technologies
Challenges and Perspectives

Edited by
Pratibha Gautam,
Vineet Kumar, and Sunil Kumar

CRC Press
Taylor & Francis Group
Boca Raton London New York

CRC Press is an imprint of the
Taylor & Francis Group, an **informa** business

Designed cover image: © Shutterstock

First edition published 2024
by CRC Press
2385 NW Executive Center Drive, Suite 320, Boca Raton FL 33431

and by CRC Press
4 Park Square, Milton Park, Abingdon, Oxon, OX14 4RN

CRC Press is an imprint of Taylor & Francis Group, LLC

ISBN: 9781032403014 (hbk)
ISBN: 9781032403021 (pbk)
ISBN: 9781003352396 (ebk)

DOI: 10.1201/9781003352396

Contents

Chapter 10 Environmental Fate, Behavior, and Risk Management
Approaches of Nanoplastics in the Environment:
Current Scenario and Future Insights 148

Shikha Gulati, Anoushka Amar, and Sheetal Olihan

About the Editors

Dr. Pratibha Gautam is Assistant Professor and Head of the Department of Environmental Science & Technology at UPL University of Sustainable Technology, Ankleshwar (Gujarat), India. Dr. Gautam is also a QCI-NABET approved Functional Area Expert (FAE) for air pollution monitoring, prevention, and control (AP). She is also designated as Technical Manager (TM) for NABL accredited Environmental Laboratory. In total, she has more than 10 years of industrial as well as academic experience. At present, she is actively involved in the environmental audit of industries and several research and consultancy projects based on environmental issues for industries. She has guided several graduate and postgraduate students on their thesis projects. She has a wide knowledge of different environmental aspects including solid waste management, air pollution control, advanced oxidation processes for wastewater treatment, environmental audits, etc. She has many publications in reputed international journals and has published several book chapters. She is a GATE and NET qualified environment professional and is the recipient of scholarships from the Indian government (MHRD) and the Netherlands government (NFP fellowship).

Dr. Vineet Kumar is presently working as a National Postdoctoral Fellow in the Department of Microbiology, School of Life Sciences at the Central University of Rajasthan, Rajasthan, India. He earned his Ph.D. (2018) in Environmental Microbiology from Babasaheb Bhimrao Ambedkar (A Central) University, Lucknow, India. Dr. Kumar's research work mainly focuses on the wastewater treatment, biofuel production, and solid waste management. He has published more than 50 articles in peer-reviewed international journals of repute, 2 books, and 55 book chapters on various aspects of science and engineering, with more than citations 2300, and h-index 28. Dr. Kumar has been serving as a guest editor and reviewer in more than 60 prestigious International Journals. In addition, he has served the editorial board of various reputed journals. He has presented several papers relevant to his research areas in national and international conferences. He is an active member of numerous scientific societies including the Microbiology Society (UK), the Indian Science Congress Association (India), the Association of Microbiologists of India (India), etc. He is the founder of the Society for Green Environment, India (www.sgeindia.org). He can be reached at drvineet.micro@gmail.com and vineet.way18@gmail.com.

Dr. Sunil Kumar is a well-rounded researcher with more than 20 years of experience in leading, supervising, and undertaking research in the broad field of environmental engineering and science with a focus on solid and hazardous waste management. His primary area of expertise is solid waste management (municipal solid waste, electronic waste, biomedical waste, etc.) over a wide range of environmental topics including contaminated sites, EIA, and wastewater treatment. He has contributed extensively to these fields and has a citation of 8825, an h-index of 44, and an i10-index of 168 (Google Scholar). His contributions since inception at CSIR-National Environmental Engineering Research Institute (NEERI), India in 2000 include 300 refereed journal publications, 5 books and 40 book chapters, 10 edited volumes, and numerous project reports to various governmental and private, local, and international academic/research bodies. He is the Associate Editor of peer-reviewed journals of the international repute, that is, *Environmental Chemistry Letter, International Journal of Environmental Science and Technology* and *ASCE Journal of Hazardous, Toxic and Radioactive Waste*. He also serves as Editorial Board in Bioresource Technology, Elsevier. He has completed more than 22 research projects as PI with 17 (7 awarded) Ph.D. and 17 M.Phil/M.Tech thesis/dissertations under his supervision. Dr. Kumar was also awarded the most prestigious award "Alexander von Humboldt-Stiftung Jean-Paul-Str.12 D-53173 Bonn, Germany" as a Senior Researcher for developing a Global Network and Excellence for more advanced research and technology innovation.

Contributors

Anoushka Amar
Department of Chemistry
Sri Venkateswara College
University of Delhi
Delhi, India

Gangagni Rao Anupoju
Academy of Scientific and Innovative
 Research (AcSIR)
New Delhi, India

Mostafa M. Besheir
Department of Environmental Science
 and Engineering
Marwadi University
Rajkot, Gujarat, India

Debleena Bhattacharya
Department of Environmental Science
 and Engineering
Marwadi University
Rajkot, Gujarat, India

Karan Chabhadiya
Department of Environmental Science
 & Technology
Shroff S R Rotary Institute of Chemical
 Technology
UPL University of Sustainable
 Technology
Vataria, Gujarat, India

Anjani Devi Chintagunta
Department of Biotechnology
Vignan's Foundation for Science,
 Technology and Research
Vadlamudi, Andhra Pradesh, India

Antônio José de Andrade Junior
Centro de Biotecnologia da
 Amazônia – CBA/SUFRAMA
Manaus, Brazil

Ana Emília M. de Freitas
Design graduate program – PPGQ
Federal University of
 Amazonas – UFAM
Manaus, Brazil

Flávio A. de Freitas
Centro de Biotecnologia da
 Amazônia – CBA/SUFRAMA
Manaus, Brazil

Silma de Sá Barros
Centro de Biotecnologia da
 Amazônia – CBA/SUFRAMA
Manaus, Brazil

Zeel Desai
Department of Environmental Science
 & Technology
Shroff S R Rotary Institute of Chemical
 Technology
UPL University of Sustainable
 Technology
Vataria, Gujarat, India

Sukhendu Dey
Department of Environmental Science
The University of Burdwan,
Burdwan, West Bengal, India

Deblina Dutta
Department of Environmental Science
SRM University
Amaravati, Andhra Pradesh, India

Vivek Gajara
Department of Environmental Science
 & Technology
Shroff S R Rotary Institute of Chemical
 Technology
UPL University of Sustainable Technology
Vataria, Gujarat, India

Alok Garg
Department of Chemical Engineering
    National Institute of Technology
Hamirpur, Himachal Pradesh, India

Arun Gautam
UCD Michael Smurfit Graduate
    Business School
Dublin, Ireland

Pratibha Gautam
Department of Environmental Science
    & Technology
Shroff S R Rotary Institute of Chemical
    Technology
UPL University of Sustainable Technology
Vataria, Gujarat, India

Shikha Gulati
Department of Chemistry
Sri Venkateswara College
University of Delhi
Delhi, India

Arijit Dutta Gupta
Department of Environmental Science
    & Technology
Shroff S R Rotary Institute of Chemical
    Technology
UPL University of Sustainable Technology
Vataria, Gujarat, India

N.S. Sampath Kumar
Department of Biotechnology
Vignan's Foundation for Science,
    Technology and Research
Vadlamudi, Andhra Pradesh, India

Sunil Kumar
Waste Re-processing Division
CSIR- National Environmental
    Engineering Research Institute
    (NEERI)
Nagpur, Maharashtra, India

Debajyoti Kundu
Department of Environmental Science
    and Engineering;
School of Engineering and Sciences;
SRM University-AP, Amaravati,
    Andhra Pradesh, India.

Wyvirlany Valente Lobo
Chemistry graduate program – PPGQ
Federal University of
    Amazonas – UFAM
Manaus, Brazil

Snehal Lokhandwala
Department of Environmental Science
    & Technology
Shroff S R Rotary Institute of Chemical
    Technology
UPL University of Sustainable
    Technology
Vataria, Gujarat, India

Rahul Mishra
Waste Re-processing Division
CSIR- National Environmental
    Engineering Research Institute
    (NEERI)
Nagpur, Maharashtra, India

Ankit Motghare
Waste Re-processing Division
CSIR-National Environmental
    Engineering Research Institute
    (NEERI)
Nagpur, Maharashtra, India

Srushti Muneshwar
Waste Re-processing Division
CSIR- National Environmental
    Engineering Research Institute
    (NEERI)
Nagpur, Maharashtra, India

Sheetal Olihan
Department of Chemistry
Sri Venkateswara College
University of Delhi
Delhi, India

Orlando A. R. L. Paes
Centro de Biotecnologia da
    Amazônia – CBA/SUFRAMA
Manaus, Brazil

Maddala Rama Krishna
Department of Energy and
    Environmental Engineering
CSIR-Indian Institute Chemical
    Technology (IICT)
Hyderabad, Andhra Pradesh, India

Rahul Rautela
Academy of Scientific and Innovative
    Research (AcSIR)
Ghaziabad, Uttar Pradesh, India

Anita Saini
Department of Microbiology
School of Basic and Applied Sciences
Maharaja Agrasen University
Solan, Himachal Pradesh, India

Darshan Salunke
Department of Environmental Science
    & Technology
Shroff S R Rotary Institute of Chemical
    Technology
UPL University of Sustainable
    Technology
Vataria, Gujarat, India

Palas Samanta
Department of Environmental Science
Sukanta Mahavidyalaya, University of
    North Bengal
Dhupguri West Bengal, India

Swati Sharma
Department of Biotechnology,
Shoolini Institute of Life Sciences and
    Business Management,
Solan, Himachal Pradesh, India

Knawang Chhunji Sherpa
CSIR-National Institute for
    Interdisciplinary Science and
    Technology (CSIR-NIST),
Trivandrum, Kerala, India

Harinder Singh
Department of Chemical Engineering
Motilal Nehru National Institute of
    Technology Allahabad,
Prayagraj, Uttar Pradesh, India

V. K. Singh
InnovatioCuris
Noida, Uttar Pradesh, India

Airi dos Santos Sousa
Centro de Biotecnologia da
    Amazônia – CBA/SUFRAMA
Manaus, Brazil

Shailaja Srinivasan
Department of Energy and
    Environmental Engineering
CSIR-Indian Institute Chemical
    Technology (IICT)
Tarnaka, Hyderabad, India

1 Waste Management
A Global Challenge

Debleena Bhattacharya and V. K. Singh

1.1 INTRODUCTION

1.1.1 WASTE

Millions of years ago, humans realised that their consumption would lead to a concentration of waste that would increase with time. It is a by-product that is most unavoidable for any human activity. The complexity of waste generation is more with the enhanced standards of living, industrial upliftment, and also the dumping of biomedical waste into the landfills and nearby streams, leading to severe environmental and human health consequences. The present precarious situation of increase in municipal solid waste (MSW) has its propensity in the growing population, increasing density in urban areas, myriad cultures, and stagnant lifestyle with low-nutrient dietary habits. The municipalities face a lot of issues pertaining to the collection, treatment, and management of solid waste. The main problem with solid waste lies in unsorted waste at the source, social taboo, the indifferent attitude of citizens towards solid waste, poor evaluation and unorganised informal waste sector, and unplanned fiscal and poor implementation of government policies. There is an upsurge in adequate treatment and strategies for recycling strategies for solid waste.

The market is laden with single-use plastic and in the recent situation, it has become a global threat as it is considered harmful and non-biodegradable. Most Indian drains are clogged due to the excess plastic waste during the monsoon season, which further leads to urban flooding (Anderson et al., 2016). Microplastics invariably come to our food chain due to the increasing use and reuse of plastic-packaged drinking water (United Nations, 1992). In this chapter, we will be discussing disposal, generation, treatment, and management of the growing waste volume as it poses challenges for both developed and developing nations.

As per the report published by the United Nations Developmental Programme (UNDP), there are 300 million tons of plastic waste generated across the globe, out of which only 9% of the plastics can be recycled, 14% is collected for recycling, and approximately the rest goes to the ocean annually (UNEP, 2019).

The most common method implemented for solid waste management is through sanitary landfills. They generate large quantities of leachate; the term leachate is defined as the contaminated liquid generated from water penetrating the landfill. The overproduction of leachate often leads to harmful effects on humans and the environment (Nakum and Bhattacharya, 2022).

DOI: 10.1201/9781003352396-1

1.1.2   GENERATION AND CHARACTERISTICS OF WASTE

The types of waste also help one to determine its treatment process. The characteristics of waste play a prominent part in the development of cost-effective solid waste management strategies. There is a lack of proper management for the waste generated as there isn't any systematic surveying of waste arising and the quantities, characteristics, and seasonal variation. The developed countries' per capita waste generation is much higher than that of developing countries. The principal sources of solid waste are residential households and the agricultural, commercial, construction, industrial, and institutional sectors. The major categories of waste are MSW, industrial waste, agricultural waste, and hazardous waste.

1.1.3   MUNICIPAL SOLID WASTE

The waste generated from households, hotels, stationery shops, other shops, institutions, and offices is termed MSW. Major components are food waste, paper, plastic, rags, metals, and glass. Wastes generated from demolition and construction activities are also included in collected waste; it also contains hazardous waste in the form of electric light bulbs, batteries, discarded expired medicines and chemicals, and also automotive parts.

The other issue related to MSWM is the production of hazardous chemical wastes by hospitals and industries, which give rise to an increase in respiratory problems and premature death (Joshi and Ahmed, 2016; Mohan, 2019). The unsystematic planning of landfill waste disposal draws rodents, mosquitoes, vultures, and scroungers to destroy the sanctity of the area (De Bercegol et al., 2017). Though many efforts were made to resolve the solid waste management–related issues, yet this waste is not managed holistically. Most cities in India collect different types of waste but the centralised or decentralised treatment augmented to them is not effective in treating them (Singh, 2020).

1.1.4   THE MANAGEMENT OF MSW (MSWM) IN INDIA

According to Kumar et al. (2017), India generates approximately 143,449 MT of MSW each day, of which approximately 35,602 MT are treated and 111,000 MT are collected. According to CPCB's 2018 annual report, city-wide waste generation shows significant variation in waste per capita/day generation at an exponential rate strictly (0.24–0.85) from 2001 to 2018, which is probably going to rise quickly (CPCB India, 2018; Kumar et al., 2017). The geographical conditions, climate, and social and economic status of the people in cities all influence the types of garbage that are disposed of there. Densely populated cities are producing more municipal waste. Municipalities face additional challenges as a result of waste variation, demonstrating that a single strategy cannot be applied to all of them. This is why municipalities are unable to handle the situation and appear to be behind, with fewer cases moving ahead.

Larger cities like Mumbai, Delhi, Kolkata, Chennai, Hyderabad, and Bangalore handle a significant portion of the country's solid waste (Figure 1.1). According to

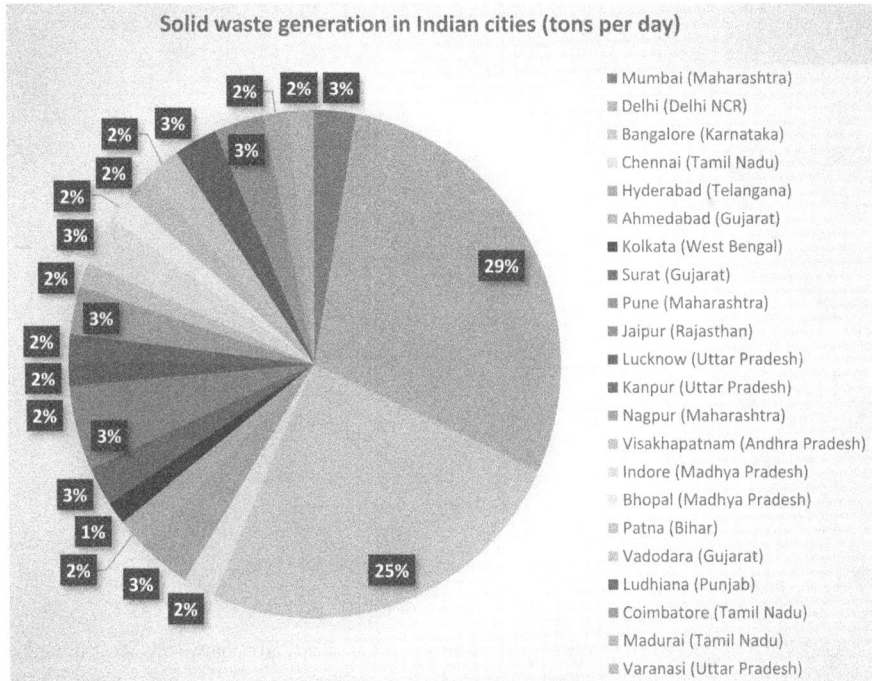

Solid waste generation in Indian cities (tons per day)

- Mumbai (Maharashtra)
- Delhi (Delhi NCR)
- Bangalore (Karnataka)
- Chennai (Tamil Nadu)
- Hyderabad (Telangana)
- Ahmedabad (Gujarat)
- Kolkata (West Bengal)
- Surat (Gujarat)
- Pune (Maharashtra)
- Jaipur (Rajasthan)
- Lucknow (Uttar Pradesh)
- Kanpur (Uttar Pradesh)
- Nagpur (Maharashtra)
- Visakhapatnam (Andhra Pradesh)
- Indore (Madhya Pradesh)
- Bhopal (Madhya Pradesh)
- Patna (Bihar)
- Vadodara (Gujarat)
- Ludhiana (Punjab)
- Coimbatore (Tamil Nadu)
- Madurai (Tamil Nadu)
- Varanasi (Uttar Pradesh)

FIGURE 1.1   Solid waste generation in tons per day in Indian cities (CPCB India, 2018).

Ministry of New and Renewable Energy (MNRE), India (2018), the highly dense population that resides in these areas generates a diverse variety of solid waste on a daily basis, accounting for approximately 70–80% of the total waste generated each day in India.

According to the 2018 MNRE report, the country produces a significant amount of waste due to overcrowding in states like Maharashtra, Tamil Nadu, Uttar Pradesh, the National Capital Region, Gujarat, Karnataka, and West Bengal.

1.2   COMPARISON OF GLOBAL SOLID WASTE GENERATION TRENDS

Twenty-seven billion tons of waste production is expected by 2050. At present, we are leading in waste generation as one-third of total waste comes from China (0–0.49 kg/capita/day) and India (0.5–0.9 kg/capita/day) (Kaza and Yao, 2018; Modak, 2011). If this situation continues, then the projected waste generation from the Indian subcontinent will be 334–661 MT/day as compared to China and east Asia Pacific, 468–714 MT/day, during 2016–2050. As per the report by CPCB (2016) and Ahluwalia and Patel (2018), India is reported to have an exponential rise in waste, i.e., 145 million tons of waste generation per year. Of which, only 117,644 MT gets collected, with approximately 49,401 MT currently being treated before disposal. Urban solid waste management faced a lot of problems during the recent pandemic

Solid waste generation tons per day in Indian cities

Regional waste generation, tons m   2016 ME 2030 forecast   2050 forecast

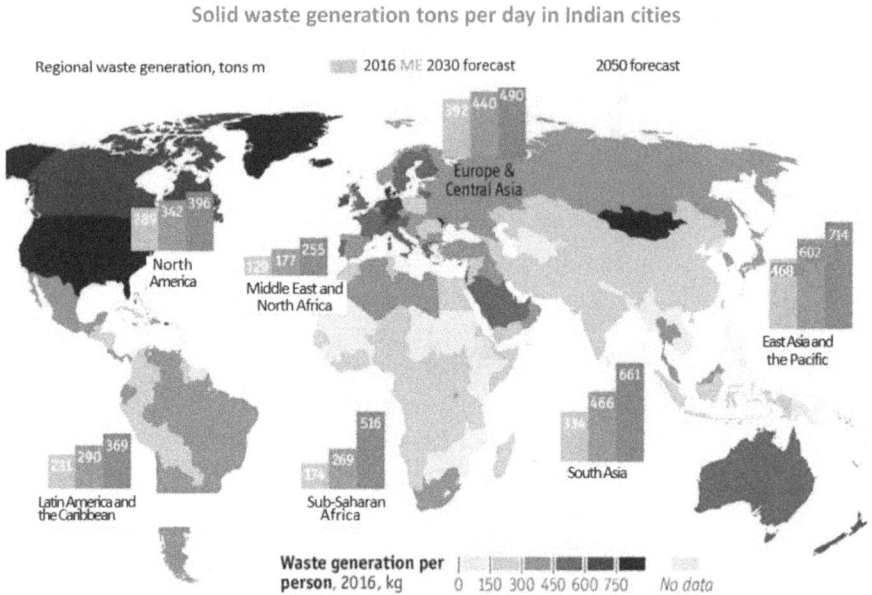

FIGURE 1.2 Global trend of waste generation and projection (daily chart – Global waste generation will nearly double by 2050|Graphic detail | The Economist (Ed.), The Economist (2018)).

as the biomedical waste accumulated around the world, especially in China, was around 240 MT on a daily basis. The Indian scenario also faced the consequences when Gujarat witnessed an increase in biomedical waste from 550–600 kg/day to around 1000 kg/day in the initial years of lockdown around the world (Bhattacharya and Singh, 2022).

1.3   COMPOSITION OF MSWM IN INDIA

The composition of waste has an important part in waste management practices. The high-income groups utilise higher volumes of paper, glass, metals, plastics, and textiles as they use more packaged products in comparison to low-income groups (Sridevi et al., 2012). Harmful wastes like paints, used medicines, pesticides, e-waste, and batteries are also part of MSW. The world scenario of solid waste is shown in Figure 1.2.

1.4   CHALLENGES FOR SOLID WASTE IN INDIA

1.4.1   HEALTH AND ENVIRONMENT

Human health is also associated with the degradation of the environment (Shukla et al., 2000). The health of the people is affected by untreated disposal of waste. The remnants of agricultural and dumpsites wastes are generally burnt off, which cause health problems like respiratory infection in the long run due to the release of fine particles and smog (Sridevi et al., 2012).

Gastrointestinal worm infestation is more common in people living near disposal sites (Giusti, 2009). The land and water are contaminated with waste. The leaching of harmful chemicals in soil and water also causes further degradation. The pollution caused due to waste disposal gives rise to epidemics in the long run (Unnikrishnan et al., 2006). The harmful gases emitted by burning solid waste lead to health complications in the form of cancer and death. These gases consist of CO, CO_2, $PM_{2.5}$, mercury and polycyclic aromatic hydrocarbon (PAHs) (Azar and Azar, 2016).

1.4.2 Operational Challenges for MSWM Processes

1.4.2.1 Collection and Segregation

Owing to the gigantic volumes of the waste generated and also to the rapidly increasing number in a densely populated areas of developing countries, the management of waste is adjourned with many difficulties. The biggest problem occurs with the collection of wet and dry waste separately. The trash that accumulates is again termed biodegradable and non-biodegradable. The non-biodegradable waste consists of plastics, aluminium cans, tyres, toxic chemicals, metals, and polystyrene. The urban waste is also laden with biomedical and electronic waste.

The composition of biomedical waste has both biodegradable and non-biodegradable elements. Apart from this, they are hazardous for open dumpsites and will often lead to harmful infection in the community. Most of the environmental issues in the health sector stem from the mismanagement of solid waste and its improper disposal. The world over the healthcare sector faces the problem of creating a sustainable environment for disposal, and studies have shown that developing nations are incurring more resistance in this area. The utmost need for handling waste lies in the availability of infrastructure, which is lacking in most of the healthcare units of developing nations like India. Figure 1.3 shows the state-level waste generation from a developing nation like India. The types of hospital wastes are given in Table 1.1.

The schematic flow chart of a typical MSW is shown in Figure 1.4. Here it is mentioned about the way the waste is collected from household and commercial areas and how it is further processed. The collection of waste plays a major role in the initial stage of segregation of waste. Door-to-door collection is still prevalent in most of the developing countries like India where 95% of the waste is collected in this manner. Local bodies in urban areas collect the domestic, trade, and institutional food/biodegradable waste from the doorstep or from local community garbage on a daily basis. There are large containers kept in fruit and vegetable markets to collect waste that are removed during non-peak hours by the local municipal bodies. Transportation of this waste is usually done in closed containers/tractors/trolleys or refuse compactors. The ragpickers still play a major role in waste segregation and help in proper recycling of wastes.

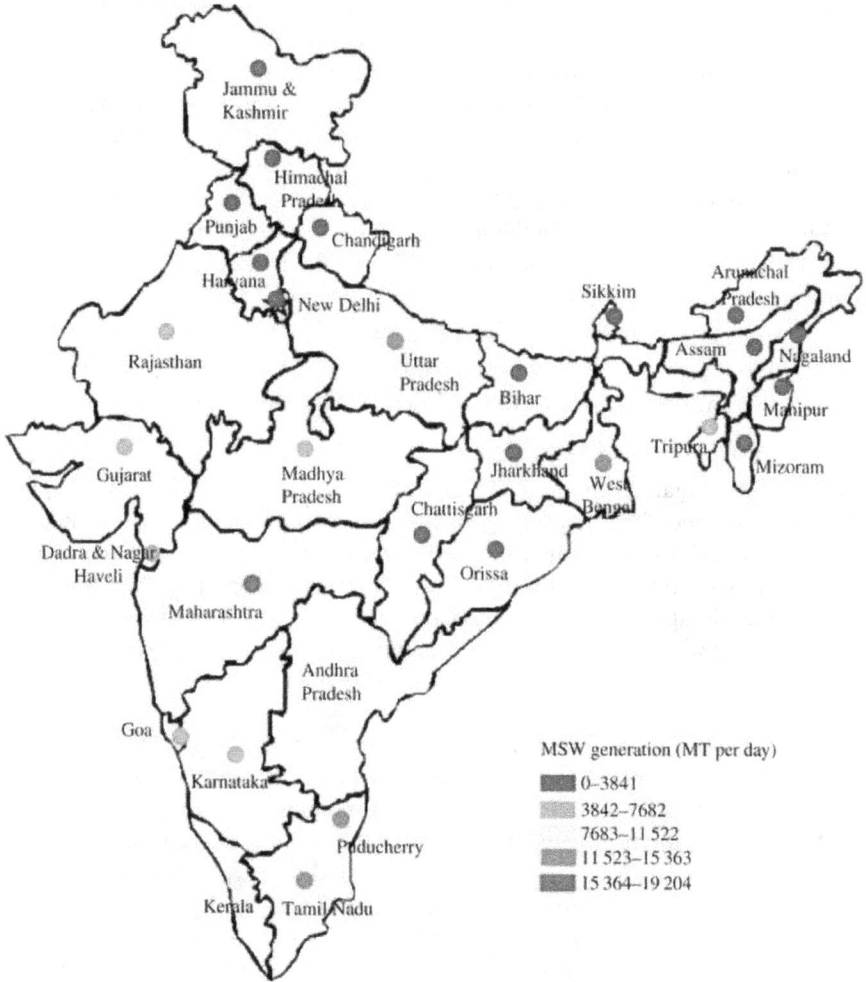

FIGURE 1.3   Statistics of state-level MSW generation from 2009 to 2012 (Kumar et al., 2017).

1.5   THE WAY FOR CONTROLLING THESE WASTES

The circular economy (CE) is a conceptual model that has been used for mini-misation of waste and better use of resources in a closed loop for waste man-agement. The biodegradable and non-biodegradable fraction of solid waste leads to the recovery of bioenergy resources and also contributes to the value added products. The best of biodegradable solid waste is it can be administered with advanced biological processes for the simultaneous production of bioenergy such as biohydrogen, biomethane, bioelectricity, and other products like butanol, etha-nol, methanol, etc.

TABLE 1.1

Types of Wastes in Hospital

Waste Category	Description and Content
Human anatomical waste	Tissue, organs, body parts (healthy or unhealthy), limb, finger, skin, muscle, toes, eyes, appendages, foetus, bones, etc. (tooth, hair, and nail are not included)
Animal waste	Tissue, organs, body parts, carcass, bleeding parts, blood and experimental animals used in research, and waste generated in animal houses
Microbiology and biotechnology waste	Laboratory culture, stocks, species of microorganisms in live or attenuated vaccines, human and animal cell culture used in research, waste from the production of biologicals, and devices used in transfer of culture
Sharps	Needles, syringes, blades, and glass capable of causing punctures or cuts
Discarded medicines	Waste consisting of contaminated, outdated, and discarded medicine, leftover medicine, and spillage
Soiled waste	Waste items contaminated with blood and body fluids including cotton dressings and linen beddings
Solid waste	Waste obtained from disposable items other than sharps such as tubings, catheters, intravenous sets, etc.
Liquid waste	Waste generated from laboratory, washing and cleaning activities, and disinfection and housekeeping activities
Incineration ash	Ash from the incinerator used for biomedical waste
Chemical waste	Chemicals used for the production of biologicals, chemicals used in disinfection, laboratory reagents, film developers used for X-rays, etc., expired medicines, and solvents
Genotoxic or cytotoxic waste	Waste containing genotoxic or cytotoxic properties, substances commonly used in therapy, e.g., antineoplastic or chemotherapeutic agents
Heavy metals	Batteries, broken thermometers, and pressure gauges
Pressurised containers	Gas cylinders, cartridges, and aerosol cans
Radioactive waste	Unused liquids from radiotherapy, lab research, contaminated glassware, packages, adsorbents, human discharges, and sealed and unsealed radionucleotides

Execution of principles of sustainable development in the context of waste management translates into the application of the hierarchy of waste management, that is, preventing waste generation; preparation for reuse, recycling, and other recovery processes; and, if necessary, disposal of the waste that cannot be recovered (Figure 1.5).

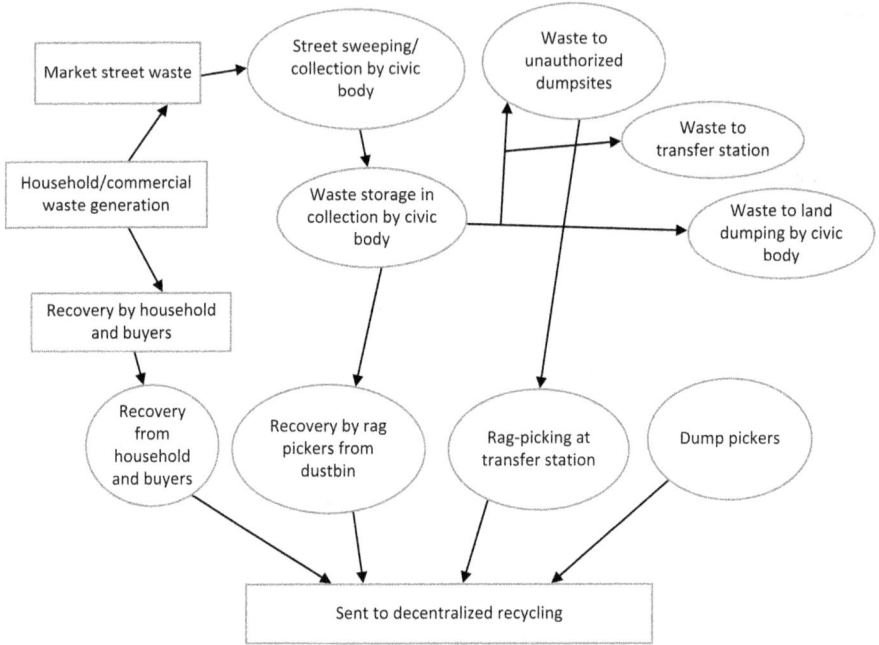

FIGURE 1.4   Schematic flow chart of typical MSW management and handling.

Most Preferred		
↕	**At Source Reduction and Reuse**	Waste minimisation and sustainable use/multiple use and reuse of products (e.g. reuse of carry bags/packaging jars)
	Recycling	Processing non-biodegradable waste to recover recycling commercially valuable materials (e.g. plastic, paper, metal, glass and e-waste recycling)
	Composting	Processing organic waste to recover compost (e.g. composting, windrow composting, in-vessel composting, vermi composting)
	Waste to Energy	Recovering energy before final disposal of waste (e.g. RDF, bio-methanation, co-processing of energy combustible non-biodegradable dry fraction of MSW, incineration)
	Landfilling	Safe disposal of inert residual waste at sanitary landfills
Least Preferred		

FIGURE 1.5   Waste management hierarchy.

1.6 CONCLUSION AND FUTURE PROSPECTS

In order to control the increased solid waste, there are many strategies taken and some of them are useful in the long run to mitigate the waste generation problem.

One of them is the generation of energy from waste. There is a requirement for lots of energy for thermal plant processing and the use of refuse-derived fuel is one such option where it generates energy due to its higher calorific value and also mitigates waste accumulation. This technology helps in using waste to generate energy and also vacate the area that would have been used for dumping.

The other way of treatment processes for combustion includes gasification, pyrolysis, production of refuse fuel, and gas plasma technology (Kumar et al., 2017).

The concept of waste management has improvised over the period. After the recent pandemic, the issues pertaining to waste management have become a global concern. Indian Government has implemented 'Swachh Bharat Abhiyan' under which many initiatives were taken to manage the solid waste in urban settlements as well as rural settlements. The present generation has seen both the wrath of untreated waste disposal and the aesthetic developed when solid waste management has been done properly. The Bowenpally agriculture market uses its rotten vegetables to generate electricity, and this form of energy is helping the vendors to avail energy resources at minimal cost. Over a 100 streetlights, 170 stalls, and administrative buildings power come from this vegetable market. These initiatives are eye-openers for the policymakers encouraging informal policy where the public is encouraged for separation of MSW at their initial stage and also creating awareness for the masses about the health hazard caused by litter accumulation beside the roads. The stringent law for littering the streets will soon mark its presence in urban and rural areas.

REFERENCES

Ahluwalia, I.J., Patel, U., 2018. Solid Waste Management in India: An Assessment of Resource Recovery and Environmental Impact. Working Paper No. 356. Indian Council for Research on International Economic Relations. https://icrier.org/pdf/Working_Paper_356.pdf

Anderson, J.C., Park, B.J., Palace, V.P., 2016. Microplastics in aquatic environments: Implications for Canadian ecosystems. Environ. Pollut. 218, 269–280. https://doi.org/10.1016/j.envpol.2016.06.074

Azar, S.K., Azar, S.S., 2016. Waste related pollutions and their potential effect on cancer incidences in Lebanon. J. Environ. Prot. 07(06), 778–783. https://doi.org/10.4236/jep.2016.76070

Bhattacharya, D., Singh, V.K., 2022. Climate Changes and Epidemiological Hotspots. CRC Press.

CPCB, and MEF&CC, 2016. CPCB, & MEF&CC Hazardous And Other Wastes Rules, 2016 Vol. 1981 (2016) Issue i https://cpcb.nic.in/displaypdf.php?id=aHdtZC9IV01fUnVsZXNfMjAxNi5wZGY.

CPCB India, 2018. Solid Waste Generation in 46 Metrocities. Retrieved from: https://cpcb.nic.in/uploads/MSW/trend_46_cities_list.pdf

De Bercegol, R., Cavé, J., Nguyen Thai Huyen, A., 2017. Waste municipal service and informal recycling sector in fast-growing Asian cities: Co-existence, opposition or integration? Resources 6(4), 70. https://doi.org/10.3390/resources6040070

Giusti, L., 2009. A review of waste management practices and their impact on human health. Waste Manag. 29(8), 2227–2239.

Global waste generation will nearly double by 2050|Graphic detail | The Economist (Ed.), The Economist (2018) https://www.economist.com/graphic-detail/2018/10/02/global-waste-generation-will-nearly-double-by-2050

Harris, N., Pisa, L., Talioaga, S., Vezeau, T., 2009. Hospitals going green: A holistic view of the issue and the critical role of the nurse leader. Holist. Nur. Pract. 23(2), 101–111. http://dx.doi.org/10.1097/HNP.0b013e3181a110fe

Joshi, R., Ahmed, S., 2016. Status and challenges of municipal solid waste management in India: A review. Cogent Environ. Sci. 2(1), 1–18. https://doi.org/10.1080/23311843.2016.1139434

Kaza, S., Yao, L., 2018. What a Waste 2.0: A Global Snapshot of Solid Waste Management to 2050. World Bank Group. https://doi.org/10.1596/978-1-4648-1329-0

Kumar, S., Smith, S.R., Fowler, G., Velis, C., Kumar, S.J., Arya, S., Rena, Kumar, R., Cheeseman, C., 2017. Challenges and opportunities associated with waste management in India. R. Soc. Open Sci. 4(3), 160764.

MNRE_India, 2018. MNRE_India Government of India Ministry of New and Renewable Energy (2018) https://mnre.gov.in/annual-reports-2018-19/

Modak, P., 2011. Municipal Solid Waste Management. Turning Waste into Resources, Shanghai Manual – A Guide for Sustainable Urban Development in the 21st Century. Retrieved from: https://www.zaragoza.es/contenidos/medioambiente/onu/1203-eng.pdf

Mohan, V., 2019 (May 10). Topic: 91% premature deaths due to air pollution in low and middle-income countries: UNEP. 1. The Economics Times. https://economictimes.indiatimes.com/news/environment/pollution/91-premature-deaths-due-to-air-pollutionin-low-and-middle-income-countries-unep/articleshow/69264417.cms

Nakum, J., Bhattacharya, D., 2022. Various green nanomaterials used for wastewater and soil treatment: A mini-review. Front. Environ. Sci. 420, Volume 9.

Shukla, A., Timblin, C., BeruBe, K., Gordon, T., McKinney, W., Driscoll, K., Vacek, P., Mossman, B.T., 2000. Inhaled particulate matter causes expression of nuclear factor (NF)-κ B–related genes and oxidant-dependent NF-κ B activation in vitro. Am. J. Respir. Cell Mol. Biol. 23(2), 182–187.

Singh, S., 2020. Decentralized Solid Waste Management in India: A Perspective on Technological Options. National Institute of Urban Affairs, New Delhi, pp. 290–304. Retrieved 31 July 2020, from https://smartnet.niua.org/sites/default/files/webform/Decentralized%20SWM%20in%20India.pdf

Sridevi V., Musalaiah M., Lakshmi M.V.V. Ch, A., Kesavarao L., (2012) A review on integrated solid waste management Int. J. Eng. Sci. Adv. Technol. Issn, 2 (5) pp. 1491–1499

UNEP (United Nation Environment Programme). 2019. Plastic recycling: An underperforming sector ripe for a remake. United Nation Environment Programme.

United Nations, 1992. Agenda 21. Report of the United Nations Conference on Environment and Development, Rio de Janeiro. United Nations publication, New York, pp. 1–486.

Unnikrishnan, H., Gowrav, B., Jathanha, S., 2006. Sustainable Decentralized Model for Solid Waste Management in Urban India. pp. 1–10.

2 Emerging Solid Wastes and Its Management

*Darshan Salunke, Arijit Dutta Gupta,
Pratibha Gautam, and Arun Gautam*

2.1 INTRODUCTION

Recent advancement in the day-to-day life cycle of the population across the world has given rise to an increase in the need and greed of people. The trust of being different from the herd has created an impression of more improvement in the mindset of industrialists. With all the development, there comes the scenario of a higher generation of by-products and waste, and some of them are even new to the world whose handling, treatment, and disposal remain a question mark for the world. These wastes are considered as emerging waste and does not fall under the common category of solid waste.

The waste discussed in this chapter are described as emerging waste. These wastes are different from the generalized solid waste categories and whose treatment facilities are not yet fully developed, products which are extensively used in present but have a tremendous load of waste in future after their end of life, which were difficult to handle during the phase of time and situation, etc. The emerging wastes are discussed separately in this chapter by describing challenges occur during the handling of these wastes as well as their possible managing opportunities.

2.2 EMERGING SOLID WASTE: CHALLENGES IN MANAGING WASTE AND THEIR MANAGEMENT OPPORTUNITIES

2.2.1 SOLAR PANEL WASTE

Energy based on solar source plays a significant role in decreasing the use of non-renewable energy. The photovoltaic solar capacity of the world has hit over 400 GW and is expected to increase ten times by 2050 (Chowdhury et al., 2020). The tremendous increase in solar power usage and production of PV cells for extraction of solar energy raises a question for the future regarding the handling and management of waste generated due to end-of-life solar panels. Generally, solar panels have a life cycle of around 20–25 years. After their end of life, these discarded waste panels need safe handling and management.

Commonly there are two types of solar panels available in the global market: crystalline silicon−based solar panels, which dominate the global market by over 90%, and thin-film solar cells, which consist of semiconductor materials like cadmium telluride or copper indium gallium. This makes an obvious impression that if these discarded solar panels, which fall under the hazardous waste category due to the presence of heavy metals like Pb and Cd, were mishandled and disposed of unobserved, it would have an adverse impact on humans as well as the environment (US EPA, 2022a).

There are various reasons for solar panel waste generation; some of which are transportation damages, defective solar cells, delaminated cells, power losses, glass or cable breakages within solar cells, loose frames, and optical failures. To understand recycling of solar panels, one must understand the construction of solar panels. Traditionally, solar panels consist of glass and aluminum frames, copper wiring, silicon solar cells, polymeric layers to protect solar cells from exposure to weather, and plastic box casing. Apart from these materials, solar panels also contain valuable metals like silver and copper components within the solar cells and trace heavy metals like cadmium and lead. Many of these materials can be processed to recycle like glass, metals like aluminum and copper, and box casing made of plastic.

Typical recycling process includes removal of fame and junction box; separation of glass and silicon wafer by thermal, mechanical, or chemical methods; chemical and/or electrochemical purification; and separation of solar cells and valuable metals like silver, copper, tin, lead, etc. (US EPA, 2022b). The physical or mechanical method basically applies to the dismantling process and is meant for external junction boxes and frames to look forward for any kinds of repair and/or replacements. Thermal and chemical treatments are methods applied for recovery and recycling purposes. A temperature of around 300–600°C is applied on the solar cells to remove the strong polymeric adhesive applied on the solar cells. Chemicals and solvents like trichloroethylene, toluene, and benzene are used for the extraction of silicon. Ultra-sonication and electrolysis methods are also used for the recovery of In and Ga (Chowdhury et al., 2020).

2.2.2 BATTERIES WASTE

In simplest terms, one can understand a battery as a form of energy storage device. Batteries make use of electrochemical reactions to provide direct current (DC) of low voltage. All batteries consist of at least one electrochemical cell, which is an arrangement of a positive electrode (cathode), a negative electrode (anode), and an electrolyte. When both the electrodes are connected through a wire, the flow of electrons is initiated. Broadly speaking, there are two types of batteries: primary batteries and secondary batteries. Primary cells/batteries cannot be recharged and have to be discarded after they are used. However, in the case of secondary batteries, when the positive electrode is made anode and the negative electrode is made cathode, upon supplying electricity, the flow of electrons reverses and batteries get recharged. Due to this, secondary batteries can go through a large number of charge-discharge cycles and be used multiple times before their use is finally discontinued. Popular examples of primary batteries are zinc/carbon cells, alkaline zinc/manganese dioxide cells, etc., and examples of secondary batteries are lead/acid, nickel/cadmium, lithium-ion batteries (LIBs), etc.

The electrode (both cathode and anode) and the electrolyte materials in batteries are composed of chemical substances and need proper handling and management after the batteries are used. Improper handling of used batteries can have an adverse impact to the environment due to the release of hazardous chemical wastes like chromium, lead, zinc, mercury, cadmium, lithium, etc. Lead/acid batteries are mostly used for household energy storage purposes because of their low cost and easy availability. After repeated use, they are discarded and often sold as scrap to small uncertified and illegal recyclers. Due to lack of professional knowledge and proper safety and disposal measures, the electrolyte (sulfuric acid) and lead components from the dismantled batteries are at often times discarded at will by the recyclers, hence impacting the nearby soil and water sources.

Almeida et al. (2006) find that small and portable household devices utilize small batteries of size AA, C, and D. These batteries contain highly alkaline waste which, during disposal, gets partially released into the atmosphere. One of the major reasons that battery wastes find their way back into the environment is that spent batteries are disposed of as municipal solid waste. Since battery waste is a part of municipal solid waste, its management is not given due importance as is the case with industrial hazardous wastes. However, a process known as vacuum metallurgical reprocessing has shown satisfactory results in the recovery of metals from batteries. This process is suitable for metals having moderate boiling points like cadmium and mercury.

For the past two decades, a battery type that has been most extensively studied, researched, and used is LIB. Its features like low self-discharge, relatively small size, high numbers of charge/discharge cycles, and high energy density make it an ideal choice for portable energy storage devices, as well as for energy-intensive uses like in electric vehicles. Until a better energy storage device is invented, the production and popularity of LIBs are only going to increase. Most of the LIB waste ends up in landfills (considered as MSW), while the remaining LIB waste is sent to waste-to-energy facilities, recycling units, or for the purpose of reusing.

Countries like the United States, Japan, and the EU block have formulated their own waste electronic and electrical equipment recycling guidelines, which are being utilized to manage spent LIBs with varying degrees of recovery. According to Zhang et al. (2017), theoretical calculations demonstrate that just in China, full recycling of the spent LIBs could lead to recovery of 75,000 tons of Li, 90,000 tons of Fe, 45,000 tons of Cu, 60,000 tons of Co, 15,000 tons of Al, and 35,000 tons of plastic.

LIBs are composed of outer shell, cathode and anode, separator, and an organic electrolyte. The outer shell is made up of metals like aluminum, stainless steel, nickel-plated steel, or plastic. According to Winslow et al. (2018), cathode makes up nearly 25–30% of the battery's total weight. It consists of a thin aluminum current collector layered with metal oxide like $LiCoO_2$, $LiMn_2O_4$, $LiNiO_2$, etc., and is called the cathode material. Cathode contains the most valuable materials used in the LIB. Anode is typically made up of copper current collector and layered with graphite sheets and makes up for nearly 20–25% of battery's total weight. Both cathode and anode are adhered to their respective metal collector sheets with the help of inert binders like polyvinylidene fluoride (PVF). The separator is made up of polyolefin materials like polyethylene and contains a lithium salt electrolyte. The separator primarily serves the purpose of avoiding physical contact between the anode and the cathode and facilitates the passage of lithium ions. Different lithium salts are used in different LIBs as the electrolyte materials also have varying impacts on the temperature, life, and stability of the battery. Some of the electrolyte salts commonly used in LIBs are $LiClO_4$, $LiBF_4$, $LiAsF_6$, $LiPF_6$, $Li(CF_3SO_3)$, $Li[N(CF_3SO_2)_2]$, etc.

Recovering and recycling the metals and salts from the spent LIBs is one of the best ways to manage the spent battery waste and to avoid the toxic chemicals present in the spent batteries from entering the environment. First, the electrode components are separated to obtain electrode materials like current collectors and metallic foils. Technologies have been devised that use optical sensors to sort LIBs based on the composition of their cathode material. Once sorted, the LIBs go through pyrometallurgical and hydrometallurgical processes and the metals are recovered. Although there is possibility of reclaiming most of the different materials used in LIBs, due to considerations of economic feasibility, mostly scarce elements like lithium and cobalt are reclaimed.

2.2.3 E-WASTE

The use of electronic devices has grown exponentially with the growth of the IT and communication sector. E-waste is added to solid waste as consumers have to dispose of outdated electronics faster due to faster technological development. The expanding e-waste problem calls for an even greater focus on recycling and better e-waste management. When electronic and electrical devices are no longer suitable for their original use or when their shelf life has passed, they become electronic waste or electrical waste. E-waste includes computers, servers, mainframes, monitors, CDs, printers, scanners, copiers, calculators, fax machines, batteries, mobile phones, transmitters, TVs, iPods, medical equipment, washing machines, refrigerators, and air conditioners (if not suitable for use). Due to the rapid development of new electronic devices and the rapid replacement of older generations, the production of e-waste has increased by leaps and bounds. People often upgrade to newer versions and products also have a shorter life span.

Metals, polymers, electron beam tubes (CRTs), circuit boards, cables, and other materials are often found in electronic waste. Valuable metals such as copper, silver, gold, and platinum can be recycled if e-waste is treated scientifically. When electronic waste is disassembled and processed raw with rudimentary techniques, it is very dangerous because of toxic substances such as liquid crystals, lithium, mercury, nickel, polychlorinated biphenyls (PCBs), selenium, arsenic, barium, brominated substances. flame retardants, cadmium, chromium, cobalt, copper, and lead. E-waste is a serious threat to people, pets, and the environment. Even small amounts of heavy metals and highly toxic compounds such as mercury, lead, beryllium, and cadmium pose a serious threat to the environment.

Consumers hold the key to better e-waste management. Initiatives such as Extended Producer Responsibility (EPR), Design for the Environment (DfE), Reduce, Reuse, Recycle (3Rs), and technology platforms for connecting markets are launched to encourage consumers to properly dispose of their electronic waste by increasing reuse and recycling, and adopting sustainable consumption habits. E-waste management is a priority in developed countries. However, in developing countries it is aggravated by the full adoption or replication of e-waste management in developed countries and many other related problems such as deficiencies, investments, and technically qualified human resources. Additional problems include inadequate infrastructure and ineffective laws, especially regarding e-waste. Also, the roles and tasks of stakeholders in the handling of e-waste, etc. are not sufficiently described. The Ministry of Environment, Forests and Climate Change (MoEFCC) issued new e-waste (Management) Rules in 2016 to replace the Indian E-Waste Rules.

International agreements like the Basel Convention regulate and reduce the movement of hazardous waste across borders. Despite the Convention, illegal shipping and disposal of electronic waste continue. In 2018, an estimated 50 million tons of electronic waste were produced globally. Half of this is made up of personal electronics like computers, TVs, displays, mobile phones, and tablets. The other half is made up of bigger home appliances like HVAC systems. Even though 66% of the world's population is covered by e-waste legislation, only 20% of the world's e-waste is recycled annually. This means that 40 million tons of e-waste are either burned for resource recovery or illegally exchanged and processed improperly. In the United States alone, less than 20% of the more than 100 million computers that are discarded are effectively recycled. In China, 160 million electronic items are thrown away each year. Previously, it was believed that China was the world's largest electronic garbage repository. Disassembling electronic waste is a skill that tens of thousands of workers have. Across the board, the amount of electronic waste is rising at a rate of 5–10% per year. According to Borthakur and Sinha (2013), India generates 146,000 tons of electronic waste annually. However, these figures only take into account national e-waste production and do not take into account illegal and legal garbage imports, which are significant in developing nations like China and India. The reason is that huge quantities of wet electrical and electronic equipment (WEEE) from other countries reach India. Switzerland is the first nation in the world to have designed and implemented a formal e-waste management system, recycling 11 kg of e-waste per person (EU), compared to the European Union's goal of 4 kg/capita.

E-waste management in developing nations is exacerbated by lax or nonexistent enforcement of existing regulatory frameworks, a low level of awareness and sensitization, and inadequate occupational safety for those involved in these processes. In contrast, the EU and Japan have well-developed initiatives aimed at changing consumer behavior at all levels. Developing nations must devise effective strategies to encourage the reuse, refurbishment, or recycling of electronic waste in specialized facilities in order to reduce environmental contamination and health risks.

The Environmental Performance Index places India 177th out of 180 countries, placing it among the bottom five, according to a study presented at the 2018 World Economic Forum. Poor results in the environmental health policy and air pollution-related death categories were linked to this. In addition, India recycles less than 2% of the electronic waste it generates annually, placing it fifth among the world's major e-waste producers, behind the United States, China, Japan, and Germany. India has imported a significant amount of electronic waste from other nations and has produced over two million tons of electronic waste annually since 2018. People frequently dump waste in open dump sites, which can cause health issues and contamination of groundwater among other issues. Computer equipment accounts for nearly 70% of all electronic waste in India, according to a study by KPMG and the Associated Chambers of Commerce and Industry of India (ASSOCHAM). The remaining e-waste comes from household sources like phones (12%), electrical equipment (8%), and medical equipment (7%).

E-waste is primarily collected, transported, processed, and recycled through the informal sector. The business is highly interconnected and unregulated. Most of the time, not all of the valuables and materials that could be retrieved. In addition, there are significant issues with worker health and safety as well as poison leaks into the environment.

Seelampur in Delhi is home to India's largest facility for the dismantling of electronic waste. Taking out the devices' functional parts, reusable parts, and precious metals like copper and gold takes between 8 and 10 hours per day for both adults and children. Open burning and acid leeching are two methods utilized by electronic waste recycling companies. This situation has the potential to be improved by raising awareness, improving the infrastructure of recycling facilities, and implementing new policies. The majority of the collected e-waste in India is managed by an unorganized sector. In addition, repair shops, dealers in used goods, and vendors on e-commerce portals collect a significant portion of the discarded electronics for recycling and reuse with the intention of reusing and cannibalizing components.

2.2.4   COVID WASTE

Novel Corona Virus, also known as COVID-19, has spread all over the world and has impacted everyone socially, economically, commercially, physically, and mentally. The virus has been identified as the greatest threat to global public health and indicates deficiency and inequity of social and commercial advancement (Chakraborty and Maity, 2020). Some of the noticeable challenges of the global medical healthcare system are lack of awareness, lack of access, workforce shortages, and lack of affordability and accountability. One of the major challenges was the management of waste. Although most of the COVID-19 waste falls under the biomedical waste category, it has been raised in this chapter to showcase the condition that the world has faced during the pandemic situation to handle the tons and tons of waste which have been generated on a daily basis while being in the procrastinating condition of pandemic.

A WHO report states that during the period of March 2020 to November 2021, over 14 crores of COVID testing kits were used, which potentially generated more than 2500 tons of non-infectious plastic-based waste. The report also highlights that nearly 1,45,000 tons of waste in the form of syringes, needles, and safety boxes was generated as an aftermath of the administration of more than 8 billion vaccination doses (WHO, 2022). As of date, roughly 30% of global healthcare facilities are facing problems in handling the waste already existing in the healthcare center due to inadequate equipped facilities. This raises the feeling of whataboutery of COVID waste load. The waste exposes the risk of burns, needle or sharps injuries, and pathogenic microbes, thereby impacting people staying near poorly managed facilities of treatment or landfill. Additionally, burning of waste contaminates the air, resulting in a serious environmental issue.

2.2.5 SANITARY WASTE

Menstruation is a physiological occurrence that both girls and women experience each month. The hygienic management of menstruation, which is connected to women's dignity and well-being, is the primary tenet of sanitation, cleanliness, and reproductive health. It may sound alarming that many communities still view it as taboo in the 21st century. Women's activities like cooking and eating certain foods are frequently restricted during menstruation, mostly to maintain antiquated social customs that have no scientific backing. Many girls decide to skip school until their periods are over due to a lack of menstrual hygiene facilities and social pressure. In fact, many schools do not have separate restrooms for girls or they are poorly maintained. This kind of social stigma is a major roadblock to achieving the lofty goal of gender equality and impedes society's overall progress.

Due to rapid urbanization, product availability and distribution, access to a wide range of options, increased mobility, awareness of menstrual hygiene, and persistent government efforts, the use of disposable sanitary napkins is expanding rapidly. According to a recent study, disposable sanitary napkins produce 44,254 cm^3 per female per year, which is more than any other hygiene product. It is estimated that 121 million women and teenage girls in India use eight sanitary napkins on average each month, producing 113,000 tons of menstrual waste annually. Due to the enormous volume of sanitary pad waste produced, the waste management industry is under pressure, and proper disposal is a concern. The handling of sanitary waste with conventional tools and methods puts the land, aquatic bodies, and public health in grave danger.

Sanitary waste disposal has become a growing problem in India as a result of the non-biodegradable plastic used in disposable sanitary napkins, which is harmful to human health and the environment. A more significant impact is caused by unorganized municipal solid waste management practices and inadequate local collection, disposal, and transportation systems in cities and villages. Sanitary waste classification, whether biological or plastic, has also long been a significant issue. Household waste, soiled napkins, diapers, condoms, tampons, and blood-soaked cotton are included in the 2016 Solid Garbage Management (SWM) Rules. After being divided into biodegradable and non-biodegradable components, these items are disposed of. However, everything contaminated with blood or bodily fluids, such as cotton, dressings, soiled plaster casts, lines, or bedding, is considered biomedical waste and must be burned, autoclaved, or microwaved to kill microorganisms, as stated in the Biomedical Waste Management Rules, 2016.

The lack of concern for proper waste management in our nation is reflected in the absence of reliable statistics on the subject. Due to the lack of waste segregation in this industry, detailed instructions for handling and managing sanitary waste are essential. According to a 2011 study titled "Sanitary Protection: Every Woman's Health Right," only 12% of the 335 million women who have menstruation had access to sanitary napkins that were disposable. The environmental website Down to Earth reports that 432 million pads are discarded each month (Down to Earth, 2022).

As a result, it is crucial to promote the use of sanitary napkins and lessen the risk posed by improper disposal of sanitary waste in landfills and other locations.

Women and girls will be able to fully participate in education and the workforce if they have easier access to sanitary waste management facilities that are socially and culturally acceptable. Typically, dirty sanitary napkins are disposed of in conjunction with standard trash cans. The waste collector must separate menstrual waste from other solid waste, which is difficult, time-consuming, and complicated. As a result, it is critical to prioritize community-level on-site disposal strategies that are quick, secure, and hygienic.

Controlled decentralized burning of used pads is the efficient method that can be found in many commercially available decentralized incinerators. If controlled burning was not used to help destroy pathogens and pad material and assist in overall sanitary waste management, sanitary pads would otherwise be mixed with municipal solid waste or disposed of. An incinerator is the option that is most commonly chosen at the local level. However, incinerators are typically known for their poor performance, energy-intensive operation, the harmful gases that are produced when plastics are burned (such as furans and dioxins), high cost, and demanding requirements for maintenance.

According to market data, only a small number of incinerators comply with environmental regulations. The sanitary pad incinerators that are currently available typically perform below average when it comes to meeting acceptable standards for the temperature of the combustion chamber and emission levels. GreenDispo is an improved electric sanitary pad incinerator that is both environmentally friendly and energy-efficient. It was developed with the intention of improving the designs of conventional incinerators by CSIR-NEERI, Sowball Aerothermics, Secunderabad, and the International Advanced Research Centre for Powder Metallurgy and New Materials, Hyderabad.

Around 660 million women live in India, the second-most populous nation. Menstruation is still frowned upon in India's conservative culture. Menstruating women were still regarded as filthy, despite advances in facilities and technology. Menstrual hygiene practices and health are negatively impacted by a variety of socioeconomic constraints in a developing nation.

According to Sinha and Paul (2018), social conventions, myths, tariffs on sanitary products, and a lack of effective disposal options make menstrual hygiene in India difficult. The authors also mentioned that during disasters like floods, earthquakes, and droughts, people received food and other necessities. However, sanitary napkins were only given to the needy a very limited number of times. Budhathoki et al. (2018) examined how women dealt with menstrual hygiene following the earthquake in Nepal. In the initial recovery zone, menstrual hygiene management activities were viewed as less important. With the readily available absorbents, women managed their periods.

Hales and Barrington (2018) conducted semi-structured interviews with the instructors in Mumbai using a questionnaire. The most practical answer, based on their ages, is to educate both boys and girls about menstruation hygiene and change their attitudes toward it. Due to the high cost of sanitary napkins, adolescent girls were only occasionally attending school in places like western Kenya, Uganda, Nepal, and Africa. According to the authors, reusable sanitary pads would be a better option to solve this issue (Parthasarthi et al., 2022).

In the beginning, women dealt with menstruation by using a variety of absorbents. The absorbents provided made menstruation more bearable. However, the napkins continue to be out of stock and expensive. Even though most people buy and use sanitary napkins, the methods for disposing of them are ineffective. A local rather than centralized approach to waste management would be preferable. As a result, incinerators might be installed in each neighborhood by government agencies. Sanitary trash is typically isolated at the point of production; it is the only household trash that is combined with it. As a result, managing that garbage separately would be straightforward.

The government's use of free contraceptive methods to implement family planning policies was successful. The same strategy should start by giving society free incinerators. In the future, we intend to construct an intelligent incinerator that is both affordable and user-friendly. By examining the data, we intend to lower the incinerator's emissions of pollutants. Smart incinerators might be made available anywhere by putting the unit inside a truck. This would help the government deal with used napkins in an environmentally friendly way.

2.2.6 MULTILAYERED AND MICROPLASTIC WASTE

Plastics have supplanted other materials in the modern economy because they are inexpensive, lightweight, and adaptable. Annual worldwide plastics output rose from 2 to 380 MT between 1950 and 2015. However, this was also accompanied by the production of an estimated 6300 MT of plastic garbage, of which only 9% was recycled, 12% was burned, and 79% was dumped in the environment or sent to landfills. Currently, 90% of plastic products are thought to only be used once before being thrown. Large amounts of plastic are released into the ocean, according to assessments of losses related to plastic waste management, losses of microplastics (5 mm in length) across the entire plastic value chain, and the transport of micro- and macro-plastics (5 mm in length) in rivers (Lebreton and Andrady, 2019). One of the most important environmental challenges now is plastic waste. Recycling has the potential to lessen environmental plastic pollution. However, it's still possible that during mechanical recycling operations, little pieces of plastic are released into the aquatic environment.

The UN Environment Program estimates that each year, 3.0 MT of microplastics and 5.3 MT of macroplastics are lost to the environment, with marine plastic pollution likely accounting for a significant portion of these amounts. As a result, it is of the utmost importance to continue expanding our understanding of plastic losses throughout the plastics value chain and developing efficient solutions to lessen the environmental harm caused by plastics. Microplastics are now a widespread pollutant that poses serious threats to human health and the environment. Because of their exposure to these contaminants, microorganisms have developed the ability to break down a variety of plastic polymers over time. Enzymes like depolymerases and lipases have been looked into for lowering the toxicity of plastic. Because plastic decomposition is a lengthy process, meta "omics" techniques have been used to determine the active microbiota and microbial dynamics involved in the mitigation of microplastic-contaminated locations. Additionally, protein engineering methods have provided novel solutions to this troubling issue. Plastic contamination is on the rise because it serves as a breeding ground and a vehicle for the growth of other persistent chlorinated pollutants.

Landfilling, recycling, and incineration are common practices for disposing of plastic. According to reports, these techniques have detrimental impacts, cause secondary pollution, and are harmful to both the environment and the human health. Due to the oxygen's low bioavailability, plastic endures longer in landfills. Additionally, it produces other poisons like xylene, toluene, and endocrine-disrupting substances like bisphenols, which have resulted in high levels of hydrogen sulfide. Additionally, the incineration process releases dangerous substances such as heavy metals, PCBs, and polyaromatic hydrocarbons. Therefore, microbial degradation has piqued researchers' interest due to the advantages it has over traditional ones. Microorganisms break down plastic waste into carbon dioxide, water, methane, biomass, and different inorganic chemicals during microbial degradation, which relies on the kind of polymer, organisms, the molecular weight of plastics, and environmental factors. Several fungi were isolated and discovered to have the capacity to break down microplastics, including *Zalerion maritimum*, *Pestalotiopsis microspora*, and *Penicillium simplicissimum*. Metagenomic investigations are necessary to discover the whole microbial variety that is capable of degrading any particular contaminant. However, culture-based methodologies only identify 1% of the total genome. Through the application of the metaproteomic methodology, omics have become a leading biotechnology method for identifying and elucidating the function of uncultivable microbes.

A recent study says that a global effort to reduce plastic waste at every stage of the plastic life cycle is needed to stop the rising trend of plastic pollution in the environment. Borrelle et al. (2020), for illustration, looked at the effects of reducing the amount of plastic waste produced, increasing the collection and recycling of plastic waste, and identifying and eliminating plastic pollution on environmental recovery. Furthermore, Lau et al. (2020) examined the effects of numerous potential future scenarios, including the use of alternative materials, an increase in plastic recycling and collection, and a reduction in plastic use. The result was the same for both groups: Cooperative global action is urgently required to reduce plastic pollution, despite the possibility of higher costs associated with the collection, sorting, quality control, and utilization of recycled plastics. Additionally, recycling plastic waste is likely to be an efficient means of combating the production of plastic pollution in the environment, including marine pollution. According to Garcia and Robertson (2017), mechanical recycling, such as milling, washing, and pelletizing, is currently the technology that is being utilized most frequently for the large-scale recycling of plastic solid waste. This method, on the other hand, generates a lot of untreated wastewater that may contain a lot of microplastics and put aquatic ecosystems in danger.

In East Asia, Vietnam is a developing country. Between 2000 and 2018, the average GDP growth rate was 6.28%, which shows significant progress. Combined with these factors, Vietnam was the fourth largest contributor to marine plastic pollution in 2015 (Geyer et al., 2017). We investigated microplastic concentrations in river discharge and surface water receiving effluent from three mechanical recycling facilities

(MRF) in Vietnam, which process different types of plastic waste in this context, to determine if MRFs are point sources of microplastic (electronic) pollution, plastic waste [EPW], polyethylene terephthalate [PET] bottle waste [PBW], and household plastic waste [HPW]). Laser confocal microscopy was used to determine the size, shape, and color of plastic flakes produced as recycled plastic from each MRF, as well as unintentionally generated plastic particles separated from sewage and downstream water. A Fourier transform infrared spectrometer was used to determine the polymer type (FT-IR). We investigated emissions and environmental impacts of accidental microplastics from these MRFs by comparing determined microplastic concentrations with published values and discussed the impact of MRFs as point sources of microplastic pollution.

2.2.7 PERSISTENT ORGANIC WASTE

POPs, or persistent organic pollutants, have a long half-life in the environment and can spread far from their source. Due to their bioaccumulative properties, these substances accumulate in the food chain after being exposed to animals. This could affect biodiversity before it reaches humans and harm the health of both current and future generations. POPs are harmful substances produced by humans and associated with the production, consumption, and disposal of particular organic compounds. Many of the chemicals were developed commercially for use in crop production, industrial applications, and pest and disease control. POPs include pesticides and PCBs, while dioxins and furans are unintended by-products of industrial processes or the result of the burning of organic molecules. POPs pose a significant threat to the world because of their persistence, transportability over long distances, capacity to bioaccumulate in fatty tissue, and extreme toxicity even at low concentrations. Due to their high stability and ability to travel great distances, POPs are found everywhere and can even be found in the Arctic. POPs can connect to the surface of solid particles or aqueous sediments and can be gaseous. Rain or ash deposition spreads POPs all over the ground.

POPs are mostly made by humans and can enter the environment through a variety of ways. Leachate from landfills, agricultural runoff, industrial effluent, urban runoff, drainage systems, and atmospheric deposition are all ways in which POPs can enter the environment. Organochlorine insecticide poisoning of the environment is common in conventional farms and rice paddy farming areas. The majority of organochlorine insecticide residues in rivers or sediments are significantly higher in agricultural areas than in other areas where farming is not practiced. Vector control, home insecticides, and structural pest management are additional factors that contribute to POP pesticide exposure. However, it is difficult to identify the actual source of POP contamination outside of agricultural areas because these studies have not been conducted so far. However, POPs also originated naturally. Volcanic activity and forest fires affected the levels of dioxin and furan in the environment. PCBs are still present in old transformers and capacitors that are still in use and contain oil contaminated with PCBs. According to Agamuthu and Narayanan (2013), activities in industrial, residential, and commercial areas may result in the presence of POPs in the waste stream. POPs can accumulate from waste disposal methods like open burning and unregulated landfilling, especially in developing nations.

In the early 1970s, POPs were outlawed in both Europe and the United States due to their harmful effects on living things, long-range transportability, tendency to bioaccumulate in fatty tissue, and resistance to chemical and biological breakdown. However, prior to the Stockholm Convention (SC)'s implementation, POPs continued to be used extensively in numerous nations. A global convention known as SC aims to protect humans and the ecosystem from POPs by regulating their release and preventing them. POPs are addressed globally by the Stockholm Convention, which was signed on May 17, 2004 and currently has 170 parties. It aims to protect the environment and human health from the harmful effects of POP exposure.

Reviewing the first ten years of the Stockholm Convention's implementation using an alternate method that measures improvements in the management of POP stocks and global POP inventories provides a clearer picture of the practical difficulties posed by POP legacies. The disposal of POP, pesticide, and PCB stockpiles has only made very modest progress, particularly in emerging nations and nations whose economies are transitioning. These nations frequently have insufficient or few demolition facilities. The end-of-life management of PCB and pesticide stocks and the rise of e-waste including PBDEs present them with enormous hurdles. Exports of POPs in trash and goods from industrial nations frequently make their issues worse (Xu et al., 2013). Even in industrialized nations, the majority of current "remediation" activities entail containment rather than the "destruction or irreversible transformation" that POPs wastes must undergo in accordance with the Stockholm Convention. This strategy violates the principles of sustainable development since it leaves pollution management in the hands of future generations.

The majority of pesticides were found to have decreased in the environment following their prohibition, with the exception of lindane and endosulfan. Numerous national rivers and sediments still contained significant concentrations of POPs, according to studies, even after pesticide use was outlawed. This demonstrated that illegal use of POP insecticides can still occur nationwide. Paddy fields and vegetable farming, according to previous research, are the primary sources of OCPs. However, there is a dearth of research on POPs that are not caused by pesticides. More research on POPs in other forms of wastewater, like agricultural farm effluent and landfill leachate, needs to be done in order to gather quantitative data for better POP management. The threat posed by POPs was not realized until significant quantities of these compounds had been widely dispersed throughout the world. In Malaysia, data on POPs, particularly POP, dioxin, and furan levels in the air, are scarce.

2.3 CONCLUSION

The chapter showcased the waste which is believed to be the emerging waste for the near future as well as present considering the waste generation scenario. It can also lead to difficult times in the future if not given importance in management. Technological advancement needs to be thought out of the box for better handling, treatment, and management of this emerging waste. Policymakers should take into account such allied types of waste generated to fulfil the current lifestyle demand and modernization. Technologies under the R&D phase need to be aided by financial as well as governmental support so that these technologies can get scaled up to handle, treat, recycle, and recover valuables from such emerging wastes.

REFERENCES

Agamuthu, P., Narayanan, K., 2013. Persistent organic pollutants in solid waste management. Waste Management & Research. 31(10), 967–968.

Almeida, M. F., Xará, S. M., Delgado, J., Costa, C. A., 2006. Caracterization of spent AA holsehold alcaline batteries. Waste Management. 26(5), 466–476. http://dx.doi.org/10.1016/j.wasman.2005.04.005. PMid:15964181.

Borrelle, S.B., Ringma, J., Law, K.L., Monnahan, C.C., Lebreton, L., McGivern, A., Murphy, E., Jambeck, J., Leonard, G.H., Hilleary, M.A., Eriksen, M., Possingham, H.P., De Frond, H., Gerber, L.R., Polidoro, B., Tahir, A., Bernard, M., Mallos, N., Barnes, M., Rochman, C.M., 2020. Predicted growth in plastic waste exceeds efforts to mitigate plastic pollution. Science. 369(6510), 1515–1518.

Borthakur, A., Sinha, K., 2013. Generation of electronic waste in India: Current scenario, dilemmas and stakeholders. African Journal of Environmental Science and Technology. 7(9), 899–910.

Budhathoki, S.S., Bhattachan, M., Castro Sánchez, E., Sagtani, R.A., Rayamajhi, R.B., Rai, P., Sharma, G., 2018. Menstrual hygiene management among women and adolescent girls in the aftermath of the earthquake in Nepal. BMC Women's Health. 18(1), 1–8.

Chakraborty, I., Maity, P., 2020. COVID-19 outbreak: migration, effects on society, global environment and prevention. Science of the Total Environment. 728. https://doi.org/10.1016/j.scitotenv.2020.138882

Chowdhury, M.S., Rahman, K.S., Chowdhury, T., Nuthammachot, N., Techato, K., Akhtaruzzaman, M., … Amin, N. 2020. An overview of solar photovoltaic panels' end-of-life material recycling. Energy Strategy Reviews. 27, 100431. https://doi.org/10.1016/j.esr.2019.100431

Down to Earth, 2022. Kicking up a stink. Retrieved from: https://www.downtoearth.org.in/coverage/kicking-up-a-stink-41036

Garcia, J.M., Robertson, M.L., 2017. The future of plastics recycling. Science. 358(6365), 870–872.

Geyer, R., Jambeck, J.R., Law, K.L., 2017. Production, use, and fate of all plastics ever made. Science Advances. 3(7), e1700782.

Hales, G., Barrington, D., 2018. Investigating the policies and practices of teaching menstrual hygiene education to schoolboys in India. The Urban World: Quarterly Publication. 11(3). 1–9.

Lau, W., Shiran, Y., Bailey, R.M., Cook, E., Stuchtey, M.R., Koskella, J., Velis, C.A., Godfrey, L., Boucher, J., Murphy, M.B., Thompson, R.C., Jankowska, E., Castillo Castillo, A., Pilditch, T.D., Dixon, B., Koerselman, L., Kosior, E., Favoino, E., Gutberlet, J., Baulch, S., Palardy, J.E., 2020. Evaluating scenarios toward zero plastic pollution. Science. 369(6510), 1455–1461.

Lebreton, L., Andrady, A., 2019. Future scenarios of global plastic waste generation and disposal. Palgrave Communications. 5, 1–11.

Parthasarthi, S., Jayaram, V., Jeganathan, S., Lakshminarayanan, A.R., 2022. Menstrual hygiene and waste management: the survey results. Materials Today Proceedings. 65(8), 3409–3416.

Sinha, R.N., Paul, B., 2018. Menstrual hygiene management in India: the concerns Indian. Journal of Public Health. 62(2), 71–74.

US EPA, 2022a. Retrieved from: https://www.epa.gov/hw/end-life-solar-panels-regulations-and-management

US EPA, 2022b. Retrieved from: https://www.epa.gov/hw/solar-panel-recycling

WHO, 2022. Tonnes of COVID-19 health care waste expose urgent need to improve waste management systems. Retrieved from: https://www.who.int/news/item/01-02-2022-tonnes-of-covid-19-health-care-waste-expose-urgent-need-to-improve-waste-management-systems

Winslow, K. M., Laux, S. J., Townsend, T. G., 2018. A review on the growing concern and potential management strategies of waste lithium-ion batteries. Resources, Conservation and Recycling. 129, 263–277. https://doi.org/10.1016/j.resconrec.2017.11.001

Xu, W., Wang, X., Cai, Z., 2013. Analytical chemistry of the persistent organic pollutants identified in the Stockholm convention: a review. Analytica Chimica Acta. 790(0), 1–13.

Zhang, W., Xu, C., He, W., Li, G., Huang, J., 2017. A review on management of spent lithium-ion batteries and strategy for resource recycling of all components from them. Waste Management & Research. 36(2), 99–112. https://doi.org/10.1177/0734242x17744655

3 Solid Waste Treatment Technologies
Thermochemical Pathway

Arijit Dutta Gupta, Pratibha Gautam, and Harinder Singh

3.1 INTRODUCTION

Thermal waste treatment is a technology that must be used for solid waste management. The global expansion of production and waste generation has been sped up by an enormous increase in urbanization and industrialization. In the pursuit of efficient and effective material recovery from various waste sources, recycling and reusing are opposed. Energy recovery is the way to go for waste streams for which material recovery isn't very effective or efficient, taking waste hierarchy into account. As a result, an integrated waste management system is required for the integration of various treatment processes: processes for recycling materials and, if necessary, biological treatments for appropriate streams, thermal treatments for energy recovery, and service landfills for disposing of treatment residues should be provided. Waste intended for or resembling a home is referred to as residual municipal solid waste (MSW). It includes household waste that is collected by local authorities as well as some commercial and industrial waste from offices, schools, and shops that may be collected by the local authority or a private business. The amount of mixed MSW that can be disposed of in landfills is limited (implicitly) by laws.

The idea of a hierarchy of waste management options, with the most desirable option being to prevent the creation of waste in the first place (waste prevention), and the least desirable option being to dispose of the waste in a landfill with no recovery of materials or energy, has been one of the guiding principles for European and UK waste management, now codified in law. Between these two extremes, there are numerous waste treatment alternatives that can be employed as part of a waste management strategy to recover resources (such as furniture reuse, glass recycling, or organic waste composting) or produce energy from the wastes (via digesting, or incineration of biodegradable wastes to produce usable gases).

To reduce the amount of MSW still needing to be disposed of in landfills, a range of alternative waste management techniques and solutions are available. The main techniques covered in this guide are gasification and pyrolysis, which together make up the broad category of technologies known as advanced thermal treatment (ATT). These technologies can help keep biodegradable municipal waste (BMW)

DOI: 10.1201/9781003352396-3

out of landfills since they are made to recover energy (in the form of heat, electricity, or fuel).

The ATT technologies have a comparatively modest track record on MSW in the UK (and a little larger track record internationally). On certain waste streams (such as biomass, industrial wastes, tyres, etc.), there are numerous instances of ATT procedures that are proved, practical, and bankable, but there are far fewer that have been tested on municipal trash. The purpose of this chapter is to increase knowledge of the technologies available and aid in the removal of obstacles to the establishment of suitable ATT procedures in England.

3.2   ADVANCED THERMAL TREATMENT – AN OVERVIEW

ATT technologies are primarily those that employ pyrolysis and/or gasification to process MSW. It excludes the incineration of wastes which has been deemed to be a mature and well-established technology in the field of thermal treatment.

A conventional technique for the thermal treatment of solid wastes is the gasification and pyrolysis of waste. It has been widely utilized to manufacture a variety of fuels, including charcoal, coke, fuel gas, and producer gas. Wood and coal are pyrolyzed to produce charcoal and coke, respectively. Producer gas is a form of flammable gas created during the coke gasification process while air (for oxygen) and high pressure steam are present.

Pyrolysis and gasification have only recently been utilized for the commercial treatment of solid waste. For some technologies in the UK, pyrolysis and gasification are becoming established for commercial and prepared municipal waste, while other technology configurations are still in the pilot-scale development stages. In North America, Europe, and Japan, additional large-scale ATT plants have been constructed and are in operation.

3.3   INCINERATION – A CONVENTIONAL FORM OF THERMAL TREATMENT

The combustion of raw or residual solid waste is known as incineration. To completely oxidize the fuel (waste), a sufficient amount of oxygen is required for combustion. Waste is transformed into carbon dioxide and water at incineration plant combustion temperatures typically exceeding 850°C. Bottom ash, which is made up of any non-combustible materials like metals and glass, remains as a solid with some carbon still in it.

Incineration also includes material combustion, which produces heat energy that can be used to generate electricity via turbines. For energy recovery from non-biodegradable sources with low moisture content, this technique is preferable. According to a literature assessment, the burning of 1 ton of garbage can provide around 500–600 kWh of power. It aids in the reduction of enormous volumes of waste and their landfill disposal. This method also aids in the reduction of methane, a by-product of open MSW disposal. However, the generation of furans, hazardous particulate particles of metals and dioxins, which are thought to be harmful to the environment, is a major constraint of incineration. Some wastes (such as clinical and biomedical wastes) benefit from incineration because the heat energy can eradicate microorganisms and their toxicity. Temperature, residence time, and turbulence are all important components in incineration. The temperature at which chemicals are oxidized is called incineration temperature. It is higher than the temperature of ignition. The better the incineration effect, the higher the incineration temperature. Incineration has several disadvantages, including the production of organic pollutants and smoke. The second factor that influences incineration is residence time, which is determined as the amount of time solid waste spends in the incinerator from start to finish. Turbulence is the third parameter, and the higher the turbulence, the more effective the combustion. The additional air coefficient has a significant impact on waste incineration. By raising the air rate, enough oxygen is delivered, and the furnace of turbulence is enhanced, which is better for the process.

Combustion and incineration aim to completely oxidize all of the feedstock material's acceptable elemental species. It is possible to directly incorporate MSW or a variety of IW into the incinerator process. In relation to MSW, its direct incineration is also referred to as "mass burn." This is typically in contrast to the incineration of mechanically pretreated waste, where the mechanical treatment (often mechanical biological treatment or MBT) aims to improve the waste's combustible quality (i.e., increasing the heating value by reducing the contents of moisture and ash) as well as the technical and environmental parameters (i.e., reducing the content of chlorine and mercury.

It is essential to keep in mind that trash with an Lower Heating Value (LHV) greater than 5–7 GJ/mg can be burned without the use of auxiliary fuels; lower LHV values are typically brought on by an excess of ash and moisture. If the waste is characterized in terms of proximate analysis, which includes combustibles, moisture, and ash, a triangular diagram (Tanner, 1965) can be used to indicate its composition. A zone of self-combustion has been recently discovered for combustibles with a mass of more than 25% and a moisture content of less than 50%, ash content of less than 60%, and both. As a reminder, gaseous combustion products, also known as flue gas, must remain at least 850°C (for non-hazardous waste) and 1100°C (for hazardous waste) for at least two seconds according to Directive 2000/76/EC. MSW takes longer to burn completely than other solid fuels due to the requirements of the largest particles it contains (Ruth, 1998).

The fact that the waste is typically solid and contains quite large particles has a significant impact on the type and design of the equipment used to burn it (i.e., the combustor) (Ruth, 1998). The majority of the waste combustors in use today fall into one of three basic categories: fluidized bed, rotary kiln, and mobile/fixed grate. According to Castaldi and Themelis (2010), mobile grate combustor technology has advanced to a significant degree. The mobile grate is the waste incineration technique that is most commonly utilized, according to data from ISWA (2012). Waste, especially MSW, is typically a complicated mixture of various materials with varying physical and chemical properties as well as a great deal of size variability. The amount of time needed for fuel combustion is specifically influenced by the size of the fuel particles. Mobile grate combustors, according to the BREF paper of the European Commission (2016), are the waste incineration equipment with the highest treatment capacity in terms of thermal power input per line (up to 120 MW on LHV basis).

Due to their internal refractory lining, rotary kilns offer the significant advantage of processing any sort of waste, including liquid waste, and can typically endure higher incineration temperatures than mobile grates (1400°C vs. 1250°C – European Commission, 2006). As a result, they are especially well suited for processing wastes like HIW that need high temperatures and pose serious corrosion risks. According to the BREF paper (European Commission, 2006), the maximum treatment capacity of these combustors is equal to 30 MW per line in terms of thermal power input on an LHV basis.

In fluidized bed combustion, a bed of fuel and inert particles are kept in suspension by an upward-moving airstream. Combustion is enhanced by the substantial turbulence, which encourages uniform mixing and efficient heat transfer. In a comprehensive study of fluidized bed combustors, Van Caneghem et al. (2012) found that the typical temperature is between 800°C and 900°C and pointed out that fluidized beds that bubble, rotate, and circulate are increasingly being used in incineration processes. However, one of the biggest drawbacks of these technologies is that they require fuel preparation, especially when working with MSW. Sludge, for example, is one type of waste that would be ideal to burn in these devices because it is readily available and has a uniform distribution of particle sizes. According to the BREF documentation (European Commission, 2006), the maximum treatment capacity for fluidized beds per line is 90 MW of LHV combustion power.

No matter what kind of device is used, solid and heterogeneous waste is burned with a little more air than the stoichiometric requirement to encourage the complete oxidation of all relevant elemental species. The European Commission (2016) stated that excess air values typically range from 40% to 150%, resulting in a significant amount of specific flue gas production (ranging from 4 to 10 mN^3/kg, with higher values for rotary kilns and lower values for grate and fluidized bed combustors). This, in turn, has some negative effects on the performance of energy recovery. In fact, the high flow rate of flue gas at the stack consumes a lot of energy and costs a lot of energy to treat.

The heat content of the combustion products is partially recovered to provide energy, often using a steam generator that is usually built into the combustor. Saturated steam is typically produced when only thermal energy is sought after, whereas superheated steam is typically produced when just electrical energy (also known as "power") or combined heat and power (CHP) are sought after. Both of these latter examples use superheated steam to create a Hirn steam cycle, which is just another term for a Rankine cycle with steam superheating. A condensing turbine is employed for the production of solely power (Figure 3.1); back pressure turbine (Figure 3.2) or an extraction-condensing turbine (Figure 3.3) is utilized for CHP.

The optimum method for recovering energy is said to be CHP (European Commission, 2006). Furthermore, CHP is the technical solution that can support

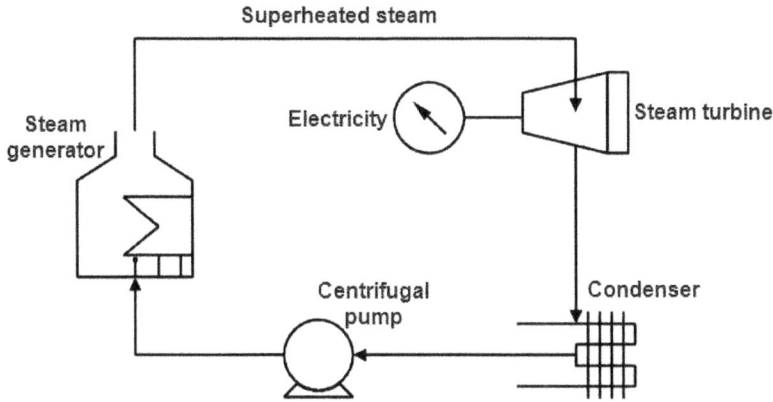

FIGURE 3.1 Schematic representation of steam cycle for conventional electricity production. (Adapted from Lombardi and Carnevale, 2015.)

high values of the energy recovery index R1 and is reported to perform better in life cycle assessment (LCA) evaluations than just electricity production (Damgaard et al., 2010). However, in the absence of thermal consumers (industrial plants or district heating) close to the WtE plant, the production of simply electricity is the only option. Reimann (2009) estimated the European average energy efficiency numbers using the following formulas to provide an overview:

$$\eta_{el} = \frac{EE_p}{E_{in}}$$

$$\eta_{th} = \frac{ET_p}{E_{in}}$$

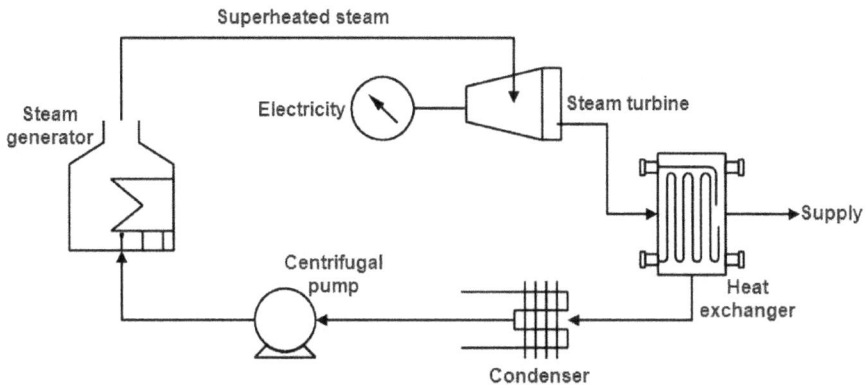

FIGURE 3.2 Schematic representation of steam cycle with cogeneration of electricity and thermal energy with back pressure turbine. (Adapted from Lombardi and Carnevale, 2015.)

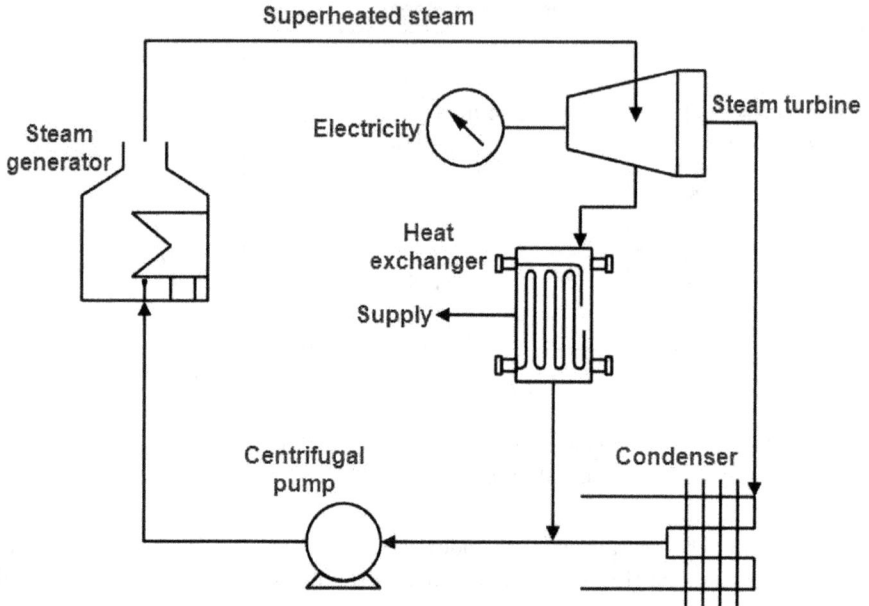

FIGURE 3.3 Schematic representation of steam cycle for cogeneration of heat and electricity with extraction-condensing turbine. (Adapted from Lombardi and Carnevale, 2015.)

$$\eta_{el,CHP} = \frac{EE_{p,CHP}}{E_{in}}$$

$$\eta_{th,CHP} = \frac{EE_{p,CHP}}{E_{in}}$$

Here, EE_p is the weighted average specific produced electricity; E_{in} is the weighted average specific energy input (including import); and ET_p is the weighted average specific heat produced and used. η_{el} is the electric energy production efficiency. When CHP is used, the electric and thermal energy production efficiencies are about 14.2% and 45.9%, respectively, when compared to the weighted average specific produced electricity in cogeneration plants $ET_{p, CHP}$, η_{el} is the electric energy production efficiency of WtE plants and is 20.7%; η_{th} is the thermal energy production efficiency of facilities with major heat production and is up to 81.3%.

3.3.1 POWER GENERATION

When simply considering the production of power (i.e., electricity), efficiency. The WtE plants seem to be fairly low in comparison with traditional fossil fuel-fired power plants. WtE plants' highest stated net electric efficiency. In the literature, the percentage is 30% on an LHV basis (Murer et al., 2011). In fact, Graus et al. (2007) determined that the weighted average LHV efficiencies were

45% for natural gas, 38% for oil, and 35% for coal-fired power generation. For coal, the maximum recorded efficiencies were around 42% LHV. According to Graus and Worrell (2009), the average efficiency of coal-fired power generation in the EU increased from 34% LHV in 1990 to 38% LHV in 2005 and was projected to reach 40% LHV by 2015.

The combined effects of economic and technical constraints are the fundamental causes of the comparatively poor performances as compared to traditional power plants. The following details can be found in particular:

i. A small plant's size
ii. Conservative steam parameters, such as superheating temperature and evaporating pressure
iii. A relatively high pressure of condensation
iv. A straightforward cycle configuration with few water preheaters and no steam reheat
v. A high stack loss with no air preheating or indirect air preheating
vi. A high rate of energy consumption within the plant

3.3.2 Small WtE Plant Size

The size of WtE plants, especially those used to process MSW, is mostly influenced by the size of the collection region that each facility serves. The size is typically quite tiny, taking into account that smaller plants could encounter less social opposition. According to Reimann (2012), small plants are those with a throughput of less than 100,000 mg/y (roughly 37.5% of the plants surveyed); medium plants are those with a throughput between 100,000 and 250,000 mg/y (roughly 39.5%); and large plants are those with a throughput greater than 250,000 mg/y (roughly 22.9%). However, it makes more sense to categorize the facilities according to their thermal input for the thermal process, which is dependent on both waste throughput and LHV. When compared to the LHV of fossil fuels, the LHV of garbage is relatively low; for example, the average LHV for MSW is 10 GJ/mg (Reimann, 2012). Therefore, WtE thermal input sizes are often low (below 35 MW LHV) or medium (below 90 MW LHV), and in any event, they are significantly smaller than conventional fossil fuel plants (1000–2000 MW LHV or even larger).

The typical small thermal input size has a number of effects on the WtE performances for a number of reasons, including the steam turbine efficiency decreasing with size (Consonni and Viganò, 2012), the combustion chamber's unfavourable volume to surface ratio imposing a significant amount of excess air, and the auxiliary devices' performance declining with size. Indirect effects of size on plant performance are also caused by economic factors. In fact, because of their small size, devices have high specific costs, necessitating the use of very straightforward configurations in order to keep overall costs under control (as will be more clearly stated in the paragraphs that follow).

In certain situations, the possibility for technological advancements that could improve performances is constrained or impractical due to financial considerations. Contrariwise, in sizable WtE plants, the utilization of financial resources is made possible by the elimination of specific costs for the performance and the plant's high level of technical sophistication and optimization. It is possible to determine the designed thermal power input per line for those plants with the design values of waste input flow rate and LHV by further analyzing the data from ISWA (2012).

3.3.3 CONSERVATIVE STEAM PARAMETERS

As the pressure and temperature of superheated steam rise, so does the Hirn cycle's efficiency. WtE boilers face severe high-temperature, acidic corrosion on their heat transfer surfaces, exacerbated by the high concentration of hydrogen chloride (HCl) in the flue gas. Furthermore, Persson et al. (2007) as well as De Greef et al. (2013) lend credence to the hypothesis that chlorine contributes significantly to the corrosion process. There will undoubtedly be metals and chlorine in the waste – except for chlorine-containing plastics like PVC, which make up a significant portion of table salt. The efficiency of high temperature acidic corrosion is significantly influenced by the surface temperature of the metal. Because the rate of corrosion increases with temperature, the surfaces must be kept at a lower temperature. Both the evaporating pressure and the superheating temperature are directly limited because both evaporating and superheating surfaces must be taken into account.

In addition, the maximum superheating temperature and the maximum moisture content at the outlet of conventional condensing turbines typically limit the maximum evaporation pressure even further. To ensure a certain steam quality at the steam turbine's output (i.e., the slope of the expansion line in the enthalpy–entropy diagram), evaporating pressure must not exceed a certain level, which is dependent on the turbine's expansion efficiency, after the superheating temperature has been determined. Steam cycles associated with WtE processes frequently employ conservative steam parameters due to the aforementioned factors. Limiting flue gas temperatures at the superheater to 650°C or less can reduce corrosion rates (De Greef et al., 2013).

According to Gohlke and Martin (2007), WtE facilities constructed in the 1980s and early 1990s in the USA employed steam specifications of 60 bar/443°C and were generally more efficient than plants constructed at the same time in Europe based on 40 bar boilers or in Japan based on 20 bar boilers. The same authors also state that, based on waste LHV of 10.44 MJ/mg (only electricity output), grate furnace WtE plants in Europe typically produce 546 kWh of electricity per mg of waste, which equates to a gross efficiency of 18%. When the average amount of electricity used inside the facility is 150 kWh per mg of trash, the amount of electricity exported is roughly 396 kWh/mg (net efficiency 13%). The majority of contemporary plants typically generate 640 kWh/mg of electricity utilizing 40 bar/400°C, which equates to a gross electric efficiency of 22% (LHV of 10.44 MJ/mg). With a net efficiency of 18% and an in-plant usage of 120 kWh/mg, this results in around 520 kWh of power exported per mg of trash.

The first heat exchanger in the series of banks used in traditional waste-fired steam generator designs typically places an evaporator at the beginning of the flue gas cooling process. With this solution, tube surfaces with high heat fluxes can maintain a low temperature. Additionally, this is the most efficient method of reducing flue gas temperature – while still recovering energy – to a level low enough to prevent corrosion on the banks that follow, particularly superheaters.

Some evaporator tubes can be inserted into the walls of the secondary combustion chamber (waterwall) above the grate furnace or within the riser of a fluidized bed combustor to recover heat through flue gas's radiation and convection heat transfer mechanisms (Van Caneghem et al., 2012). Additionally, this arrangement provides sufficient cooling of the flue gases to the point where they can safely enter the superheater section downstream. Compared to older furnaces without integrated heat recovery in the walls, this type of combustor requires less surplus air and also reduces the sensible heat losses from the exhaust. The available surface for this integrated heat recovery is the bed or grate that surrounds the firing apparatus. It continues to the exit plane from the top of the stoker or bed until the next surface, usually a screen or superheater, starts (Jegla et al., 2010). Exhausts come into contact with the superheater tubes when the temperature drops to at least 650°C.

Lee et al. (2007) found a way to extend the boiler tube's lifespan and provide a variety of methods for safeguarding them against corrosion at high temperatures. The Prism is a technological boiler feature that optimizes secondary air injection and combustion management by installing a prism-shaped body in the first empty boiler pass. De Greef et al. (2013) mention the Prism as one of the primary methods for reducing high temperature corrosion. One of the secondary methods used to combat high temperature corrosion is the use of various corrosion-resistant material systems, such as Inconel 625 (21Cr-9Mo-3.5Nb-Ni base), which can be layered over waterwall areas and superheater tubes to protect them from the attack of HCl/Cl_2 (De Greef et al., 2013). In numerous WtE plants, this technology has been utilized successfully in the waterwall tubes and a portion of the superheater's tube bundles.

Adaptation of the water-steam cycle, the use of corrosion-resistant material for the coating of the waterwall tubes, and improved boiler design all make it possible to conveniently raise the temperature of the superheated steam and, as a result, the pressure. Recent WtE plants operate at pressures and temperatures of up to 600°C and 60 bar, respectively. A trade-off between investments, anticipated declines in plant availability, and profit from increased efficiency and electricity sales results in high steam characteristics (Pavlas et al., 2011). Only WtE facilities with a high throughput can withstand the rising costs of investment and operation.

3.3.4 RELATIVELY HIGH CONDENSING PRESSURE

Lowering the pressure at the condenser also increases the efficiency of the Hirn cycle. The capacity of the plant, the cooling medium's temperature, and the condenser pressure all have an impact. The temperature of the surrounding air has a considerable impact on the pressure for air-cooled condensers. A temperature of 20°C has a negative effect, increasing the condensing pressure to 0.17–0.12 bar, while a temperature of 10°C allows for pressure of 0.1–0.08 bar (European Commission, 2006). Using water-cooled condensers in a closed loop with an atmospheric cooling tower can reduce condenser pressure (Pavlas et al., 2011). However, because they are less expensive than water-cooled condensers, air-cooled condensers are frequently employed in classic WtE facilities. Larger air-cooled condenser surfaces or the use of water-cooled condensers may reduce the condensing pressure for big installations, enhancing efficiency. The average value is close to 0.15 bar.

A specific layout of a WtE cogeneration facility with an air-cooled condenser in parallel with a water-cooled condenser is the subject of a 2011 study by Barigozzi et al. (2011). They show that the air cooled is the optimum approach to reject heat if the ambient temperature is lower than 15°C in order to increase the power production while taking into account the energy requirements for the two types of condensers. When the ambient temperature is higher, the condensation should use as much of the water-cooled condenser's cooling capacity as possible while letting the air-cooled condenser's residual heat escape.

3.3.5 STRAIGHFOREWARD CYCLE CONFIGURATION

In fossil fuel steam power plants, specific arrangements like internal regeneration and reheating are also used to improve efficiency. The advantage of reheating is that it permits work at extremely high pressure levels even when the temperature of superheating is controlled. In fact, steam is heated once more at an intermediate expansion pressure, causing the expansion curve to deviate from the saturation line and preventing excessive steam condensing in the turbine's final stage, which could cause the blades to be eroded by liquid droplets. Using external superheaters to superheat live steam from lower temperatures to higher ones is another option (Main and Maghon, 2010).

In high-capacity WtE installations, these kinds of measures can be utilized to significantly boost net electrical efficiency (Gohlke and Martin, 2007). However, according to Pavlas et al. (2011), the potential for the aforementioned measures is limited and/or not financially feasible for facilities with low processing capacities (100,000–150,000 mg/y).

3.3.6 HIGH STACK LOSS WITH NO OR INDIRECT AIR PREHEATING

Temperatures over 200°C are not uncommon, and exhausts are cooled down to between 150°C and 380°C throughout the heat recovery process. Since different processes require different temperatures in order to function properly, the temperature at the steam generator's outlet is a trade-off between maximum boiler efficiency (equivalent to 150°C) and the type of flue gas treatment (FGT) that is being used (De Greef et al., 2013).

Thus, the temperature and volume of departing exhausts, or thermal losses by sensible heat, affect the boiler's performance. Additionally, feedstock moisture affects stack losses. However, the impact of feedstock moisture is already taken into account when the LHV of waste is taken into account for losses and boiler efficiency calculation, rather than HHV. Depending on the exhaust temperature and air excess, these energy losses can account for up to 25% of the total process losses. Other losses in the overall process, such as those caused by radiation and convection, incomplete combustion, and thermal losses in unburned fuel, are less significant (between 3% and 4%) (Pavlas et al., 2011). According to Tabasová et al. (2012), boiler efficiency, which is determined by comparing the thermal energy entering with waste to the energy acquired by the steam, ranges between 75% and 85%. According to Pavlas et al. (2011), if a boiler exiting temperature of 250°C and 6% oxygen content after the last air supply are assumed, boiler efficiency reaches 81%. However, a significant increase in efficiency depends on lowering flue gas temperature, which on the other hand places more demands on the FGT system.

The lowering of the flue gas flow rate offers one way to lower stack losses. This can be accomplished by improving the combustor design with heat recovery integration (as previously stated) and lowering the extra air from higher values, such as 1.75–1.9 (Gohlke, 2009; Main and Maghon, 2010), to lower values, such as 1.39 (Main and Maghon, 2010). Flue gas recirculation (FGR) can also lessen thermal losses through sensible heat at the boiler outflow. The fresh air flow rate is chosen to bring the oxygen content of the exhausts leaving the combustion chamber as close to 6% as possible, which is the reference value to ensure that the combustion reactions are complete, and it determines the corresponding minimum flue gas flow rate. The recirculated flue gas flow rate is used to control the combustor temperature. According to Liuzzo et al. (2007), the application of FGR is expected to boost net electric efficiency by 1–3% points.

Additionally, because the installed capacity of the FGT is smaller when FGR is used, there are lower capital expenses involved. Castaldi and Themelis (2010) point out that FGR improves combustion performance by raising the reaction zone's temperature and lowering emissions levels, enabling more complete combustion in a short amount of time without the addition of high-grade fuel. However, they also point out that FGR causes systemic concerns with complexity and robustness.

In addition to controlling NO_x emissions, effective manipulation of the FGR improves the performance of the WtE under waste uncertainty, enabling increased throughput at the required temperature and within the permitted FGR residence time range. Preheating air or water utilizing low temperature streams present in the plant is another approach to increase energy recovery and conversion efficiency. This improves the efficiency of the heat recovery system. For example, combustion air preheating could be advantageous for the enthalpy balance in the combustion chamber. One low temperature stream that can be utilized for air preheating is the water from the grate cooling circuit, which has a temperature range of 70–80°C (grate cooling is typically used by some industries when the LHV of waste is higher than 12.5–14.5 GJ/mg). Preheating of air or water may also be accomplished by further heat recovery from exhausts at the steam generator's output.

3.3.7 HIGH RATE OF IN-PLANT ENERGY CONSUMPTION

A sizable portion of the gross power comes from the in-plant consumption, which is significantly influenced by the FGT. Energy consumption in the plant ranges from

120 to 150 kWh per mg of incoming garbage, according to Gohlke and Martin (2007). The amount of electricity required by FGT ranges from 10 to 80 kWh per mg of trash, depending on the technology used. A bag house filter, selective non-catalytic reduction (SNCR), semi-dry flue gas cleaning, activated carbon injection to remove dioxin, and a FGT use approximately 45 kWh/mg of energy. If flue gases are condensed and selective catalytic reduction (SCR) is utilized in place of SNCR, the consumption rises to 75 kWh/mg. A bag house filter with activated carbon injection and SNCR is followed by a 70 kWh/mg consumption in the case of wet flue gas cleaning. SCR can be used in place of SNCR, increasing consumption to 80 kWh/mg, and flue gas condensation can be added. When waste pretreatment is required to feed fluidized beds, the percentage of electrical power used for in-plant purposes can reach up to 21%, depending on the plant's size.

3.4 GASIFICATION

By partially oxidizing the solid fuel in the presence of an oxidant amount less than that required for stoichiometric combustion, a process known as gasification aims to transform a solid fuel—in this case, solid waste—into a gaseous fuel known as "producer gas" or "syngas." Additionally, char (carbonaceous chemicals) and ashes make up the solid product. The gasification of solid waste generally necessitates temperatures greater than 600°C, a complicated process that is largely determined by the kind of reactor used and the characteristics of the waste. The operating and performance process parameters, as well as the various process phases, are combined by Arena.

Both an external source and partial combustion of the fuel being used (autothermal gasification) can provide the heat needed for the gas formation process (allo-thermal gasification). The gasification medium that is utilized has a significant impact on the LHV of the syngas. Partial oxidation occurs when air, oxygen-enriched air, or pure oxygen is utilized; the LHV of the syngas varies from 4 to 7 MJ/mN^3 (when air is used) to 10–15 MJ/mN^3 (when pure oxygen is used). A high hydrogen concentration syngas with an LHV of 15–20 MJ/mN^3 is created when steam is the only gasification medium employed; in this situation, no exothermic reaction occurs, necessitating an external heat source, such as thermal plasma.

Gasification reactors include fixed beds (updraft and downdraft), fluidized beds (bubbling, circulating, and internally circulated fluidized beds), entrained beds, vertical shafts, moving grate furnaces, rotary kilns, and plasma reactors. Their explanation and a list of companies that make plants based on gasification offer a design tool that has been proven to work in the real world.

Pretreatment is typically required for MSW, particularly when fluidized beds are utilized, even though some gasification methods can handle MSW that has not been pretreated. If MSW pretreatment is necessary, it should be taken into account that MBT processes use more energy than was previously noted, and this should also be taken into account in the overall energy balance.

Partially oxidized products, primarily hydrogen, carbon monoxide, and methane in smaller amounts, are abundant in the leaving syngas. As a result, it has a lot of energy and can be used as an energy carrier in a variety of energy conversion devices. The main problem with using syngas, especially syngas made from waste, is that it often contains undesirable substances like particles, tar, alkali metals, chloride, and sulphide. The cooling of syngas and the installation of cleaning equipment to remove its tarry compounds and particle solid matter are required because tars, in particular, pose problems with fouling in transportation pipeline routes and moving machinery parts (Bebar et al., 2005). The use of the generated syngas in subsequent processes is the primary concern, rather than the potential for energy recovery from waste gasification. When syngas is burned in a boiler, there are no specific restrictions on the amount of tar, dust, alkalis, heavy metals, and hydrogen sulphide. This eliminates the problems caused by cooling the products and condensation of their tarry components (Arena, 2012). With the exception of the fact that oxidation is divided into two processes in this instance (two-step oxidation), the plant in question is essentially comparable to an incineration plant. If potential improvements in energy efficiency are sought, it would be more intriguing to use the syngas in an internally fired cycle, which typically has higher conversion efficiencies than steam cycles, a synthesis process, or a system to produce hydrogen. The difficulty lies in properly cleaning the syngas in these circumstances.

3.4.1 GASIFICATION: COMBUSTION OF SYNGAS IN BOILER

The combustion of syngas, which is a homogeneous gas-phase one, can be performed under better conditions than in the case of MSW, so processes based on waste gasification and syngas combustion in a boiler have some advantages, according to several authors (Consonni and Viganò, 2012). Because a significantly smaller volume of fuel is required, smaller equipment for heat recovery and FGT may be utilized (Bebar et al., 2005). Plants, on the other hand, typically have higher operating costs, are more difficult to manage and maintain, cost more to do so, and are less reliable (Consonni and Viganò, 2012).

For instance, one of the earliest examples of these plants in Italy, the gasification plant in Greve in Chianti (Italy), was forced to shut down due to operational issues in July 2001, after three periods of operation over a period of five years. Based on two atmospheric pressure circulating fluidized beds that worked with air at 850°C and each had a 15 MW thermal power input, the plant was put into operation in 1992 to process 200 mg/day of RDF pellets.

Thermoselect is one example of a gasification technology that garnered a lot of interest in the past. The Thermoselect process begins with a degassing channel (where water is evaporated and organic compounds are partially degasified using external heating in an airless environment). This is followed by oxygen, high-temperature gasification, and a melting stage. A quencher, a gas scrubber, a desulfurization stage, a gas drying system, and an activated carbon filter typically make up the syngas cleaning system (Richers et al., 1999).

The development of this procedure began in 1989. A demonstration plant with a licensed capacity of 4.2 mg/h (33,000 mg/y) was put into operation in Fondotoce (Italy) in 1992 and continued to operate until 1998. Malkow (2004) claims that this plant generated between 200 and 500 kWh, or 12 GJ, of energy per mg of LHV waste. The facility for producing syngas, which was then burned in two boilers to produce steam for a back pressure turbine with a planned output of 12.5 MW and 50 MW of heat for district heating, was built in Karlsruhe, Germany, with a nominal capacity of 225,000 mg/y of MSW and industrial waste (three lines, 10 mg/h each). Based on the design parameters, the electric efficiency and cogeneration efficiency are estimated to be approximately 12.5% and 62.5%, respectively. The facility was put into operation in January 2002, after two years of commissioning, and it was taken down in November 2004.

Thermoselect technology is utilized in a small number of Japanese manu- facturing facilities, the most well-known of which is located in Chiba (Japan). Construction on it began in 1998 with a theoretical capacity of 300 mg/day. This technology was initially licensed to the JFE Group, which is now Kawasaki Steel (two lines). In 1999, the plant successfully completed operation of MSW treatment for approximately 15,000 mg of MSW with 8.5 GJ/mg LHV for a continuous period of 93 days and a total of more than 130 days. In 2000, the plant established a busi- ness that combines fuel production with industrial waste treatment. In addition to other by-product gases (blast furnace gas, coke oven gas, etc.) produced by the steel works, the purified syngas produced during this process is transported to the nearby steel works for use as part of the fuel for the combined cycle power plant. The Chiba factory installed a 1.5 MW gas engine generator and ran a demonstration test of gas engine power production at full and partial load with some of the fuel gas provided to the steel plant to demonstrate that syngas could potentially be used in other devices.

Consonni and Viganò (2012) describe and simulate two power plants that pro- duce steam for feeding a Hirn cycle using gasification and syngas combustion in a boiler, replicating as closely as possible two commercial gasification technolo- gies: a gasifier with a high temperature grate that is similar to the technology that Ener-G suggests; and a fluidized-bed gasifier that operates at low temperatures and is based on the technology proposed by the Japanese company Ebara. In the case of a plant with 300 MW of input thermal power, steam parameters being adjusted up to 70 bar/450°C, the gross efficiency may reach 31%, with a corresponding net efficiency of roughly 27% for the two scenarios, assuming a waste LHV of 10.34 GJ/mg and steam characteristics at 40 bar/400°C. In the case of a plant with 50 MW of thermal power input, the gross efficiency is approximately 23.7–23.8%. The authors concluded that the energy performance is largely influenced by the configuration and operational parameters of the steam cycle in traditional incin- eration facilities, as well as the utilization of auxiliary power. Vigan et al. (2010) used ASR with 19.2 GJ/mg LHV for sequential gasification and syngas combustion with a thermal power input of 100 MW (approximately 140,000 mg/y) and steam cycle parameters equal to 30 bar/325°C. The estimate of net electric efficiency was 22.5% and a gross electric efficiency of 25.7% after accounting for internal consumption.

Lombardi et al. developed a power plant that was based on the high-temperature gasification of RDF in pure oxygen with ash melting and a steam cycle with steam characteristics of 56 bar/405°C, mimicking an existing Italian facility (Lombardi et al., 2012). The air separation unit consumes roughly 20% of the total power required to produce pure oxygen, resulting in a calculated net efficiency of approximately 13% in this instance. Similar to waste incineration, the steam cycle can be configured to operate in cogeneration mode, with the same beneficial effects that were previously highlighted, and the discussion about how to increase the electric efficiency of the steam cycle when gasification with syngas combustion in a boiler is evidently analogous.

For instance, the Metso waste gasification plant in Lahti, Finland, which uses a circulating fluidized bed operating at 850°C, can use improved steam parameters of 121 bar/540°C without running the risk of high temperature corrosion because of a fairly demanding treatment of the syngas based on its cooling to 400°C and hot gas filtration by ceramic filters, which allows reaching a net electric efficiency of about 31%. The facility converts 250,000 mg of SRF into 50 MW of electricity and 90 MW of district heating every year.

The operation and performance of waste gasification plants on a commercial scale are rarely documented in detail because many gasification technologies are still in the development or pilot stages. Also, while a lot of studies focus primarily on the gasification of various inputs, very few go into detail about how syngas is used downstream and how the gasification and power production sections are integrated.

3.4.2 GASIFICATION: APPLICATION OF SYNGAS IN INTERNALLY FIRED FURNACES/DEVICES

As an alternative to what was depicted in the preceding paragraph, a highly efficient internally fired cycle, such as the GT and GTCC or internal combustion/reciprocating engines, may also be fed the syngas after adequate treatment (ICE). The use of a gaseous fuel in highly efficient equipment opens up the possibility of increasing the process's total energy efficiency. According to Consonni and Viganò (2012), the efficiency of GT-based systems at the small scale typical of waste treatment plants is limited to values similar to those previously observed for incineration or gasification due to the need for demanding syngas treatment, atmospheric pressure gasification, and the relatively low efficiency of GT-based systems.

When syngas is fed to GT, particular restrictions on the amount of pollutants present apply: the following are the limits for ICE: for tar: for dust: 10 mg/m_N^3 for alkalis, 5 mg/m_N^3: for heavy metals, 0.1 ppm in weight: 0.1 parts per million in weight, and for hydrogen sulphide: volume: 20 ppm. According to Arena (2012), the volume contains 100 mg/m_N^3 of tar, 50 mg/m_N^3 of dust, 0.1 ppm of alkalis, 0.1 ppm of heavy metals, and 20 ppm of hydrogen sulphide. Simulations of how syngas is used in highly efficient internally fired cycles are sometimes used instead of the actual syngas pretreatment that is required to produce syngas of a sufficient quality for feeding such devices in order to assess the potential amount of energy produced.

Yassin et al. (2009) report that when ICE or GTCC is combined with fluidized bed gasification, it compensates for RDF's mass and energy for gasification (LHV 16.7 GJ/mg). The authors show that when a GTCC is used, the net electric efficiency is between 24% and 27%, while when an ICE is used, it is between 23% and 25%

(lower values for 50,000 mg/y throughput or 29.4 MW thermal power input; higher values for 58.8 MW thermal power input or 100,000 mg/y throughput).

Lombardi et al. (2012) also look at the energy efficiency of using syngas from RDF high-temperature gasification (64 MW of thermal power input) in a GTCC with pure oxygen and ash melting by means of simulation. In this instance, the final net electric efficiency is 14.7%. This is primarily due to internal consumptions, which consume approximately 69% of the generated power, specifically the power required for air compression (approximately 52%). The same authors show that one way to improve overall efficiency is to improve integration between the power section and the gasification section. The cooling of the syngas prior to the cleaning process produces heat that can be recovered and used to generate additional steam for the bottoming steam cycle (17.1%) to be injected into the GT combustion chamber (net electric efficiency: 18.7%).

In addition, gasification at pressures higher than atmospheric pressure may increase opportunities to increase energy conversion efficiency in GT-based cycles and reduce costs; however, a process of this kind would be extremely challenging and is not yet offered by any company (Consonni and Viganò, 2012). Giugliano et al. (2008) offer syngas co-combustion and RDF gasification as specific examples for replacing some of the natural gas in a natural gas combined cycle power plant that has undergone extensive cleaning (including the removal of tar, particulates, hydrogen sulphide, ammonia, and chlorine and fluorine compounds).

Di Gregorio and Zaccariello (2012) investigate the bubbling fluidized bed air blown gasification of fuel obtained from packaging (LHV 23.2 GJ/mg) on a small industrial scale application, and they recommend the use of an ICE following suitable syngas cleaning. It is conceivable to estimate a net electric efficiency of around 15% for a modest system with a net electric power output of about 500 kW and a waste entering flow of about 500 kg/h, or about 4000 mg/y.

3.4.3 PLASMA GASIFICATION

The process of thermal plasma gasification creates syngas and solid residue by partially thermal oxidizing waste. An oxidizing agent can be added to the process as steam, air, or oxygen. The plasma arc's high temperature considerably reduces the possibility of unwelcome compounds in the syngas, and the solid by-products are created in the form of vitrified slag.

The availability of thermal plasma production torches that are more reliable and effective than in the past has likely prompted recent research on the viability of implementing the gasification process using plasma technology. Over the course of the past three decades, a variety of wastes, including ASR, sludge, asbestos fibers, medical waste, and MSW, have been the subject of extensive research and the construction of small-scale plants utilizing plasma technology for the purpose of disposing of organic waste. However, the most important application of the thermal plasma process is the elimination of hazardous pollutants, not energy recovery.

Spraying a plasma jet onto solid waste to accelerate its rapid heating, release volatile materials, and shattering was the initial approach to using thermal plasma to treat waste. A thermochemical model can be developed to determine the syngas composition and required energy for the gasification processes of RDF (12.9 GJ/mg, or roughly 70 MW of thermal power input with waste). After that, the syngas goes to a GTCC, which converts it into electricity with a net efficiency of 31%. Galeno et al. (2011) modelled the application of syngas from plasma gasification of RDF (12.9 GJ/mg, 12.9 MW thermal power input with waste) to a solid oxide fuel cell (SOFC). The net electric efficiency in this instance is 32.7% after accounting for the torch's consumption and air separation.

Based on the use of plasma in a two-stage thermal process, a second method was suggested: traditional gasification of waste is followed by plasma jet refinement of syngas to decompose and convert tar to improve its quality for engine feeding. The plasma gasification of the entire solid waste needed less electricity, which was the goal of this method. The plasma jet can be used to vitrify the gasification of solid leftovers in the future. A system based on the conventional RDF gasification (LHV 18 GJ/mg, around 40 MW of thermal power input with waste), syngas plasma refinement, and use of the refined syngas in an ICE was determined to have a net electric efficiency of 25.3%. The estimated net electric efficiency for a process using pulper leftovers (LHV 21.4 GJ/mg) was 27.4%.

3.5   ADVANCED THERMAL TREATMENT – PYROLYSIS

Pyrolysis is a thermal process that occurs at temperatures higher than 400°C and typically lower than 800°C, heating the incoming feed by an external heat source while there is absolutely no oxygen present. It generates three output streams, all of which have flammable properties: gas, liquid (oil), and solid (char). Since there is no oxygen present, there is no oxidation; instead, the organic substance being fed undergoes thermal decomposition, sometimes with the aid of catalysts. The process temperature and heating rate have a significant impact on the yields of pyrolysis gas, oil, and char. In general, the gaseous fraction increases with temperature. While lower temperatures and rapid heating rates can produce higher oil yields, oil with a high yield is generally of interest because it is portable and easy to store.

While waste pyrolysis has made some progress recently and is primarily at the research/pilot stage, biomass pyrolysis has lagged behind. In particular, several studies on the pyrolysis of specific types of waste (e.g., various plastic wastes, scrap tyres, wood waste, waste electric and electronic equipment RDF) are available, but they primarily concentrate on the pyrolysis process itself rather than on potential future uses of the fuels produced and on an overall energy balance.

Solid waste is decomposed by heating, a process known as pyrolysis, which is anoxic and converts solid waste to liquid bio-oil, gas, and carbon. The pyrolysis process and reaction conditions determine the final product. At lower temperatures, a slow process (500°C) produces a lot of carbon black; at higher temperatures, a high-speed process (700–1100°C) produces primary combustible gas; and at middle temperatures, quick pyrolysis (500–700°C) produces liquid bio-oil. As the moisture level of the feed increases, the amount of solid and gaseous products decreases, allowing more liquid products to be produced.

Pyrolysis is categorized into three types based on operating conditions: conventional, rapid, and flash pyrolysis. Pyrolysis takes place at low temperatures

(300–600°C) with extended residence durations and modest heating rates (0.1–10°C/s). Maximum char generation is favoured in this method of pyrolysis. Fast pyrolysis occurs at high temperatures (500–700°C), fast heating rates (100–1000°C/s), and a short residence time (0.5–5.0 s). Fast pyrolysis converts biomass into vapours, which are then condensed to make bio-oil. For a few seconds, flash pyrolysis occurs at high temperatures (>1000°C) at very rapid heating rates (500–1000°C/s). The main by-products of flash pyrolysis are flue gases.

Temperature, residence time, pressure, and heating rate are all factors that influence pyrolysis. The temperature is important in the pyrolysis and distribution of organic wastes, as well as the formation of burning gas. Low-temperature and long-term maintenance are required for pyrolysis slowly process for efficient carbon generation; in this scenario, mass and energy yields will approach 30% and 50%, respectively. The temperature was lower than 600°C, and the reaction ratio was in the middle; the formation of non-condensable gas, carbon, and bio-oil was nearly equal. The temperature of flash pyrolysis was 500–650°C, with the goal of increasing bio-oil production to 80%. When the temperature exceeds 700°C, the product is mostly gaseous and can approach 80%.

The second most critical factor is the residence time. To totally change the materials at the desired size and reaction temperature, a modest solids hold-up was required. The volatile chemicals created during the pyrolysis process leave the reactor fast, reducing tar cracking time and maximizing bio-oil production. As a result, the gaseous stage retention time was crucial in achieving the highest bio-oil output. The size, shape, and dispersion of biomass are critical for product creation after pyrolysis. When the particle size is less than 1 mm, the reaction kinetic ratio can regulate the process; when the size is greater than 1 mm, the reaction kinetic ratio can control the process, and particles play a vital role in heat conversion. Pressure is the third significant component that affects pyrolysis. Secondary cracking and gas stage sustain time under pressure have influenced the yield distribution. The heating rate is the fourth and most important element in pyrolysis. The low heating ratio encourages the production of char but has drawbacks in the production of tar. As a result, flash pyrolysis is utilized to accelerate the heating rate in order to boost bio-oil output. The generation of tar in addition to syngas is a key flaw in this process, limiting its use in internal combustion engines.

In fact, only one MSW pyrolysis plant is in operation in Europe, and it has been in Burgau, Germany, since 1986. The technology for the Burgau plant was provided by WasteGen UK Ltd. Each year, the facility processes 38,000 mg of MSW and sewage sludge. To begin, MSW must be shredded to a maximum of 300 mm in size. From its initial value of 8.5 GJ/mg, the feedstock's heating value increased over time to around 10.2 GJ/mg. A rotary kiln, where pyrolysis takes place at a temperature of approximately 500°C, is indirectly heated by using about 20% of the flue gases from the combustion of syngas. The syngas, which has a volume of 636 m^3/mg and contains hydrogen, carbon monoxide, carbon dioxide, methane, and other hydrocarbons, is burned alongside landfill gas that is taken from a nearby landfill at a temperature of more than 1250°C. It has an LHV content of 10–14 GJ/m^3.

About 80% of the flue gases that are fed into the boiler, which was replaced in 2003 to handle the increasing heating value of the MSW, drive a 2.2 MW steam turbine generator. A nearby greenhouse receives pumped condensate and residual steam. However, assuming 38,000 mg/h, 8000 h/y of operation, 10.2 GJ/mg LHV, and a gross power output of 2.2 MW, the gross electric conversion efficiency is approximately 16%. Given that landfill gas adds additional thermal input to the combustion chamber, this figure is obviously exaggerated. There is no information about the efficiency or effective production of electric energy in the cited reference.

Pyrolysis, which is also used to treat MSW, has found great success in Japan, where at the start of the 2000s, multiple plants based on Mitsui Recycling 21 technology were constructed (Wang et al., 2020). This method is based on the processing of garbage in a pyrolysis drum after it has been shredded. In a combustion chamber, carbon and pyrolysis gas are burned. In a steam cycle, the boiler generates steam for power production. As an alternative, cement factories might receive the pyrolysis carbon. Aluminium and steel can also be recovered using this method.

No further MSW pyrolysis facilities are currently planned, other than these two instances, which are based on a traditional steam cycle for the energy recovery phase and do not take advantage of the potential of employing the gaseous fuel in extremely efficient internally fired cycles (Du et al., 2021). When combined with specific waste fractions like plastic, rubber tyres, sewage sludge, and waste wood, pyrolysis is still intriguing.

However, there is still a great deal of research being conducted in this field. Considering start-up time, scrap tyre pyrolysis can supply gases with LHV greater than 40 MJ/mN3 that can fully satisfy the needs of the pyrolysis plant. The solid char can be gasified to produce fuel gases or used as a smokeless fuel, carbon black, or activated carbon. Pyrolytic gas can be burned by itself or in a predetermined proportion with natural gas or propane. The pyrolysis oils from used tires can reach 44 MJ/kg of LHV, but this is contingent on the composition of the waste tires and the conditions of the pyrolysis process. They can be added to the feedstock for petroleum refineries or used directly as liquid fuels, particularly as an alternative fuel in compression ignition engines (Martinez et al., 2013). Frigo et al. (2014) determined whether leftover tire oil can be blended with diesel fuel to power diesel engines. When 20% scrap tyre oil and 80% diesel are combined (on a volume basis), engine performance does not change.

In a similar vein, Kumar et al. (2013) investigated the performance of diesel-oil blends produced by catalytic pyrolysis of used high-density polyethylene. Singhabhandhu and Tezuka (2010) looked at the integrated co-processing of waste plastics, waste lubricating oil, and waste cooking oil by pyrolysis in order to create pyrolysis oil for energy recovery.

A novel system based on the pyrolysis of MSW (thermal input 34.8 MW, 7200 kg/h) that yields solid char, oil, and gas fuels used in a GT combined cycle for the production of electricity was simulated by Baggio et al. (2008). The combustion of some of the char and oil produces the thermal energy needed for pyrolysis. After compression and filtering, the gas feeds two GTs. After post-combustion with char and oil, the exhaust from the GTs powers a steam turbine cycle. According to the authors, the plant's net electric efficiency is between 28% and 30%.

3.6 CASE STUDIES FOR WASTE-TO-ENERGY RECOVERY

3.6.1 USA

WtE plants are running in 25 states in the USA. They are fed by 26.3 million mg/y of MSW or 7.4% of the total amount of MSW produced. The majority of WtE facilities, or 66% of the country's total WtE capacity, are located on the East Coast. Fifteen WtE plants (5.71×10^6 mg/y) are fed by RDF, while 65 WtE plants (20.05×10^6 mg/y) are mass-burn facilities where the MSW is fed as collected into enormous furnaces (Psomopoulos et al., 2009).

According to a 2009 report, the typical specific net amount of electricity produced by MSW in a contemporary WtE power plant is 600 kWh/mg. The same authors add that, even if a stricter definition of "renewable" is used, which excludes only material from non-fossil sources, approximately 64% of US MSW, after material recovery for recycling and composting, comes from renewable sources. WtE plants are considered to be a kind of biomass plant by the US Department of Energy (US DOE). Psomopoulos et al. (2009) estimate that, excluding hydropower, WtE accounts for 28% of renewable electric energy in the United States.

3.6.2 INDIA

According to Nixon et al. (2013), even though the MSW Management and Handling Rules were implemented in India in 2000, the collection efficiency of MSW is still only around 70%, and the majority of landfills are unsatisfactory, accounting for about 90% of them. According to Sharholy et al. (2008), landfilling, composting (aerobic and vermicomposting), and a relatively small number of waste-to-energy programmes make up the majority of MSW treatment and disposal options in India (incineration, RDF, and biomethanation). According to Unnikrishnan and Singh (2010), the calorific value ranges between 3.3 and 4.2 GJ/mg. According to Singh et al. (2011), a WtE facility in India – which includes biogas generation in the term as well – is not only feasible but also essential to meet the demands of expanding cities and to enhance the environment.

3.6.3 JAPAN

Castaldi and Themelis (2010) say that Japan is the biggest user of this method in the world, processing 40 million mg/y of MSW through thermal treatment. Grate incineration of untreated MSW is the most common method. According to Gohlke and Martin (2007), Japan has not utilized landfills in the past due to a lack of available space. As a result, the majority of garbage is burned in a number of small plants. Even though over 80% of MSW is burned in Japan, only 24.5% of MSW incineration plants use energy recovery, according to Tabata (2013). When energy recovery plants are taken into consideration, the national average power generation efficiency is rather low at 11.3%, producing an average of 200 kWh/mg, with higher values for large installations (300–390 kWh/mg). According to the same author, district heating is only used to heat pools or hot water in public facilities, and heat production is relatively low (on average 0.76 kWh/mg) despite its brief use.

Tanigaki et al. (2012) pointed out that one of the primary objectives of MSW management in Japan is to reduce landfill use, which is a key issue and the reason for the heavy reliance on thermal treatment even in the absence of energy recovery. Tanigaki et al. (2012) also point out that Japan's stringent waste management regulations typically necessitate melting MSW incinerator bottom ash prior to landfilling it. Processes that produce vitrified residues (ash melting) as an addition to conventional incineration, as major modifications to the incineration process (like oxygen enrichment), or as integrated treatments in high-temperature gasification processes have all been developed as a result of these circumstances. This is the main reason why there are so many commercial MSW gasification facilities operating in Japan for the purpose of recovering materials and energy

3.7 FUTURE PROSPECTS AND CONCLUSIONS

Thermal treatment of waste is an essential component of the integrated waste management system, so it must be included. The opportunity to recover the valuable energy component of waste, which is in any case created every day by local activities and is, in part, essentially renewable, is now combined with the fundamental capability of thermal treatment to reduce waste mass/volume and putrescibility. In this way, garbage is a local, continuously produced energy source that can be found in all regions. It has the potential to grow, improving plant performance and more than making up for the lack of thermal treatment plants in many nations, even though its contribution to global energy production and consumption is currently quite low.

In thermal treatment, incineration and energy recovery in a steam cycle dominate. The mobile grate, the most advanced and widely used furnace for trash incineration, has developed significantly. When only talking about the production of power, it is true that the technological level of the designed steam cycle has a direct impact on the energy performances of both plants that use incineration or gasification with the use of syngas in a boiler. This includes maximum steam pressure and temperature as well as complex cycle arrangements. Because investments and profits from selling energy and increasing efficiency must be balanced, large-scale plants are the only ones capable of keeping the steam cycle at its highest technological level. Given this point of view, it is reasonable to assert that, when operated solely in power mode, large-scale incineration or gasification plants can achieve net electric efficiency of between 30% and 31%. Small- to medium-sized incineration or gasification plants typically run on steam at 40–50 bar and 400°C, with a maximum net electric efficiency of around 24%.

However, the gasification process is less dependable than incineration at the equivalent size and steam cycle technology since it is more complicated, expensive, difficult to operate, and maintain. Additionally, a variety of feeding streams for gasification reactors must be fed with pretreated waste. Therefore, pretreatment consumptions and losses must be taken into account in the total energy balance.

For small- to medium-sized plants based on incineration or gasification, cogeneration is the only practical and effective way to increase energy recovery. It also meets some regulatory requirements (the R1 criterion in the EU) and has additional positive environmental effects. Estimates of efficiency based on the use of syngas from gasification in internally fired devices are comparable to those from conventional incineration-based systems according to studies. The use of thermal plasma in a two-step process that involves conventional gasification and syngas plasma refinement to produce clean syngas seems promising, but the scientific community needs more information about pilot operations.

Pyrolysis is only used to recover energy from a select few waste streams. Pure and uniform waste streams are especially necessary for the production of high-quality oil that can be used in extremely efficient energy conversion technologies. Additionally, it is essential to investigate the possibilities of incorporating the produced fuels into such devices.

It was discovered that there was a general lack of published information regarding the performances of plants. The authors of this chapter focused on the scientific literature on energy recovery from waste which could be found in scientific journals and conferences, omitting information that could be found in other sources (plant technical reports, personal communications, etc.). With a few exceptions, the authors of this chapter did not include any information that could be found in other sources.

REFERENCES

Arena, U., 2012. Process and technological aspects of municipal solid waste gasification. A review. Waste Manage. 32, 625–639.

Baggio, P., Baratieri, M., Gasparella, A., Longo, G.A., 2008. Energy and environmental analysis of an innovative system based on municipal solid waste (MSW) pyrolysis and combined cycle. Appl. Therm. Eng. 28, 136–144.

Barigozzi, G., Perdichizzi, A., Ravelli, S., 2011. Wet and dry cooling systems optimization applied to a modern waste-to-energy cogeneration heat and power plant. Appl. Energy. 88, 1366–1376.

Bebar, L., Stehlik, P., Havlen, L., Oral, J., 2005. Analysis of using gasification and incineration for thermal processing of wastes. Appl. Therm. Eng. 25, 1045–1055.

Castaldi, M.J., Themelis, N.J., 2010. The case for increasing the global capacity for waste to energy (WTE). Waste Biomass Valor. 1, 91–105.

Consonni, S., Viganò, F., 2012. Waste gasification vs. conventional waste-to-energy: a comparative evaluation of two commercial technologies. Waste Manage. 32, 653–666.

Damgaard, A., Riber, C., Fruergaard, T., Hulgaard, T., Christensen, T.H., 2010. Life-cycle-assessment of the historical development of air pollution control and energy recovery in waste incineration. Waste Manage. 30, 1244–1250.

De Greef, J., Villani, K., Goethals, J., Van Belle, H., Van Caneghem, J., Vandecasteele, C., 2013. Optimising energy recovery and use of chemicals, resources and materials in modern waste-to- energy plants. Waste Manage. 33, 2416–2424.

Di Gregorio, F., Zaccariello, L., 2012. Fluidized bed gasification of a packaging derived fuel: energetic, environmental and economic performances comparison for waste-to-energy plants. Energy. 42, 331–341.

Du, Y., Ju, T., Meng, Y., Lan, T., Han, S., Jiang, J., 2021. A review on municipal solid waste pyrolysis of different composition for gas production. Fuel Process Technol. 224, 107026.

European Commission (2016) ICCS-NTUA, IIASA, Eurocare. EU reference scenario 2016. Energy, transport and GHG emission trends to 2050. ICCS-NTUA, IIASA, Eurocare (2016) report for the European commission. 2016. Retrieved from: https://energy.ec.europa.eu/system/files/2016-07/refweb_1_0.pdf.

Frigo, S., Seggiani, M., Puccini, M., Vitolo, S., 2014. Liquid fuel production from waste tyre pyrolysis and its utilisation in a diesel engine. Fuel. 116, 399–408.

Galeno, G., Minutillo, M., Perna, A., 2011. From waste to electricity through integrated plasma gasification/fuel cell (IPGFC) system. Int. J. Hydrogen Energy. 36, 1692–1701.

Giugliano, M., Grosso, M., Rigamonti, L., 2008. Energy recovery from municipal waste: a case study for a middle-sized Italian district. Waste Manage. 28, 39–50.

Gohlke, O., 2009. Efficiency of energy recovery from municipal solid waste and the resultant effect on the greenhouse gas balance. Waste Manage. Res. 27, 894–906.

Gohlke, O., Martin, J., 2007. Drivers for innovation in waste-to-energy technology. Waste Manage. Res. 25, 214–219.

Graus, W., Voogt, M., Worrell, M., 2007. International comparison of energy efficiency of fossil power generation. Energy Policy. 35, 3936–3951.

Graus WHJ, Worrell E (2009). Trend in efficiency and capacity of fossil power generation in the EU. Energy Policy. volume 37 pp 2147–2160. DOI: 10.1016/j.enpol.2009.01.034.

ISWA, 2012 "Globalization and waste management final report from the ISWA task force," International Solid Waste Association, Rotterdam. Retrieved from: https://www. researchgate.net/profile/Antonis-Mavropoulos/publication/275017171_Globalisation_ and_Waste_Management_-_Final_Report/links/552e39980cf22d43716def32/ Globalisation-and-Waste-Management-Final-Report.pdf.

Jegla, Z., Bebar, L., Pavlas, M., Kropac, J., Stehlik, P., 2010. Secondary combustion chamber with inbuilt heat transfer area – thermal model for improved waste to- energy systems modelling. Chem. Eng. Trans. 21, 859–864.

Komilis, D., Kissas, K., Symeonidis, A., 2013. Effect of organic matter and moisture on the calorific value of solid wastes: an update of the tanner diagram. Waste Manage. 34, 249–255. http://dx.doi.org/10.1016/j.wasman.2013.09.023

Kumar, S., Prakash, R., Murugan, S., Singh, R.K., 2013. Performance and emission analysis of blends of waste plastic oil obtained by catalytic pyrolysis of waste HDPE with diesel in a CI engine. Energy Convers. Manage. 74, 323–331.

Lee, S.H., Themelis, N.J., Castaldi, M.J., 2007. High-temperature corrosion in waste to-energy boilers. J. Therm. Spray Tech. 16, 104–110.

Liuzzo, G., Verdone, N., Bravi, M., 2007. The benefits of flue gas recirculation in waste incineration. Waste Manage. 27, 106–116.

Lombardi, L., Carnevale, E., 2015. A review of technologies and performances of thermal treatment systems for energy recovery from waste. Waste Manage. 37, 26–44.

Lombardi, L., Carnevale, E., Corti, A., 2012. Analysis of energy recovery potential using innovative technologies of waste gasification. Waste Manage. 32, 640–652.

Main, A., Maghon, T., 2010. Concepts and experiences for higher plant efficiency with modern advanced boiler and incineration technology. In: Proceedings of the 18th Annual North American Waste-to-Energy Conference NAWTEC18 May 11–13, 2010, Orlando, Florida, USA.

Malkow, T., 2004. Novel and innovative pyrolysis and gasification technologies for energy efficient and environmentally sound MSW disposal. Waste Manage. 24, 53–79.

Martinez, J.D., Puy, N., Murillo, R., Garcia, T., Navarro, M.V., Mastral, A.M., 2013. Waste tyre pyrolysis – a review. Renew. Sustain. Energy Rev. 23(2013), 179–213.

Murer, M.J., Spliethoff, H., De Waal, C.M.W., Wilpshaar, S., Berkhout, B., Van Berlo, M.A.J., Gohlke, O., Martin, J.J.E., 2011. High efficient waste-to-energy in Amsterdam: getting ready for the next steps. Waste Manage. Res. 29, 20–29.

Nixon, J.D., Dey, P.K., Ghosh, S.K., Davies, P.A., 2013. Evaluation of options for energy recovery from municipal solid waste in India using the hierarchical analytical network process. Energy. 59, 215–223.

Pavlas, M., Tous, M., Klimek, P., Bebar, L., 2011. Waste incineration with production of clean and reliable energy. Clean Technol. Environ. Policy. 13, 595–605.

Persson, K., Broström, M., Carlsson, J., Nordin, A., Backman, R., 2007. High temperature corrosion in a 65 MW waste to energy plant. Fuel Process.Technol. 88, 1178–1182.

Psomopoulos, C.S., Bourka, A., Themelis, N.J., 2009. Waste-to-energy: a review of the status and benefits in USA. Waste Manage. 29, 1718–1724.

Reimann, C., Filzmoser, P., Fabian, K., Hron, K., Birke, M., Demetriades, A., Dinelli, E., Ladenberger, A., Albanese, S., Andersson, M., Arnoldussen, A., Baritz, R., Batista, M. J., Bel-lan, A., Cicchella, D., De Vivo, B., De Vos, W., Duris, M., Dusza-Dobek, A., Zomeni, Z. (2012). The concept of compositional data analysis in practice — Total major element concentrations in agricultural and grazing land soils of Europe. Science of The Total Environment, 426, 196–210. DOI: 10.1016/j.scitotenv.2012.02.032

Reimann, D.O., 2009. CEWEP Energy Report II (Status 2004–2007). https://www.cewep. eu/wp-content/uploads/2013/01/401_09_04_29_final_version_CEWEP-Report.pdf

Richers, U., Vehlow, J., Seifert, H., 1999. Evaluation Program for Municipal Solid Waste Incineration Plants. FZKA 6298. Institut für Technische Chemie – Bereich thermische Abfallbehandlung. Forschungszentrum Karlsruhe GmbH, Karlsruhe.

Ruth, A.L., 1998. Energy from municipal solid waste: a comparison with coal combustion technology. Prog. Energy Combust. Sci. 24, 545–564.

Sharholy, M., Ahmad, K., Mahmood, G., Trivedi, R.C., 2008. Municipal solid waste management in Indian cities – a review. Waste Manage. 28, 459–467.

Singhabhandhu, A., Tezuka, T., 2010. The waste-to-energy framework for integrated multi-waste utilization: waste cooking oil, waste lubricating oil, and waste plastics. Energy. 35, 2544–2551.

Singh, R.P., Tyagi, V.V., Allen, T., Ibrahim, M.H., Kothari, R., 2011. An overview for exploring the possibilities of energy generation from municipal solid waste (MSW) in Indian scenario. Renew. Sustain. Energy Rev. 15, 4797–4808.

Tabasová, A., Kropác, J., Kermes, V., Nemet, A., Stehlík, P., 2012. Waste-to-energy technologies: impact on environment. Energy. 44, 146–155.

Tabata, T., 2013. Waste-to-energy incineration plants as greenhouse gas reducers: a case study of seven Japanese metropolises. Waste Manage. Res. 31, 1110–1117.

Tanigaki, N., Manako, K., Osada, M., 2012. Co-gasification of municipal solid waste and material recovery in a large-scale gasification and melting system. Waste Manage. 32, 667–675.

Tanner, V.R., 1965. Die Entwicklung der Von-Roll-MÃ1=4llverbrennungsanlagen (the development of the Von-Roll incinerators). Schweiz Bauzeitung 83, 251–260. http://dx.doi. org/10.5169/seals-68135 (last accessed May 2014) (in German).

Unnikrishnan, S., Singh, A., 2010. Energy recovery in solid waste management through CDM in India and other countries. Resour. Conserv. Recy. 54, 630–640.

Van Caneghem, J., Brems, A., Lievens, P., Block, C., Billen, P., Vermeulen, I., Dewil, R., Baeyens, J., Vandecasteele, C., 2012. Fluidized bed waste incinerators: design, operational and environmental issues. Prog. Energy Combust. 38, 551–582.

Vigan, A., Mopntou, C., Langlois, M., Allard, F., Bocalettu, A., Carbillet, M., Moiullet, D., Smith, I., 2010. Photometric characterization of exoplanets using angular and spectral differential imaging. Monthly Notices of the Royal Astronomical Society. Volume 407, 1–13.

Wang, G., Dai, Y., Yang, H., Xiong, Q., Wang, K., Zhou, J., Li, Y., Wang, S., 2020. A review of recent advances in biomass pyrolysis. Energy Fuels. 34, 15557–15578.

Yassin, L., Lettieri, P., Simons, S.J.R., Germanà, A., 2009. Techno-economic performance of energy-from-waste fluidized bed combustion and gasification processes in the UK context. Chem. Eng. J. 146, 315–327.

4 Solid Waste Treatment Technologies
Biochemical Pathway

Mostafa M. Besheir and Debleena Bhattacharya

4.1 INTRODUCTION

Sewage sludge is a by-product of wastewater treatment, which poses an environmental problem, especially with the production of increasing quantities of wastewater, which results in more and more production of sewage sludge. There are various sources of wastewater and its content of sewage sludge, which includes many anthropogenic activities such as agricultural, industrial, and domestic (Ahmad et al., 2016). The chemical composition of sewage sludge constitutes of about 70% organic and 30% inorganic compounds, respectively (Templeton and Butler, 2011). Sewage sludge also contains many organic and inorganic pollutants, which vary in their severity and impact, such as heavy metals, polychlorinated biphenyls (PCBs), and polycyclic aromatic hydrocarbons (PAHs), which may cause serious damage to the environment and public health (Delibacak et al., 2020). Recently sewage sludge has shifted from a burden on the environment to a potential resource for many applications. This approach helps in safe disposal of sewage sludge and at the same time optimizing its utilization as a renewable resource. In this chapter, the use of sewage sludge in agriculture and energy production and the pros and cons of using sludge in these applications will be addressed along with the biochemical pathway taken by the sludge.

4.2 SEWAGE SLUDGE PRODUCTION

Sewage sludge is the by-product of a number of wastewater treatments that aim to reduce the level of unwanted substances and bring them to the limits that allow for reuse, and the values of these limits depend on the nature of the intended use (Ghazy et al., 2009; Rorat et al., 2019). There are more than 50,000 wastewater treatment plants in the European Union (EU), and they produce more than 8 million tons of sewage sludge per year (Eriksson et al., 2008). Recent statistics indicate that the annual production of sewage sludge in the European Union countries amounted to 10.9 million dry tons, and the contribution of the old members to it reaches 89.5%, while the new members contribute 10.5%, and this increase is not only due to the number of the population and due to the substantial contribution of the five major industrial countries in Europe, Germany, England, France, Italy and Spain. Together, these countries contributes to more than 65% of the total sewage sludge produced by the European Union countries combined (Kelessidis and Stasinakis, 2012). It was expected that the production of sewage sludge would increase to 13 million tons in the year 2020 compared to the production of 11 million tons in the year 2010, with an increased rate of about 18.2% (Kominko et al., 2017). However, contrary to what is expected, the amount of sewage sludge produced in the European Union increased by 45 million tons, which is equivalent to 2.5 times the expected even before the year 2020 (Zhang et al., 2017).

In developing countries such as India and Egypt with an increasing population and industrial growth there is a huge amount of wastewater generation., India has a daily municipal wastewater treatment capacity of about 11,786 MLD, while the daily production is up to 38,354 MLD; therefore, even using the maximum capacity of the treatment plants, it will not exceed 30.7% of the daily production amount (Kaur et al., 2012). Another study indicated the effect of population growth. The amount of wastewater produced per day reached 42,301 MLD, which means a decrease in the proportion of treated wastewater to produced wastewater per day, reaching 27.9% (Mateo-Sagasta et al., 2015). With regard to industrial wastewater, 60% of it is treated, while the remaining 40% is left untreated, which raises concerns about the release of a number of dangerous pollutants and pathogens into the surface or groundwater systems alike, especially if municipal and industrial wastewater outputs are integrated into a unified drainage system (Kaur et al., 2012; Saha et al., 2017). Reports issued by the United Nations indicated that more than 80% of the wastewater released to surface waters such as rivers and lakes are untreated and further leads to surface water contamination and harmful consumption. Pollution causes the death of 297,000 children every year (UN, 2019; Wu et al., 2021).

In Egypt, there are 303 wastewater treatment plants capable of treating 11,850 MLD, which produce about 1.2 million tons of sewage sludge per year (Ghazy et al., 2009; Wahaab et al., 2020). With the rapid increase in population and industrial growth and the accompanying expansion of the infrastructure to accommodate human and industrial growth there is a steep growth in the daily flow of wastewater, which requires planning to improve the absorptive capacity of the treatment plants already operating or the establishment of more treatment plants, taking into account the emergence of a new challenge represented in the production of more sewage sludge, which necessitates developing future plans for managing sewage sludge and achieving maximum benefit as it being a non-traditional renewable resource (Ayoub et al., 2016).

Biological treatment of wastewater containing several pollutants has been found to be very effective. Biological treatment can be classified into two categories: (i) aerobic treatment and (ii) anaerobic treatment

Aerobic treatment involves using aerobic bacteria which in the presence of oxygen degrade pollutants. The air is diffused into the bioreactor mechanically which is used by the aerobic bacteria. The sludge generated is rich in aerobic microbes.

In anaerobic treatment, the anaerobic bacteria in the absence of oxygen degrade the pollutants. The sludge bed can be divided into three parts. The bottom zone of the sludge bed is known as the active zone where the concentration of short-chain fatty acids is increased to a maximum level because of acidification. High hydrogen partial pressure is generated in this zone because hydrogen is produced during acidification. Under such high hydrogen partial pressure conditions, methanogens, presumably *methanobacterium* strains which use hydrogen only as the energy source, produce all of their amino acids for cell synthesis. The excess production of amino acids induces the organisms to form extracellular long chain polypeptides, which bind the species in the form of granules. In the second zone, called the upper active zone, the partial hydrogen pressure is reduced to a minimum by the action of *hydrogenotrophs*. In the third zone, called the upper inactive zone, there are no observable biological microorganisms.

4.3 CHEMICAL COMPOSITION OF SEWAGE SLUDGE

Sewage sludge is formed as a result of physical, chemical, and biological treatments, where suspended and dissolved compounds are deposited in sewage in a semi-solid form. Sewage sludge constitutes only 0.1% of the wastewater composition, of which organic matter represents 70%, which in turn is further divided into a mixture of organic compounds of biological origin (carbohydrates, proteins, and fats), and the inorganic part of sewage sludge represents 30%, distributed between sediments, minerals, and salts (Templeton and Butler, 2011). The organic portion of sewage sludge rich in carbon (47% by weight) represents the largest proportion of sewage sludge, which in turn contains many chemical elements (Vouk et al., 2017). Some of these are important as elements necessary for plant nutrition, such as nitrogen (N), phosphorous (P), and potassium (K). In general, the organic part of the sewage sludge constitutes more than 50%, which in turn contains the largest part of the carbon (96–99%), while the inorganic carbon in the sewage sludge constitutes the remainder of the total carbon content (1–4%), and the organic nitrogen constitutes the majority sewage sludge nitrogen, due to the fact that nitrogen, along with carbon and hydrogen, forms the basic structure from which organic compounds are built (Delibacak et al., 2020). In addition to NPK and C, there are a number of micronutrients that must be available to the plant in trace concentrations such as iron (Fe), copper (Cu), and zinc (Zn), so sewage sludge is a good source that provides these elements to the plant, and due to the presence of organic matter, which decays very slowly, it provides these nutrients over a long period of time; however, due to the presence of microelements in concentrations higher than the required levels, which are chemically considered heavy metals, in addition to other heavy metals hazardous trace elements such as cadmium (Cd), lead (Pb), and rare earth elements such as scandium (Sc), there is a negative effect on the environment due to their toxicity, which necessitates the management of sewage sludge in a right way (Hopcroft, 2015; Iticescu et al., 2018). The sewage sludge content of heavy and trace metals varies between 0.5% and 2%, which may reach 4% in some cases (Cai et al., 2007). Some older studies indicated that the sewage sludge content of the elements that make up the basic fertilizer elements is relatively constant, while the trace minerals, which include micronutrients, vary in their concentrations in the sewage sludge. This discrepancy appeared clearly in cadmium, copper, nickel, lead, and zinc. The total concentrations of these elements in the sewage sludge showed a positive correlation with their concentrations in the soil solution if the sewage sludge was applied as an organic fertilizer. This may be due to the release of organic acids in the sewage sludge, which may lead to lowering the acidity of the soil and increasing the availability of these elements to the plant (Bradford et al., 1975; Sommers et al., 1976).

The danger of these elements is not only due to their toxicity but also to the fact that they are not degradable and therefore have a long residence period. The content of sewage sludge from dangerous materials is not limited to heavy metals but includes carcinogens such as aromatic hydrocarbons (PAHs) and pathogens (Fijalkowski et al., 2017). PAHs are characterized by their high toxicity, long stay period, and cumulative effect; they are chemically inert and are hardly affected by acids and oxidizing agents. Many research results have shown that there are a number of compounds belonging to this group, which have been listed as dangerous pollutants in many countries, especially with evidence that it is a carcinogen (Zhai et al., 2011). The combustion of fossil fuels and their derivatives and materials of biological origin such as wood is the primary source of PAHs (Soclo et al., 2000). The concentration of PAH depends on more than one factor, including wastewater sources and treatment methods; a variation in the concentration of these substances is noted according to the type of treatment. In the case of primary treatment, the concentrations of PAHs were lower than the concentrations in the case of secondary treatment (Zhai et al., 2011).

4.3.1 Type of Sludge

The presence of bacterial culture and carrier materials for bacterial attachment is essential for the inhibition and stimulation of sludge aggregation (Lettinga and Pol, 1986). Hulshoff Pol et al. (1983) have reported that well digested sludge is characterized by low residual methanogenic activity (<0.6 kg CH_4-COD/m^3.d) and good settleability (SVI, 50 ml/g VSS). The digested sewage sludge with total solids less than 40 kg/m^3 usually has higher methanogenic activity than the thicker type with total solids greater than 60 kg/m^3. The sludge with a concentration of 30–40 kg/m^3 exhibits the highest methanogenic activity.

Activated sludge also has enough methanogenic bacteria and can be used as an alternative to digested sewage sludge (Guyot et al., 1990; Wu et al., 1987). Successful cultivation of granules using digested sludge, activated sludge, and cow manure has been reported for a variety of wastewater over a wide temperature range from 20°C to 55°C (Hulshoff Pol et al., 1983; Wu et al., 1987).

The schematic of the process in accordance with the above description is depicted in Figure 4.1. Stronach et al. (1986) have given the details of various groups of bacteria mediating these reactions during the biological conversion of organic matter to its final end products. It should be noted that the active metabolism of each group of bacteria is interdependent, and hence the whole process is an integrated one.

4.3.1.1 Hydrolysis

As bacteria, in general, can only take up organic matter in soluble form, microbial assimilation of heterogeneous particulate biopolymer requires breakdown or hydrolysis as the first step. This process is mediated by extracellular enzymes like hydrolase, lipase, protease, and cellulose (Stronach et al., 1986).

The reaction rates of the hydrolytic enzymes are influenced by pH, mean cell residence time (MCRT), and composition of the reacting substrate. At constant pH and temperature, the solubilization of particulate organic carbon in domestic sludge during the acidogenic phase of digestion is approximately the first-order reaction in respect of the remaining biodegradable particulate substrate, in keeping with Eastman and Ferguson function as an empirical one that reflects the cumulative effect of all the bacterial reactions occurring when the sludge is being digested. However, the first-order kinetics may be suitable for complex substrates (Gujer and Zehnder, 1983; Stronach et al., 1986) Proteolytic bacteria reportedly fulfill an important role in the hydrolysis of raw sewage sludge. Most of the proteolytic organisms isolated from digesters have been clostridia and anaerobic cocci. The proteolytic enzymes of the hydrolytic bacteria are stable in the pH range of 5.0–11.0 and their actions could be highly specific or broad ranging.

4.3.1.2   Fundamentals of Anaerobic Digestion Process

The anaerobic biological conversion of organic waste matter to methane is a complex process involving several groups of bacteria carrying out rather specific reactions. This process is described as a multi-step process. According to Gujer and Zehnder (1983), the anaerobic bio-conversion involves the following steps (Figure 4.1):

   i. Hydrolysis of proteins, lipids, and carbohydrates
   ii. Fermentation of sugars and amino acids
  iii. Anaerobic beta-oxidation of long chain fatty acids and alcohols
  iv. Anaerobic beta-oxidation of intermediates such as volatile fatty acids (with the exception of acetate)
   v. Conversion of acetate to methane
  vi. Conversion of H_2 to methane

Fermentation and Anaerobic Beta-Oxidation

The breakdown products of the hydrolysis phase of biopolymer degradation form the substrates of the intermediary stages of the anaerobic digestion process. The acid-forming bacteria are the predominant microflora and acetate production is the principal result of their activity although higher volatile acids such as propionate, butyrate, iso-butyrate, valerate, and iso-valerate may also be formed. The major routes of product fermentation are either fermentation or anaerobic beta-oxidation process (Stronach et al., 1986).

4.3.1.2.1   Fermentation

Fermentation in the present context has been defined by Gujer and Zehnder (1983) as a microbial process in which organic compounds act as both electron donors and electron acceptors. During fermentation, amino acids and sugars are converted usually to short-chain fatty acids, alcohols, ammonia, carbon dioxide, and hydrogen. The source of hydrogen is reported to be the metabolism of sporulating clostridia and many nonsporulating anaerobes, hydrogen being generated from a reduced non-heme iron protein called ferredoxin by mediation of enzyme hydrogenase (Gujer and Zehnder, 1983).

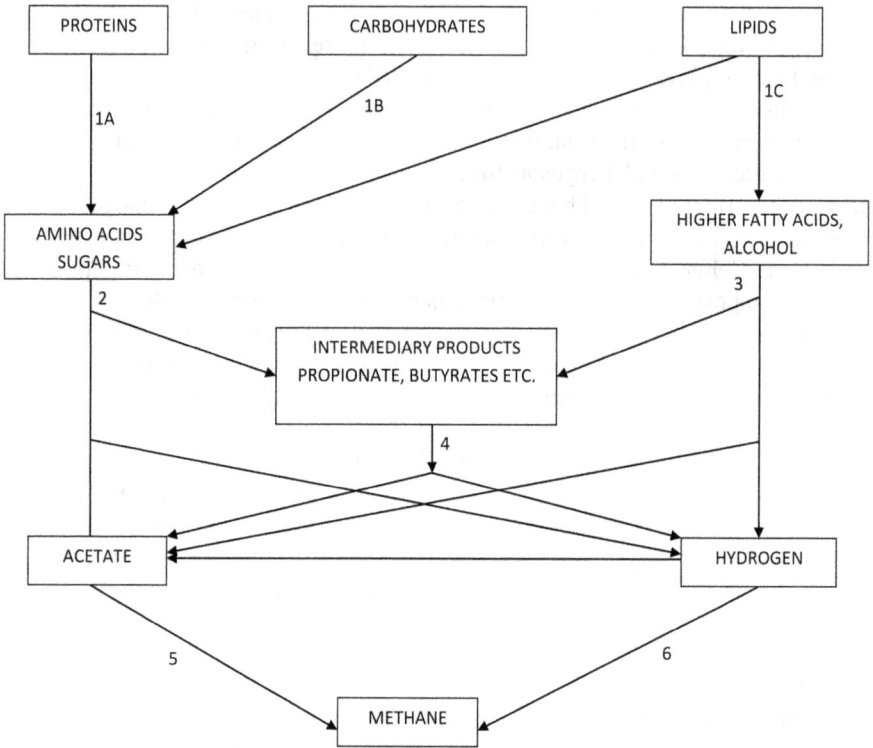

FIGURE 4.1 Pathway of anaerobic biodegradation.

1. HYDROLYSIS
2. FERMENTATION
3. ANAEROBIC BETA-OXIDATION
4. ANAEROBIC OXIDATION
5. DECARBOXYLATION
6. HYDROGEN OXIDATION

4.3.1.2.2 Anaerobic Beta-oxidation

Gujer and Zehnder (1983) defined anaerobic beta-oxidation as a microbial process in which molecular H_2 is the major sink for electrons.

The beta-oxidation process results in successive removal of acetate units from long chain fatty acid. The primary step in the reaction is the activation of a fatty acid by its transformation into corresponding CoA thioester, an ATP-dependent process. Beta-oxidation is a cyclic process with one mole of acetyl-CoA released at every completed turn of the cycle. In contrast to hydrogen fermentation, this process is inhibited by the elevated partial pressure of H_2 (Stronach et al., 1986).

Mosey (1983) presented H_2 regulated catabolic pathways possible for the conversion of glucose in an anaerobic process. Accordingly, during the course of oxidation, glucose are transferred first to the carrier molecule NAD^+, converting it to ($NADH^+$ H^+) and then released into solution as dissolved H_2 gas. It was hypothesized that the relative availabilities of the reduced (NADH) and oxidized forms (NAD^+) of carrier nicotinamide adenine dinucleotide control both the overall rate of conversion and the composition of the mixture of acids formed. Thus, in order for catabolism to proceed continuously in a constant pool of NAD^+, NADH produced during substrate-level phosphorylation of glyceraldehydes-3-phosphate and oxidative decarboxylation of pyruvic acid to acetyl-CoA must be regenerated. This function is achieved by the reduction of protons to form H_2 gas, which is subsequently removed by H_2 consuming bacteria (Harper and Pohland, 1986).

The importance of H_2 concentration in process control could be well appreciated by considering energy of the beta-oxidation.

Groups of bacteria responsible for anaerobic beta-oxidation called obligate hydrogen producing acetogens (OHPA) are thus syntrophically associated with methanogenic bacteria (or other hydrogen consuming bacteria) allowing the latter to assimilate the H_2 produced by the former. H_2 and acetate synthesized by OHPA have been estimated to provide the substrate 54% of the total methane produced in anaerobic systems. OHPA are intolerant of pH fluctuations (optimal pH = 7.0) and their doubling times are of the order of 2–6 days.

4.3.1.3 Methane Formation

The last two steps of anaerobic digestion consist of methane formation by decarboxylation of acetate and reduction of CO_2 by H_2. CH_4 produced escapes the system as it is insoluble in water whereas CO_2 either escapes as gas or is converted to bicarbonate alkalinity. Thus, CH_4 and CO_2 form principal constituents of gas produced during these steps.

The composition of gas depends mainly on the mean oxidation state of carbon in the organic matter, the CO_2 saturation of the reacting liquor, and the nitrogen content of the organic matter (Gujer and Zehnder, 1983). The presence of sulfate in the reacting liquor may give rise to H_2S in evolved gas.

Thus sewage sludge plays an eminent role in most of the toxicant biodegradation, but there are certain legislative rules to control the disposal of such useful resource.

4.4 SEWAGE SLUDGE LEGISLATIONS

During recent decades, there has been an increase in the generation of sewage sludge especially with the upward increase in the number of beneficiaries of wastewater treatment plants (Stylianou et al., 2008). Data indicates in this regard that the annual production in the EU27 countries is estimated at about 10 million tons (Inglezakis et al., 2014).

In view of the chemical composition of sewage sludge which mainly consists of organic matter, in addition to the possibility of relatively high levels of organic and inorganic pollutants, there are also a number of pathogens; despite this the sewage slidge is still considered as a good raw material for a number of promising applications. Among the available alternatives for sludge management is the production of sludge compost and/or energy production. The process of producing sludge compost in addition to other traditional methods such as landfill are of the least expensive methods, but they are less efficient as there is a possibility of pollution, which is a disadvantage for these methods. In order to avoid these issues that may result from this, countries, especially the European Union, have developed environmental legislations and directives, which represent the legal framework for sludge management in various potential applications. These directives establish a set of determinants that govern sludge management to achieve the maximum possible benefit and reduce the damage resulting from the use of sludge.

In the second half of the last century, several countries paid attention to the role played by sludge as a source of environmental pollution. On July 15, 1975, Directive 75/442/EEC was issued, in which no reference was made to the application of sewage sludge in agriculture as fertilizer, which is a clear shortcoming in the relevant environmental legislation. this has been remedied legislative shortcomings enacted a set of environmental directives and laws since then (Tomlinson, P. 1986)). There are many directives related to wastewater treatment and sludge management among these legal frameworks: Directive 86/278/EEC for the use of sewage sludge in agriculture; Directive 91/271/EEC for urban wastewater treatment; Directive 96/61/EC of reducing pollutants; Directive 99/31/EC of landfilling; and Directive 2000/60/EC for the protection of surface and groundwater (Inglezakis et al., 2014). Council Directive 91/271/EEC aims to protect surface and ground natural waters from pollutants of all kinds as well as pathogens, and the possibility of using sludge in crop production, taking into account preventing or at least reducing direct and indirect damages that could be caused to the soil, the directive works likewise to establish a clear concept and to form a mental image of the concept of sludge, as article 14 of the directive indicated that "the sludge arising from wastewater treatment shall be reused whenever appropriate; disposal routes shall minimize the adverse effects on the environment" (Directive, E.U.W., 1991.)). In Directive 99/31/EC, it was indicated that the landfill for biodegradable waste in Article 2 is degradable organic waste in case it is subjected to aerobic and anaerobic decomposition, and this waste extends to include food residues and garden waste in addition to sewage sludge; through the strict application of Directive 99, the landfill contribution was reduced to gradually to reach 35% by the year 2016 Inglezakis et al., 2014).

Among all the previous directives, Directives 86/278/EEC and 2000/60/EC are those that deal directly with sewage sludge. Article 14 of Directive 91/271/EEC referred to the establishment of the concept of sludge, and Directive 86/278/EEC, article 2, paragraph A, previously referred to sludge as "residual sludge from sewage plants treating domestic or urban wastewaters and from other sewage treating wastewaters of a composition similar to domestic and urban wastewaters," and according to Directive 86/278/EEC, limit values for heavy metals have been established (Table 4.1), where the ANNEX IB of the directive shows these limit values for a selected number of heavy metals (EEC, 1986). If the concentrations of heavy metals exceed the maximum limit indicated in Table 4.1, they become unsuitable for use in agriculture (Kelessidis and Stasinakis, 2012). With reference to heavy metals as potential pollutants in sludge, no reference was made to organic pollutants (Salado et al., 2009).

TABLE 4.1
Limit Values of Heavy Metal Concentrations in
Sewage Sludge for Use in Agriculture (mg/kg DM)

Parameter	Limit Values
Cd	20–40
Cu	1000–1750
Ni	300–400
Pb	750–1200
Zn	2500–4000
Hg	16–25
Cr	–

4.5 APPLICATIONS OF SEWAGE SLUDGE

4.5.1 AGRICULTURE

Soil is the natural environment for plant growth, which is characterized by a set of physical, chemical, and biological properties in a way that makes it able to provide the appropriate medium for the healthy growth of crops (Weil and Brady, 2017). Soil organic matter is considered one of the most important factors that have a great impact on soil properties such as porosity and the ability to hold ground moisture and provide an incubating media for the growth of beneficial soil microbes (Osman, 2012). The level of organic matter is proportional to the soil's ability to provide the ground moisture in an appropriate manner, thereby allowing the supply of agricultural crops with quantities of water to suit their needs. Table 4.2 shows that the increase in soil organic matter from 0.12% to 4.33% leads to an increase in the soil's ability to hold ground moisture approximately seven times compared to the initial level (Hartmann and De Boodt, 1974). Because the sewage sludge is very rich in organic matter, it behaves in the same way that works to modify the physical properties of the soil, which leads to improving its ability to retain water (Ramulu, 2002). Sewage sludge

TABLE 4.2
Effect of Organic Matter Level on Critical Moisture Content
and Bulk Density (Hartmann and De Boodt 1974)

Texture	Organic Matter (%)	Critical Moisture Content W (%)	Bulk Density (g cm^{-3})
Sandy soil	0.12	1.9	1.480
	1.62	7.4	1.230
	2.10	8.7	1.237
	4.33	12.6	0.979

greatly affects the accumulative pattern and cohesion of the sandy soil, which has a coarse texture, thus makes the accumulative pattern closer to the sandy loam soil, to make it more cohesive and less affected by erosion factors (Delibacak et al., 2020; Hartmann and De Boodt, 1974).

Several studies showed an inverse relationship between soil organic matter content and soil bulk density (Ojeda et al., 2003). The increase in organic matter from 0.12% to 4.33% result in a clear decrease of 33.8% (Hartmann and De Boodt, 1974). In a study of the sewage sludge application conducted in different regions of Sweden (Table 4.3), the results showed that by adding sludge as an organic matter by 4, 8, and 12 mg, the values of the bulk density of the soil decreased by 28.7%, 5.8%, and 4.2%, respectively, compared to the control in each case (Kirchmann et al., 2017).

TABLE 4.3

Yield, N Use Efficiency, Soil Balances of N and P, and Bulk Density in the Four Swedish Long-Term Field Experiments with Sewage Sludge (Kirchmann et al., 2017)

Site, Years, and Soil Treatments	Mean Yield (kg h^{-1})	Nutrient Application (kg ha^{-1} y^{-1}) N P		Nutrient Removal (kg ha^{-1} y^{-1}) N P		ΔN in Soil	ΔP in Soil	Bulk Density (2009)
Ultuna 2002–2009								
Control	3329	0	20	35	8	−35	+12	1.43
Mineral fertilized	7176	80	20	95	15	−15	+5	1.28
Sewage sludge treated, 4 mg dry matter ha^{-1} every 2nd year	9719	276	233	146	19	+131	+215	1.02
Lanna 1996–2009								
Control	1316	0	0	19	4	−19	−4	1.38
Mineral fertilized	3407	80	20	57	11	+23	+9	1.36
Sewage sludge treated, 8 mg dry matter ha^{-1} every 2nd year	3450	236	194	63	12	+173	+183	1.30
IgelÖsa 2006–2009								
Control	4038	0	0	96	26	−96	−26	NA
Mineral fertilized	8010	128	19	154	38	−26	−19	NA
Sewage sludge treated, 12 mg dry matter ha^{-1} every 4th year	4838	105	171	110	29	−5	+141	NA
Sewage sludge + N treated, 12 mg dry matter ha^{-1} every 4th year	7698	232	190	161	38	+71	+152	NA
Petersborg 2006–2009								
Control	3308	0	0	79	21	−79	−21	1.68
Mineral fertilized	6850	139	23	167	41	−28	−18	1.59
Sewage sludge treated, 12 mg dry matter ha^{-1} every 4th year	4235	132	105	98	25	+75	+80	1.61
Sewage sludge + N treated, 12 mg dry matter ha^{-1} every 4th year	7330	271	128	196	43	+34	+84	1.56

Fertility is the most important determinant of the productivity of agricultural lands, but the shift to intensive agriculture leads to the depletion of nutrients from the soil, and this leads to a decline in crop yields and the consequent deficit in the food supply, especially with the steady increase in the global population. All this requires the supply of plant nutrients to the soil to compensate for the deficit in the natural fertility of the soil through manufactured fertilizers (Logan and Harrison, 1995). In addition to the role of drainage sludge in improving the physical properties, it plays an additional role by increasing the soil content of macronutrients and micronutrients, which gives importance to sewage sludge as a fertilizer rich in the essential nutrients. Currently, 34% of the drainage sludge production is consumed to improve soil fertility and provide part of plant needs (Ekane et al., 2021). The level of basic fertilizer elements NPK in 100 kg of sewage sludge reaches 1.4 kg, 1.3 kg, and 0.95 kg, respectively, while micronutrient concentrations such as Fe, Cu, Zn, and Mn reach 49.56, 13.50, 15.24, and 21.47 mg kg^{-1} (Jatav et al., 2018).

Although the yield increases with the increase in the amount of sewage sludge added to the soil (Table 4.3), the rate of increase decreases with the increase in the amount of sewage sludge. This inverse relationship between the yield and the rate of increase may be due to the increase in heavy metals and other pollutants, which negatively affects plants and leads to reduced productivity per unit area of agricultural land. The residual phosphorous and nitrogen in the soil after harvesting in the case of using sewage sludge is greater if compared to the residuals in the case of applying mineral fertilizers, and this is mostly due to the slow decomposition of the organic matter over a long time. Thus, the phosphorous and nitrogen associated with the organic matter slowly separate (Kirchmann et al., 2017).

4.5.2 BIOFUELS PRODUCTION

With the industrial development during the recent period, the demand for energy has increased, which represents the heart of any development, and with it the search for new alternatives to reduce dependence on fossil fuels, which affects the environment, has increased the fossil fuels release 15 billion tons of carbon dioxide (CO_2) into the atmosphere annually (Chandra et al., 2012). Carbon dioxide, along with other gases such as nitrous oxide (N_2O) or methane (CH_4), is dangerous to the atmosphere and the environment as a whole, the totally of these gases are known as greenhouse gases (GHGs) (Rutz and Janssen, 2007). It is primarily responsible for climate extremism, which poses a real threat to the environment and requires measures to be taken to reduce it (Ahmed and Huddersman, 2022; Djandja et al., 2020). Sewage sludge is one of the forms of biomass that enhances energy economics and mitigates the effects of GHGs (Hsu et al., 2011). It consists of a group of organic compounds of biological origin, such as carbohydrates, proteins, and fats, and a wide range of microorganisms used in the production of a number of types of biofuels, which can be considered potential alternatives to fossil fuels and their derivatives, in order to distinguish sludge sewage has latent energy between its molecules, which is released by the oxidation of sewage sludge, as one gram of it is capable of producing energy estimated at 8300 joules (Bharathiraja et al., 2014b; Rodríguez et al., 2013).

Biodiesel is a renewable alternative to fossil diesel oil (Tarigan et al., 2019). There are a set of characteristics that give biodiesel an advantage, such as high flash point and biodegradability (Centi and Perathoner, 2008). This is in addition to other advantages such as low carbon emissions on the one hand and on the other hand being low toxicity and safe for workers in the production of biodiesel during handling and storage operations (Bharathiraja et al., 2014a). Biodiesel fuel is a substance similar in nature to conventional diesel oil; biodiesel oil is chemically formed through the transesterification process by the interaction of fatty acids or triglycerides with alcohols in the presence of acidic or acidic catalysts to form alkyl esters of fatty acids such as fatty acid methyl esters (FAMEs) (Qi et al., 2016). It is preferable to use acidic catalysts instead of basic catalysts to avoid the saponification process to reduce the free fatty acids (FFAs) content and, in turn, increase the percentage of triglycerides as a substrate for the production of biodiesel; in addition, the separation of the base catalysts is difficult and uneconomical as it increases production costs (McNeff et al., 2008).

Bioethanol is another form of biofuel produced from sewage sludge biomass resulting from the effect of microorganisms such as yeasts on a group of simple or complex sugars (Hsu et al., 2011; Mtui, 2009). Lignocellulose as represented in sewage sludge, consists of cellulose 30–50%, hemicellulose 15–35%, and lignin 10–20%. Bioethanol is produced from a successive series of treatments such as hydrolysis, fermentation, and distillation (Knauf and Moniruzzaman, 2004; Limayem and Ricke, 2012). For the purpose of biological stimulation, a number of microorganisms have been used that have the ability to produce the desired effect, including *Saccharomyces cerevisiae, S. Diastatitus, Kluyveromyces marxianus, E. coli, Zymomonas mobilis, Klebsiella oxytoca*, and *Pichia kudriavzevii* (Yu et al., 2009). Several studies have shown that *S. cerevisiae* is the most effective of them in the production of ethanol (Taouda et al., 2017). In addition to the use of carbohydrates, a trend has emerged to produce ethanol from the fatty substances present in the oil wastewater sludge (OWS), to maximize the use of oily wastewater sludge (OWS), bioethanol was produced from oil extracted (OWS) from the production and refining of edible oil from oil crop seeds, using ethanol and n-hexane as solvents at boiling. Points for obtaining the residue after extraction (RAE) from which the dried pre-treated substrate was obtained, which is the main substrate used in the production of bioethanol, in which a successive series of treatments (hydrolysis, fermentation, and distillation) was carried out to extract bioethanol in its final form; bioethanol, according to the production mechanism, again enters the biodiesel manufacturing line (Figure 4.2), which has resulted in the optimization of the production of two types of biofuels from one raw material (Ngoie et al., 2020).

One of the growing trends in the field of environmentally friendly energy from sewage sludge biomass is the production of biogas (Simpson-Holley et al., 2007).

Chandra et al. (2012) reported that gaseous biofuels are characterized by the following features:

1. Biogas is produced from renewable resources.
2. It produces GHGs at much lower levels than natural gas.
3. Production takes place locally without relying on external supplies, as is the case with natural gas.
4. A successful method in the management of solid waste of biological origin

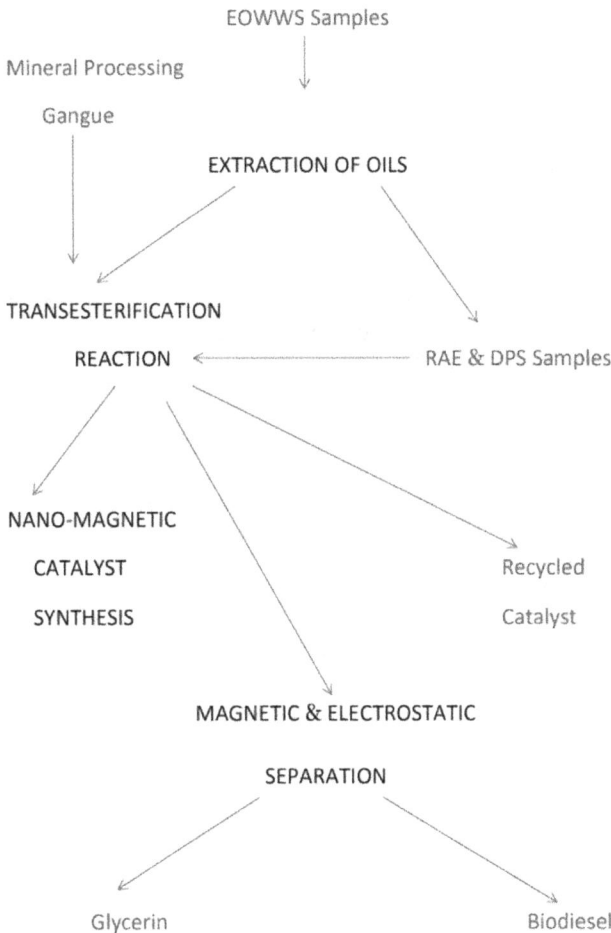

FIGURE 4.2 Bioethanol and biodiesel production from edible oil wastewater sludge EOWW.

4.6 CONCLUSION AND FUTURE PROSPECTS

The global production of biomass is 220 billion dry tons per year. Part of this biomass contributes. Sewage sludge used in energy production represents 24% of the total biomass (Chandra et al., 2012; Hall and Rosillo-Calle, 1998). Biomass produces gases such as 92 of CH_4 and 5.6 kg of hydrogen per ton of biomass, which encourages work on the production of sewage sludge (Neto et al., 2009). Natural emissions of methane are generated by microbial activity on sewage sludge (Rutz and Janssen, 2007).

The perception of sewage sludge has recently shifted from being a mere waste for wastewater treatment to being a potential source for a number of applications through which two benefits are achieved. The first is the disposal of sewage sludge and the second is to take advantage of the characteristics of sewage sludge.

One of the advantages of sewage sludge is that it is rich in its content of organic matter, which improves the natural properties of the soil, which has a great impact on the growth medium of the plant in a way that allows the support of the plant, which makes it grow in a healthy way throughout production season. It is noted that the increase in the percentage of organic matter in the soil is accompanied by an increase in the ability of the soil to hold water, and this is noticed more clearly when the sewage sludge is applied to the coarse sandy soil, which by its nature is unable to retain water, and in this case the organic matter of sewage sludge makes a significant change in this regard; also, other physical properties such as soil porosity and permeability are affected by the soil content of organic matter, which in turn depends on the aggregate model of the soil particles, and bulk density of soil is also affected; therefore, the organic matter in the sewage sludge has an exaggerated effect on these properties.

Returning to the chemical composition of sewage sludge, we find that it contains, in addition to organic matter, a number of elements, some of which represent essential elements for plants, such as nitrogen, phosphorous, and potassium, along with a number of micronutrients. This is in addition to the fact that the sewage sludge contains hazardous elements, such as cadmium, which affect the environment and public health, even if they are found in trace concentrations; the last group, which is considered toxic substances such as cadmium and lead, may accumulate in the soil in the long term due to the repeated application of untreated sewage sludge to the soil.

Among the applications that were used to manage sewage sludge during the last two decades, and more research was conducted on it, is the production of waste types of biofuels, depending on the fact that sewage sludge is one of the forms of biomass that can be exploited in energy production, and the reason is due to the latent chemical energy within molecules of organic matter. Recently, different types of biofuels have been produced depending on the biomass of sewage sludge, such as biodiesel, bioethanol, and biogas, which represents one of the potential alternatives to fossil fuels, as it has characteristics similar to fossil fuels, but it is characterized also by low emissions of greenhouse gases, which gives biofuels produced from biomass, in general, a relative advantage compared to fossil fuels. The biodegradation pathway of the sludge also gives a clear picture of the utilization of the sludge to curtail the generation of solid waste. The degraded sludge acts as an enrichment medium for the soil.

REFERENCES

Ahmad, H.R., Aziz, T., Zia-ur-Rehman, M., Sabir, M. and Khalid, H., 2016. Sources and Composition of Waste Water: Threats to Plants and Soil Health. In Soil Science: Agricultural and Environmental Prospectives (pp. 349–370). Springer, Cham.

Ahmed, R. and Huddersman, K., 2022. Review of biodiesel production by the esterification of wastewater containing fats oils and grease (FOGs). Journal of Industrial and Engineering Chemistry, 110, pp.1–14

Ayoub, M., Rashed, I.G.A.A. and El-Morsy, A., 2016. Energy production from sewage sludge in a proposed wastewater treatment plant. Civil Engineering Journal, 2(12), pp. 637–645.

Bharathiraja, B., Chakravarthy, M., Kumar, R.R., Yuvaraj, D., Jayamuthunagai, J., Kumar, R.P. and Palani, S., 2014a. Biodiesel production using chemical and biological methods—A review of process, catalyst, acyl acceptor, source and process variables. Renewable and Sustainable Energy Reviews, 38, pp. 368–382.

Bharathiraja, B., Yogendran, D., Ranjith Kumar, R., Chakravarthy, M. and Palani, S., 2014b. Biofuels from sewage sludge—A review. International Journal of ChemTech Research, 6(9), pp. 4417–4427.

Bradford, G.R., Page, A.L., Lund, L.J. and Olmstead, W., 1975. Trace element concentrations of sewage treatment plant effluents and sludges; their interactions with soils and uptake by plants. American Society of Agronomy, Crop Science Society of America, and Soil Science Society of America, 4(1), pp. 123–127.

Cai, Q.Y., Mo, C.H., Wu, Q.T., Zeng, Q.Y. and Katsoyiannis, A., 2007. Concentration and speciation of heavy metals in six different sewage sludge-composts. Journal of Hazardous Materials, 147(3), pp. 1063–1072.

Centi, G. and Perathoner, S., 2008. Catalysis by layered materials: A review. Microporous and Mesoporous Materials, 107(1–2), pp. 3–15.

Chandra, R., Takeuchi, H. and Hasegawa, T., 2012. Methane production from lignocellulosic agricultural crop wastes: A review in context to second generation of biofuel production. Renewable and Sustainable Energy Reviews, 16(3), pp. 1462–1476.

Delibacak, S., Voronina, L. and Morachevskaya, E., 2020. Use of sewage sludge in agricultural soils: Useful or harmful. Eurasian Journal of Soil Science, 9(2), pp. 126–139.

Djandja, O.S., Wang, Z.C., Wang, F., Xu, Y.P. and Duan, P.G., 2020. Pyrolysis of municipal sewage sludge for biofuel production: a review. Industrial & Engineering Chemistry Research, 59(39), pp.16939–16956.

Directive, E.U.W., 1991. Council Directive of 21. May 1991 concerning urban waste water treatment (91/271/EEC). European Journal of Communication, 34, p.40.

Ekane, N., Barquet, K. and Rosemarin, A., 2021. Resources and risks: Perceptions on the application of sewage sludge on agricultural land in Sweden, a case study. Frontiers in Sustainable Food Systems, 5, p. 647780.

Eriksson, E., Christensen, N., Schmidt, J.E. and Ledin, A., 2008. Potential priority pollutants in sewage sludge. Desalination, 226(1–3), pp. 371–388.

Fijalkowski, K., Rorat, A., Grobelak, A. and Kacprzak, M.J., 2017. The presence of contaminations in sewage sludge—The current situation. Journal of Environmental Management, 203, pp. 1126–1136.

Ghazy, M., Dockhorn, T. and Dichtl, N., 2009. Sewage sludge management in Egypt: Current status and perspectives towards a sustainable agricultural use. International Journal of Environmental and Ecological Engineering, 3(9), pp. 270–278.

Gujer, W. and Zehnder, A.J., 1983. Conversion processes in anaerobic digestion. Water Science and Technology, 15(8–9), pp. 127–167.

Guyot, J.P., Macarie, H. and Noyola, A., 1990. Anaerobic digestion of a petrochemical wastewater using the UASB process. Applied Biochemistry and Biotechnology, 24(1), pp. 579–589.

Hall, D.O. and Rosillo-Calle, F., 1998. Biomass Resources Other than Wood. World Energy Council, London.

Harper, S.R. and Pohland, F.G., 1986. Recent developments in hydrogen management during anaerobic biological wastewater treatment. Biotechnology and Bioengineering, 28(4), pp. 585–602.

Hartmann, R. and De Boodt, M., 1974. The influence of the moisture content, texture and organic matter on the aggregation of sandy and loamy soils. Geoderma, 11(1), pp. 53–62.

Hsu, C.L., Chang, K.S., Lai, M.Z., Chang, T.C., Chang, Y.H. and Jang, H.D., 2011. Pretreatment and hydrolysis of cellulosic agricultural wastes with cellulase-producing streptomyces for bioethanol production. Biomass and Bioenergy, 35(5), pp. 1878–1884.

Hulshoff Pol, L.W., De Zeeuw, W.J., Velzeboer, C.T.M. and Lettinga, G., 1983. Granulation in UASB-reactors. Water Science and Technology, 15(8–9), pp. 291–304.

Inglezakis, V.J., Zorpas, A.A., Karagiannidis, A., Samaras, P., Voukkali, I. and Sklari, S., 2014. European Union legislation on sewage sludge management. Fresenius Environmental Bulletin, 23(2), pp.635–639.

Iticescu, C., Georgescu, L.P., Murariu, G., Circiumaru, A. and Timofti, M., 2018, November. The characteristics of sewage sludge used on agricultural lands. In AIP conference proceedings (Vol. 2022, No. 1). AIP Publishing.

Jatav, H.S., Singh, S.K., Singh, Y. and Kumar, O., 2018. Biochar and sewage sludge application increases yield and micronutrient uptake in rice (Oryza sativa L.). Communications in Soil Science and Plant Analysis, 49(13), pp. 1617–1628.

Kaur, R., Wani, S.P., Singh, A.K. and Lal, K., 2012, May. Wastewater production, treatment and use in India. In National Report presented at the 2nd regional workshop on Safe Use of Wastewater in Agriculture (pp. 1–13).

Kelessidis, A. and Stasinakis, A.S., 2012. Comparative study of the methods used for treatment and final disposal of sewage sludge in European countries. Waste Management, 32(6), pp. 1186–1195.

Kirchmann, H., Börjesson, G., Kätterer, T. and Cohen, Y., 2017. From agricultural use of sewage sludge to nutrient extraction: A soil science outlook. Ambio, 46(2), pp. 143–154.

Knauf, M. and Moniruzzaman, M., 2004. Lignocellulosic biomass processing: A perspective. International Sugar Journal, 106(1263), pp. 147–150.

Kominko, H., Gorazda, K. and Wzorek, Z., 2017. The possibility of organo-mineral fertilizer production from sewage sludge. Waste and Biomass Valorization, 8(5), pp. 1781–1791.

Lettinga, G. and Pol, L.H., 1986. Advanced reactor design, operation and economy. Water Science and Technology, 18(12), pp. 99–108.

Limayem, A. and Ricke, S.C., 2012. Lignocellulosic biomass for bioethanol production: Current perspectives, potential issues and future prospects. Progress in Energy and Combustion Science, 38(4), pp. 449–467.

Logan, T.J. and Harrison, B.J., 1995. Physical characteristics of alkaline stabilized sewage sludge (N-Viro Soil) and their effects on soil physical properties. American Society of Agronomy, Crop Science Society of America, and Soil Science Society of America, 24(1), pp. 153–164).

Mateo-Sagasta, J., Raschid-Sally, L. and Thebo, A., 2015. Global Wastewater and Sludge Production, Treatment and Use. In Wastewater (pp. 15–38). Springer, Dordrecht.

McNeff, C.V., McNeff, L.C., Yan, B., Nowlan, D.T., Rasmussen, M., Gyberg, A.E., Krohn, B.J., Fedie, R.L. and Hoye, T.R., 2008. A continuous catalytic system for biodiesel production. Applied Catalysis A: General, 343(1–2), pp. 39–48.

Mosey, F.E., 1983. Mathematical modelling of the anaerobic digestion process: Regulatory mechanisms for the formation of short-chain volatile acids from glucose. Water Science and Technology, 15(8–9), pp. 209–232.

Mtui, G.Y., 2009. Recent advances in pretreatment of lignocellulosic wastes and production of value added products. African Journal of Biotechnology, 8(8), pp. 1398–1415.

Neto, T.S., Carvalho, J.A. Jr, Veras, C.A.G., Alvarado, E.C., Gielow, R., Lincoln, E.N., Christian, T.J., Yokelson, R.J. and Santos, J.C., 2009. Biomass consumption and CO_2, CO and main hydrocarbon gas emissions in an Amazonian forest clearing fire. Atmospheric Environment, 43(2), pp. 438–446.

Ngoie, W.I., Oyekola, O.O., Ikhu-Omoregbe, D. and Welz, P.J., 2020. Valorisation of edible oil wastewater sludge: Bioethanol and biodiesel production. Waste and Biomass Valorization, 11(6), pp. 2431–2440.

Ojeda, G., Alcañiz, J.M. and Ortiz, O., 2003. Runoff and losses by erosion in soils amended with sewage sludge. Land Degradation & Development, 14(6), pp. 563–573.

Osman, K.T., 2012. Soils: Principles, Properties and Management. Springer Science & Business Media, Dordrecht, Netherlands.

Qi, J., Zhu, F., Wei, X., Zhao, L., Xiong, Y., Wu, X. and Yan, F., 2016. Comparison of biodiesel production from sewage sludge obtained from the A2/O and MBR processes by in situ transesterification. Waste Management, 49, pp. 212–220.

Ramulu, U.S.S., 2002. Reuse of Municipal Sewage and Sludge in Agriculture. Scientific Publishers, Jodhpur.

Rodríguez, N.H., Martínez-Ramírez, S., Blanco-Varela, M.T., Donatello, S., Guillem, M., Puig, J., Fos, C., Larrotcha, E. and Flores, J., 2013. The effect of using thermally dried sewage sludge as an alternative fuel on Portland cement clinker production. Journal of Cleaner Production, 52, pp. 94–102.

Rorat, A., Courtois, P., Vandenbulcke, F. and Lemiere, S., 2019. Sanitary and environmental aspects of sewage sludge management. In Industrial and Municipal Sludge (pp. 155–180). Butterworth-Heinemann, Oxford.

Rutz, D. and Janssen, R., 2007. Biofuel technology handbook. WIP Renewable energies, München

Saha, S., Saha, B.N., Pati, S., Pal, B. and Hazra, G.C., 2017. Agricultural use of sewage sludge in India: Benefits and potential risk of heavy metals contamination and possible remediation options—A review. International Journal of Environmental Technology and Management, 20(3–4), pp. 183–199.

Salado, R., Vencovsky, D., Daly, E., Zamparutti, T. and Palfrey, R., 2009. Environmental, economic and social impacts of the use of sewage sludge on land. Consultation Report on Options and Impacts. Interim Report.

Simpson-Holley, M., Higson, A. and Evans, G., 2007. Bring on the biorefinery. Chemical Engineer, 795, pp. 46–48.

Soclo, H.H., Garrigues, P.H. and Ewald, M., 2000. Origin of polycyclic aromatic hydrocarbons (PAHs) in coastal marine sediments: Case studies in Cotonou (Benin) and Aquitaine (France) areas. Marine Pollution Bulletin, 40(5), pp. 387–396.

Sommers, L.E., Nelson, D.W. and Yost, K.J., 1976. Variable nature of chemical composition of sewage sludges. American Society of Agronomy, Crop Science Society of America, and Soil Science Society of America, 5(3), pp. 303–306.

Stronach, S.M., Rudd, T. and Lester, J.N., 1986. The Microbiology of Anaerobic Digestion. In Anaerobic Digestion Processes in Industrial Wastewater Treatment (pp. 21–38). Springer, Berlin, Heidelberg.

Stylianou, M.A., Inglezakis, V.J., Moustakas, K.G. and Loizidou, M.D., 2008. Improvement of the quality of sewage sludge compost by adding natural clinoptilolite. *Desalination*, 224(1-3), pp.240–249.

Taouda, H., Chabir, R., Aarab, L., Miyah, Y. and Errachich, F., 2017. Biomass and bioethanol production from date extract. Journal of Materials and Environmental Science, 8(9), pp. 3391–3396.

Tarigan, J.B., Ginting, M., Mubarokah, S.N., Sebayang, F., Karo-Karo, J., Nguyen, T.T., Ginting, J. and Sitepu, E.K., 2019. Direct biodiesel production from wet spent coffee grounds. RSC Advances, 9(60), pp. 35109–35116.

Templeton, M.R. and Butler, D., 2011. Introduction to wastewater treatment. Bookboon.

Tomlinson, P., 1986. Environmental Assessment in the UK: Implementation of the EEC Directive. *The Town Planning Review*, pp.458–486.

UN, 2019. The United Nations World Water Development Report. Leaving No One behind. Paris, France. https://www.un.org/

Vouk, D., Nakic, D., Stirmer, N. and Cheeseman, C.R., 2017. Use of sewage sludge ash in cementitious materials. Reviews on advanced materials science, 49(2), pp. 158–170.

Wahaab, R.A., Mahmoud, M. and van Lier, J.B., 2020. Toward achieving sustainable management of municipal wastewater sludge in Egypt: The current status and future prospective. Renewable and Sustainable Energy Reviews, 127, p. 109880.

Weil, R.R. and Brady, N.C., 2017. The Nature and Properties of Soils (Global Edition). Pearson, Harlow.

Wu, G., Hong, J., Tian, Z., Zeng, Z. and Sun, C., 2021. Assessing the total factor performance of wastewater treatment in China: A city-level analysis. Science of The Total Environment, 758, p. 143324.

Wu, W., Hu, J., Gu, X., Zhao, Y., Zhang, H. and Gu, G., 1987. Cultivation of anaerobic granular sludge in UASB reactors with aerobic activated sludge as seed. Water Research, 21(7), pp. 789–799.

Yu, J., Zhang, X. and Tan, T., 2009. Optimization of media conditions for the production of ethanol from sweet sorghum juice by immobilized Saccharomyces cerevisiae. Biomass and Bioenergy, 33(3), pp. 521–526.

Zehnder, A.J. and Brock, T.D., 1979. Methane formation and methane oxidation by methanogenic bacteria. Journal of Bacteriology, 137(1), pp. 420–432.

Zhai, J., Tian, W. and Liu, K., 2011. Quantitative assessment of polycyclic aromatic hydrocarbons in sewage sludge from wastewater treatment plants in Qingdao, China. Environmental Monitoring and Assessment, 180(1), pp. 303–311.

Zhang, Q., Hu, J., Lee, D.J., Chang, Y. and Lee, Y.J., 2017. Sludge treatment: Current research trends. Bioresource Technology, 243, pp. 1159–1172.

5 Solid Waste Treatment Technologies
Physicochemical Pathways

Karan Chabhadiya and Darshan Salunke

5.1 INTRODUCTION

Growth in economic development can be witnessed easily by observing rapid urbanization but embracingly it also showcases how mankind is generating an abundance of solid waste in the thrust of achieving a better and better lifestyle. Due to advancements in technology, developed countries are producing a higher rate of solid waste than developing or underdeveloped countries; likewise, high-income groups tend to generate more solid waste than the lower income ones. It can be summarized that advanced technology, higher population, lifestyle, and urbanization are the major factors of waste generation.

Having a look at the global waste generation scenario, municipal solid waste (MSW), industrial and commercial waste, hazardous waste, and biomedical waste contribute the major fraction and are under limelight in terms of their management. Various environmental issues are likely to be faced if these wastes are not properly managed. To name a few, damage to the ecosystem, changes in climate, deteriorated health and environment, etc. are the common issues that catch the attention. Open dumping of solid waste as well as its open burning gives rise to climate change problems and releases around 1,200 g CO_{2eq} of greenhouse gases (GHGs) into the atmosphere. Thus, it makes clear sense that formal management of solid waste needs to be looked upon which would include a proper collection of solid waste from the site, onsite and offsite segregation, transportation of waste in a closed container system, and technological innovations in recycling and recovery of material and energy from the waste. Regrettably, it has been witnessed that these methods of treatment are not executed effectively in many developing countries and exhibit huge losses to the economy and environment.

5.2 MUNICIPAL SOLID WASTE: INDIAN SCENARIO

India is a rapidly emerging developing nation in the world. It houses 1.4 billion people and several industrial complexes. Solid waste is the by-product of rapid growth. The municipalities and other commercials are accountable for environmental disruption and are mandated to work on principles like "sustainable development" and "polluters pay" (Down to earth, 2022). India has adopted the public-private partnership (PPP) model for waste management which promotes waste utilization as a

DOI: 10.1201/9781003352396-5

TABLE 5.1
Per Capita Waste Generation of Indian Cities (2010–2011)

City	*Population (2011)×10^6	#Total waste generated in tons per day	Waste generation (kg per capita per day)
Ahmedabad	6.3	2300	0.36
Hyderabad	7.7	4200	0.54
Bangalore	8.4	3700	0.44
Chennai	8.6	4500	0.52
Kolkata	14.1	3670	0.26
Delhi	16.3	5800	0.41
Mumbai	18.4	6500	0.35

Sources: *Census of India 2011, #CPCB Report 2011 (Census India, 2022).

resource. Table 5.1 shows the rapid growth and SW generation of megacities in India. The Government of India calculated about 65 million tons of waste generation every year, of which 62 million tons is MSW; only 20–28% of MSW is processed and treated and the remaining is dumped into landfill sites (Teri, 2022). Furthermore, the degradable and moist SW in India is around 60–65%, which is more as compared to the developed nations. Because of high moisture content, the calorific value of waste in India is 1400–2100 kcal/kg as compared to 1900–3800 kcal/kg in the USA, Norway, and Germany. Non-recyclable plastics, papers, and cardboards would undergo thermal treatment like pyrolysis, incineration, or gasification or can be used as a source of energy as refuse derived fuel (RDF) (Pujara et al., 2019). In India, 11% of GHGs is produced from informal waste management (Ramachandra et al., 2018). The informal sector plays an important role in reusing and processing of MSW but causes harm to humans and their surroundings (Sharma and Chandel, 2017). Presently, ragpickers are trained to collect the organic and inorganic waste separately where organic waste is diverted to composting and inorganic waste goes to mechanical sorting for recycling. The six-year data of per capita waste generation, treatment, and disposal from 2015 to 2021 is given in Table 5.2 (CPCB, 2022).

TABLE 5.2
Six-Year Solid Waste Data of India

Year	Solid waste generation per capita (kg/day)	Solid waste treated (%)	Solid waste landfill (%)
2015–2016	0.12	19.9	39.8
2016–2017	0.13	20.1	42.1
2017–2018	0.09	36.9	54.5
2018–2019	0.12	37.2	32.8
2019–2020	0.12	48.1	28.7
2020–2021	0.12	50.1	18.7

TABLE 5.3
Different Types of Pyrolysis

Types of pyrolysis	Retention time	Temperature (°C)	Products
Fast	<2 s	500	Bio-oil
Flash	<1 s	<650	Bio-oil, chemicals, and gas
Ultra-rapid	<0.5 s	1000	Chemical and gas
Vacuum	2–30 s	400	Bio-oil
Hydro-pyrolysis	<10 s	<500	Bio-oil
Carbonization	Days	400	Charcoal
Conventional	5–30 min	600	Char, bio-oil, and gas

As per the MNRE report, the total conceivable waste to energy (WtE) generation in India stands at 5690 MW (MNRE, 2022). WtE is a promising technology for waste utilization but it has a high initial cost, and out of a total of 16 WtE plants 07 plants are completely closed.

5.3 PHYSICOCHEMICAL TREATMENT

5.3.1 Pyrolysis

Pyrolysis indicates the thermal decomposition of waste at an excessive temperature in the absence of oxygen. The distinctive types of pyrolysis process are based on the yields of products like gas, liquid fuel, and charcoal. Based on the product, the pyrolysis classification is given in Table 5.3. The selection of the particle size for thermogravimetric analysis (TGA) was carefully carried out as there was a variation of product for different temperatures of solid waste (Buah et al., 2007).

Case Study

Plastic has the most versatile properties. Due to its numerous applications, it is useful for packaging currency notes and as a fuel. The non-biodegradability of this material makes it vulnerable to living beings and the surroundings. The idea of utilizing plastic waste for producing fuel, syn gas, and charcoal is established at Rudra Environmental Solutions, Pune. The equipment implied here is Thermo Catalytic Depolymerization (TCD). The process is shown in Figure 5.1.

To optimize the fuel consistency, initially it requires LPG, natural gas, or diesel. The batch process takes 3–7 hours at 280–430°C. Long-chain polymer cracks into short-chain polymers in the absence of oxygen. The syn gas produced during the process is recirculated to the reactor itself and vapors

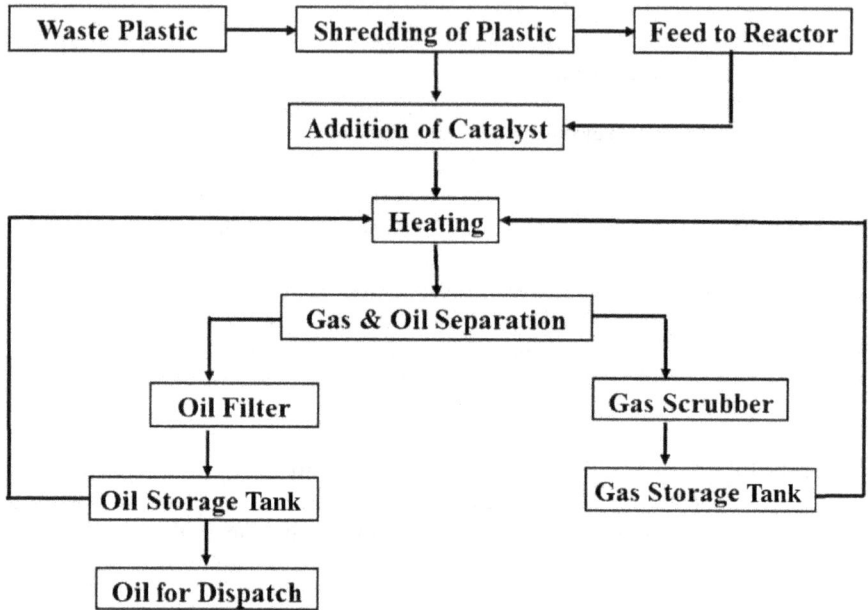

FIGURE 5.1 Transformation of waste plastic to fuel (Rudra envsolution, 2022).

are condensed to make fuel, making the process zero emission. The gas and fuels are then purified through filtration and scrubber system. The char collected from the overall process is used for road construction or landfilling. For every single ton of plastic, the process yields 20–25% of syn gas (reused in the process), 600–650 L of polyfuel, and moisture and residual char of 5–10%.

5.3.2 REFUSE DERIVED FUEL

Solid Waste Management (2016) rule obligates municipalities and private bodies to use recycled solid waste for various purposes like construction and WtE facilities. RDF is classified based on the type of material, structure, volume, and appearance of palletization. The particle size of 10–300 mm and bulk density of 120–300 kg/m^3 with 10–30% moist content are some of the properties of an ideal RDF. Excess calorific and non-recyclable segment of MSW is processed and used as energy for generating electrical energy or used as a substitution for kiln and boilers. The specific industrial and MSW such as plastics, textile, agriculture waste, expended oil, and wood chips, along with RDF, is used with the feed to increase the calorific value for power generation. Segregation of non-combustibles like stone, sand, glass, metals, etc. should be carried out for higher efficiency

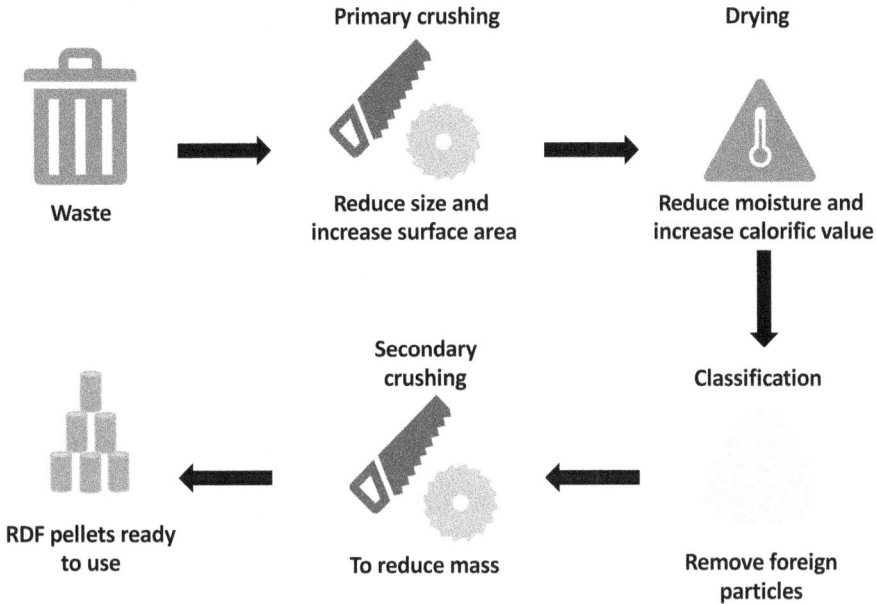

FIGURE 5.2   Synthesis of RDF from waste.

along with wet waste like food, garden, peel of fruits, etc. RDF can be formed in pellets or used directly as feed into the kilns or boilers. The process of RDF is shown in Figure 5.2 (cd waste, 2022).

5.3.3   MECHANICAL RECYCLING

In 1970, mechanical reprocessing was established to retrieve the plastic matter from waste plastic. Polymer technologists made mechanical recycling a simple and consistent system to utilize the waste plastic. Though the system is promising, the recycling plastic lacks the quality of virgin plastic. During mechanical recycling, the quality degrades because of several processes resulting in poor quality of polymer (Panda et al., 2010; Singh and Ruj, 2015). Extruder and palletizer play a major role in altering the polymer properties and transforming to a new product with near virgin product quality. Moreover, the quality of recycled plastic is brittle as compared to virgin plastic because of several processes that break the long-chain polymers and make them less durable. Recycling is the dawning trend in India with a plastic recycling marketplace of 6 million tons in the year 2021 and is projected to grow 7% annually (Businessware, 2022) Furthermore, downcycling process degrades the product quality, but replaces the virgin plastic to a certain extent when it comes to recycled Polyfibre. Therefore, it is not economical to recycle all classes of plastic, but it can be opted for several gradients of products as described in Figure 5.3 (Safar Polyfibre, 2022).

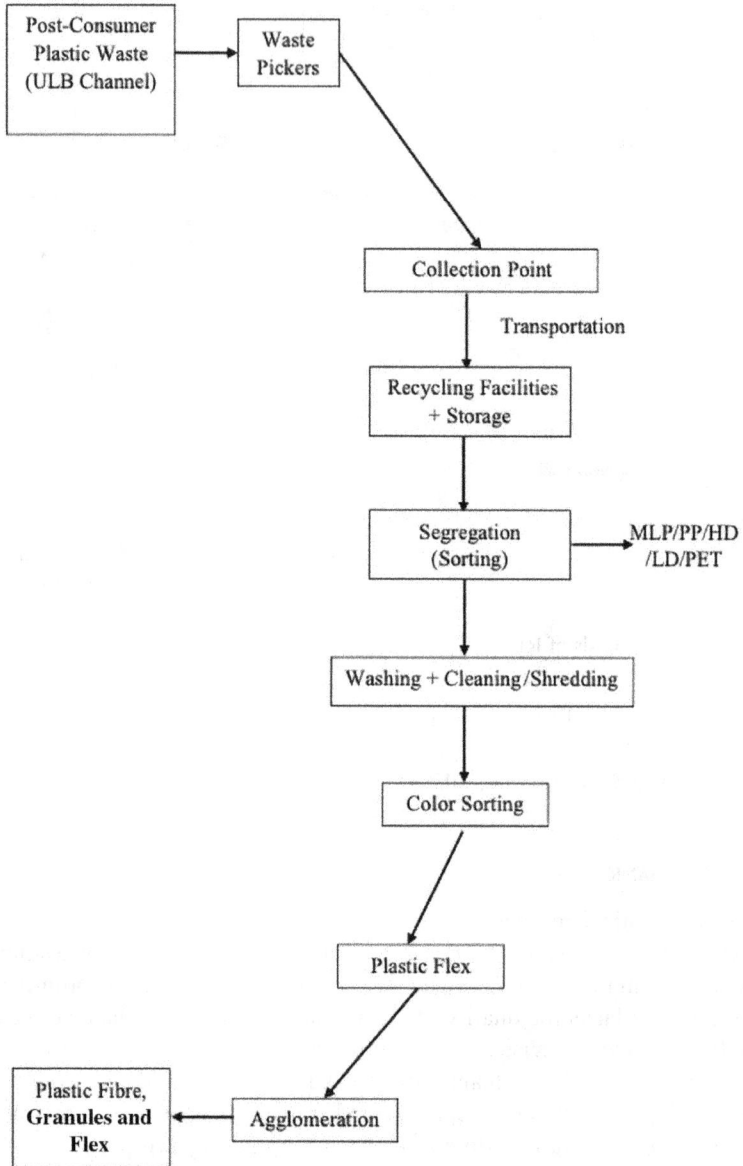

FIGURE 5.3 Mechanical recycling process.

5.4 CONCLUSION

The management of solid waste is one of the prime concerns if the world wants to achieve its targeted goal of attaining sustainability by 2050. The intention can be obtained by managing solid waste by applying appropriate treatment methodology to them. Treatment technologies should be designed in such a way that the by-product formation is very low. It should provide a robust solution to the problem. Treatment of solid waste should be aided by policy remakes and financial support so that the systems can be more trustworthy to the stakeholders and corporate management. Many technologies are found very reliable at the R&D stage and have been proven to be a key element in solving this burning issue. These technologies should be given economic and legislative backup so that technology transfer can be done at an easier rate to overcome this problem at a larger scale.

REFERENCES

Buah, W.K., Cunliffe, A.M., Williams, P.T. (2007). Characterization of products from the pyrolysis of municipal solid waste. Process safety and environmental protection. 85(5), 450–457.

Businessware, 2022. Retrieved from: https://www.businesswire.com/news/home/20221109005572/en/India-Plastic-Recycling-Market-Report-2022-Favourable-Government-Initiatives-Promise-to-Boost-Growth—ResearchAndMarkets.com

cd waste, 2022. Retrieved from: http://cdwaste.co.uk/refuse-derived-fuel-rdf/

Census India, 2022. Retrieved from: https://censusindia.gov.in/census.website/

CPCB, 2022. Retrieved from: https://cpcb.nic.in/uploads/MSW/MSW_AnnualReport_2020-21.pdf

Down to earth, 2022. Retrieved from: https://www.downtoearth.org.in/blog/waste/india-s-challenges-in-waste-management-56753

MNRE, 2022. Retrieved from: https://mnre.gov.in/waste-to-energy-overview/

Panda, A.K., Singh, R.K., Mishra, D.K., 2010. Thermolysis of waste plastics to liquid fuel: a suitable method for plastic waste management and manufacture of value added products – a world prospective. Renewable and Sustainable Energy Reviews. 14(1), 233–248.

Pujara, Y., Pathak, P., Sharma, A., Govani, J., 2019. Review on Indian municipal solid waste management practices for reduction of environmental impacts to achieve sustainable development goals. Journal of environmental management. 248, 109238.

Ramachandra, T.V., Bharath, H.A., Kulkarni, G., Han, S.S., 2018. Municipal solid waste: generation, composition and GHG emissions in Bangalore, India. Renewable and Sustainable Energy Reviews. 82, 1122–1136.

Rudra envsolution, 2022. Retrieved from: http://rudraenvsolution.com/productplantdetails.html

Safar Polyfibre, 2022. Retrieved from: https://safarpolyfibre.com

Sharma, B.K., Chandel, M.K., 2017. Life cycle assessment of potential municipal solid waste management strategies for Mumbai, India. Waste Management & Research. 35(1), 79–91.

Singh, R.K., Ruj, B., 2015. Plastic waste management and disposal techniques – Indian scenario. International Journal of Plastics Technology. 19(2), 211–226.

Teri, 2022. Retrieved from: https://www.teriin.org/article/msw-management-pitiable-situation-municipal-solid-waste-management

6 Waste Disposal and Its Environmental Impact

Snehal Lokhandwala, Pratibha Gautam,
Zeel Desai, and Vivek Gajara

6.1 INTRODUCTION

The existence of mankind solely depends on the resources provided by Mother Earth which was earlier thought to be directly proportional to the food requirements and basic amenities. But the waste generated in any form, the extent of resource utilization, and even the technologies adopted for varying applications contribute to the balance between the environment and human beings (Wong et al., 2016). Improper solid waste management has been the major contributing factor to environmental degradation and now even developing countries are facing problems with solid waste disposal and its management. Ever since the Industrial Revolution, solid waste has been a major environmental issue. Different kinds of waste are generated on a day-to-day purpose from different kinds of sectors. An approximate annual generation of 200 million tons of municipal solid waste (MSW) is calculated, constituting approximately 29% of the worldwide total MSW production generated annually is municipal solid waste whereas the other major contributors are industrial waste and construction/demolition waste. Therefore, the treatment and disposal of solid waste becomes increasingly important (Chen and Zhang, 2013). The unscientific municipal solid waste management results not only in air, soil, and water pollution but its indiscriminate dumping contaminates both ground and surface water. The solid waste in urban areas with dense population clogs the drains which accumulates the drain water and invites insect breeding and floods in monsoon. Greenhouse gases are generated from decomposition of organic waste in landfills and not only this but there also are gaseous emissions from activities like uncontrolled burning and improper incineration of MSW. Landfills are also responsible for generating the leachate which pollutes surrounding water bodies. The polluted environment thus arising from improper MSW management results in severe health and safety issues (Alam and Ahmade, 2013).

The useless output of human activities as non-liquid waste, non-gaseous waste, or their by-products can be termed as solid wastes. Such waste needs to be properly managed and disposed. Otherwise, it can create problems for the environment. Scientific and sustainable waste management techniques give economic benefits, better health, and a reduction in vulnerability. In this chapter, we have discussed waste management technologies that are helpful in the case of managing such waste. It will also help us to understand the hierarchy of waste management, that is, the procedure to be followed while managing waste. There are different conventional and

DOI: 10.1201/9781003352396-6

alternative technologies that can be followed for the disposal of waste. This chapter also focuses on the different future technologies that will be implemented to increase the efficiency of solid waste management systems and convert the waste-to-energy. We have also discussed the environmental and health impacts caused by such waste and during the handling of waste.

6.2 HIERARCHY OF SOLID WASTE DISPOSAL

The tool to evaluate the processes that protect the environment including its energy consumption and resources from most sustainable to least sustainable options is termed as solid waste hierarchy. Its main priority is to achieve sustainability.

The waste hierarchy works in the direction to reduce and manage solid waste which can be represented in the form of a pyramid. The aim of this hierarchy is to maximize the resource utilization and minimize the waste generation. It helps to reduce pollutants, save energy, and prevent greenhouse gas emissions into the environment and also emphasizes on implementation of greener technologies. Thus, there is a need for modern integrated waste management which infuses sustainability in all its components along with consideration of demand and supply resources (Wong et al., 2016).

The solid waste hierarchy is ranked from most preferable to least preferable options as shown in Figure 6.1.

This hierarchy consists of the following:

- Waste minimization
- Reuse
- Recycle
- Energy recovery
- Disposal

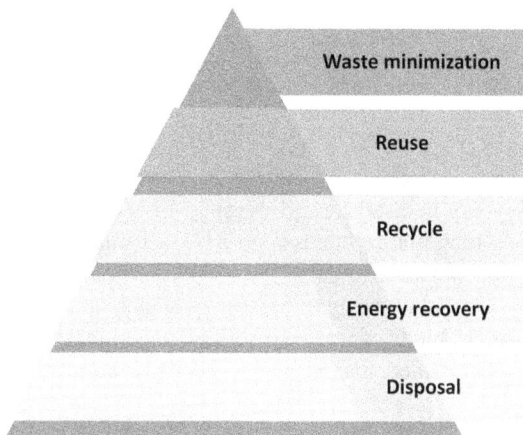

FIGURE 6.1 Hierarchy of solid waste.

6.2.1 Waste Minimization

Waste minimization basically focuses on reducing the amount and harmful effects of waste generated. It is an approach towards preventing waste formation. It is the first step of the waste hierarchy and is a most preferable option to control waste formation right from the source, to cancel its major long-term effects. Reducing waste from both the finished product and the manufacturing process can reduce the stress and challenges during the time of disposal.

Waste minimization does not include the treatment processes like compacting, pyrolysis, incineration, etc. Instead it goes in a hierarchical approach by reusing, recycling, and recovering followed by the final disposal. The main benefit of this technique is that it protects the environment by reducing emissions and also by conserving the natural resources that are linked with raw material extraction and waste disposal. The main four functional components of waste management include storage and collection, transportation, treatment, and disposal of waste. In order to minimize waste, these elements should be optimized (Jain and Singhal, 2014). Waste minimization of hazardous waste leads to the reduction of toxins during production and final output, which can improve the overall health of the workplace by fewer exposures to such harmful toxins.

More the work done on the minimization during producing and finishing of the product, less is the time spent during the time of disposal.

6.2.2 Reuse

Reuse is an approach involving conventional reuse (original purpose) or creative reuse (different purpose) of an item used products again saves time, money, energy, and resources and generates less waste which eventually protects our environment. Reusing eventually leads to waste minimization. For example, reusing a plastic bottle by refilling it or creating some useful art out of it instead of throwing it out eventually reduces the plastic waste and also decreases the load during the time of disposal.

6.2.3 Recycle

Recycling is a process wherein new materials or products are obtained from the waste and now they can be used for a different purpose. It is the third component of waste hierarchy which mainly focuses on environmental sustainability by using waste in an economic way. It also helps us to avoid unnecessary carbon emissions. There is a recovery of solid waste up to 65% with the help of recycling and composting processes. Different kinds of waste like glass, paper, plastics, textiles, electronics, etc. can be sorted, crushed, and further sent for the recycling process.

6.2.4 Energy Recovery

The process often termed waste to energy (WTE) is the conversion of non-recyclable waste materials into usable electricity, heat, or fuel through a variety of processes, including gasification, combustion, etc. This energy recovery component of waste hierarchy helps in reducing the amount of waste that is to be sent to the landfills to produce useful energy in the form of heat and electricity. This component of the hierarchy provides a system that simultaneously solves the issues of energy demand, waste management, and emissions of greenhouse gases and achieves the circular economy system. Another benefit it provides is a reduction in the demand of land for landfilling and also it reduces the transportation cost of transporting waste to the disposal sites.

6.2.5 DISPOSAL

Disposal means removal, discarding, or destruction of undesired materials (waste) that are produced from different sectors like domestic, commercial, industrial, etc. The correct process followed with the proper method of treatment and disposal will protect the environment from pollution and other hazards. It is the final stage of waste hierarchy which only includes waste materials. There are different types of conventional disposal techniques like landfill, incineration, and composting for solid waste management. The most convenient and cheapest technique among all these is landfilling.

6.3 SOLID WASTE DISPOSAL TECHNOLOGIES

Solid waste can create a big nuisance to the environment if it is not disposed of properly. The main disposal method for solid waste is either open dumping or landfill. Uncontrolled dump sites and improper disposal nature can cause a lot of leachate generation with other such environmental issues (Chandrappa and Das, 2012). Figure 6.2

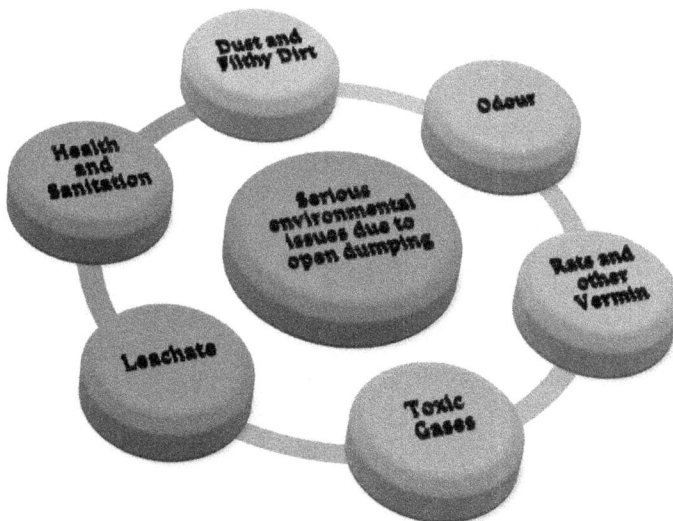

FIGURE 6.2 Environmental issues due to open dumping.

shows different environmental issues that resulted from open dumping. There are different conventional technologies like landfilling, incineration, and composting and other alternating technologies like gasification, pyrolysis, autoclaving, etc. which helps to deal with the management of solid waste disposal.

6.3.1 CONVENTIONAL DISPOSAL TECHNOLOGIES

6.3.1.1 Landfill

Landfills are the final and preferred destination for waste selected after all other waste management options are taken care of. Landfills are classified as open dumps, controlled dumps, or sanitary landfills (integrated waste management disposal systems). Disposal in landfills provides long-term containment of waste materials. The landfill is the physical facility specifically designed, constructed, and operated for the disposal of waste including the residual waste generated from well-established waste reduction, recycling, and recovering programmes.

The lifecycle of a landfill involves the following steps:

- Planning and execution
- Site selection and preparation
- Landfill bed construction including leachate and gas collection system
- Landfilling (scientifically)
- Monitoring
- Closure of landfill and post-closure monitoring.

The typical landfill process during operation involves waste dumping at properly planned site, waste spreading, shredding, and compaction followed by waste covering and monitoring (Chandrappa and Das, 2012).

6.3.1.1.1 Process of Landfill

The main process of landfilling involves decomposing of the organic matter and other such waste that is present in the disposal. Organic matter in solids is composed of mainly proteins, carbohydrates, lignin, etc. which are easily degradable and other organic matter like lignin and cellulose as recalcitrant. The landfill ecosystem is diverse in nature and hence promotes stability. However, environmental conditions like pH, temperature, moisture content, etc. influence the processes involved in landfills (Chandrappa and Das, 2012).

Table 6.1 shows the different phases involved in municipal solid waste landfills.

6.3.1.1.2 Controlling and Monitoring of Leachate and Gas

Leachate generated from waste is the complex liquid that has percolated through the wastes, which is a source of soil and groundwater pollution and gas is produced by the fermentation of organic matter. The major reason for generation of leachate is precipitation. The percolating water combines with both soluble and suspended components of the biodegrading waste through complex reactions (physical and chemical) resulting in leachate which results in surface water run-off,

TABLE 6.1
Phases of MSW Landfill

Phase	Phase Name	Particulate
I	Initial Adjustment Phase	Microbes acclimatize to the landfill condition
II	Transition Phase	Transformation from an aerobic to an anaerobic environment occurs
III	Acid Formation Phase	The continuous hydrolysis of solid waste followed by the microbial action on biodegradable organic fraction results in the generation of intermediate volatile organic acids
IV	Fermentation Phase	Intermediate acids are converted into methane, carbon dioxide, hydrogen sulphides, and ammonia by microbial action
V	Maturation Phase	Nutrient becomes scarce and the biological activity shifts to dormancy resulting in a drop in gas production and leachate strength will be at lower concentrations. Slow decomposition of resistant organic matter may continue resulting in humic-like substances

groundwater inflow, and biological degradation. The leachate from a stabilized landfill can be treated by coagulation-flocculation whereas chemical precipitation can be implemented to pre-treat leachate so as to eliminate ammonical nitrogen. However, stripping can always be used to reduce ammonical nitrogen in leachate. Advanced oxidation techniques are to be used to treat leachate as it is highly non-biodegradable.

Regular monitoring of the landfill is carried out for groundwater quality, liner leakages (using a lysimeter), gas absorption in surrounding soil, quality and quantity of landfill gas, leachate quantity and quality, and stability of final cover. Groundwater monitoring is done by drilling monitoring wells, both upstream and downstream of the landfill. Landfill gas is collected from gas extraction wells (Chandrappa and Das, 2012).

6.3.1.1.3   Closure of Landfills

Scientific closure of the landfilled waste involves daily covers to form cells including liners to protect the waste heap from rain. Layers of geo-membrane, clay, gravel, geotextile, and topsoil are laid down once the landfill attains the capping limit. Landscaping, run-off control, gas and leachate collection/treatment, erosion control, and environmental monitoring are the essential components of the closure plan. Safety and environment are taken care for 30–50 years post closure of landfill involving routine inspection and mitigation measures in case of uncontrolled gas emissions and degraded leachate quality. The final cover on the landfill helps in minimizing infiltration of rainwater, avoiding fugitive emissions, isolating waste from the environment, minimizing frost and soil erosion, resisting penetration or roots and burrowing animals, etc. (Chandrappa and Das, 2012).

6.3.1.2 Incineration

Incineration is one of the most acceptable techniques to dispose of all combustible waste (waste that can be burned). The incineration technology has been developed to reduce waste volumes and solve the hygienic problems related to modern waste-to-energy plants and emission control systems. Waste is burnt in a specialized engineered setup. Incineration is a thermal treatment converting waste to ash, flue gas, and heat through combustion (Chandrappa and Das, 2012). The flue gases must be monitored and controlled with control equipment before they are dispersed to the atmosphere. The solid mass of the original waste is reduced by 80–85% and the volume by 95–96% after incineration which depends on the composition and degree of recovery of materials (metals from the ash) for recycling. Thus, incineration does not replace landfill, but helps in reduction of volume for disposal in such dumping sites.

Figure 6.3 shows the block diagram of a conventional incineration plant. Incineration results in combustion of waste to release heat and other environmentally friendly gases. Further, the moisture content of the waste is reduced at the initial stage of the incineration process, and the incombustible parts of the waste form solid residues (bottom ash, fly ash) (Christensen, 2011). This residue is then sent to the dumping sites and the gases that are released are treated with the control equipment in order to avoid the problem of environmental pollution.

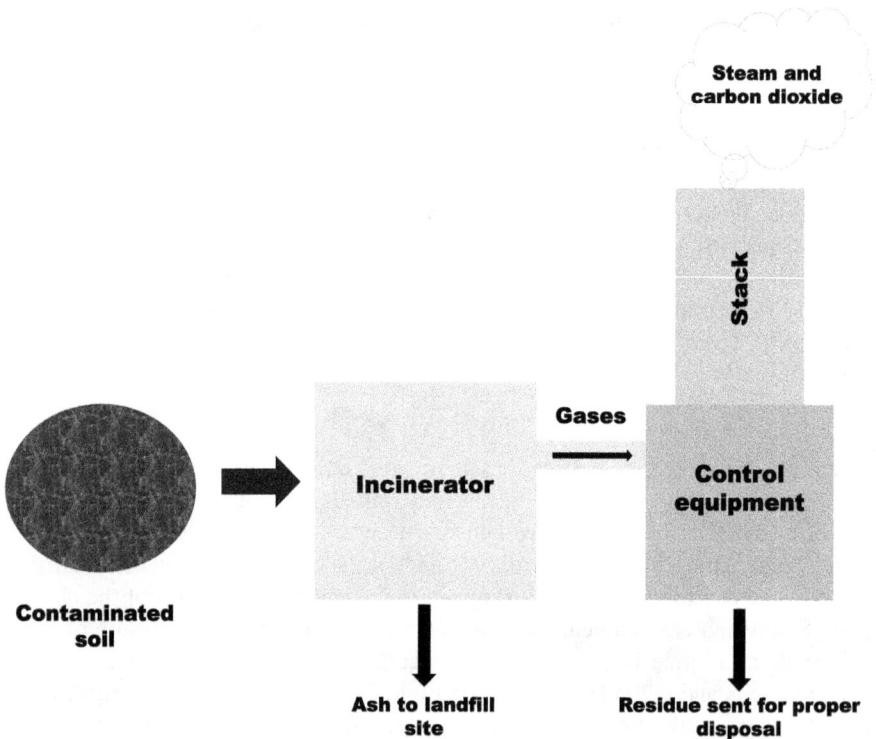

FIGURE 6.3 Block diagram of incineration plant.

6.3.1.3   Recycling to Useful Products

Recyclable waste is the reusable waste material which renovates into useful materials and objects.

Recycling is one of the best alternatives to waste disposal as it preserves natural resources and lowers greenhouse gas emissions. The wastage of potentially useful resources is extensively prevented which indirectly reduces fresh raw material consumption and also helps in energy saving, water pollution (from landfilling), and air pollution (incineration). Recycling is a major component of the most modern waste reduction strategy of 3Rs (Reduce, Reuse, and Recycle) aiming at a sustainable environment by substituting raw material inputs. Recycling has proved to be the remedy for environmental issues like climate change, resource depletion, and environmental pollution which are the results of overpopulation. The three main types of recycling are as follows (Farooqi et al., 2021):

Primary recycling: Waste is recycled again to the original thing (e.g., paper to paper).

Secondary recycling: Waste is modified for some different use (e.g., cutting waste bottles and using them in garden).

Tertiary recycling: Complete transformation of waste to a new product by chemical/physical processes.

6.3.1.4   Anaerobic Digestion

Anaerobic digestion (AD) involves sequential processes of breaking down biodegradable material by microorganisms in the absence of oxygen. The process is applicable both for domestic and industrial purposes to manage waste and for fuel production. The digestion process in AD generates biogas which has methane in maximum concentration and thus can be used as an alternative source of energy. Sewage sludge, cow manure, and even waste activated sludge can be treated with AD.

The four major stages involved in anaerobic digestion are as follows:

- **Hydrolysis**: Macromolecules are converted to smaller components.
- **Acidogenesis**: Acidogenic bacteria utilize the smaller components through cell membranes to produce intermediate volatile fatty acids (VFAs) and other products.
- **Acetogenesis:** VFAs and other intermediates are converted into acetate along with the generation of hydrogen.
- **Methanogenesis:** Accessible intermediates are consumed by methanogenic microorganisms to produce methane.

6.3.1.5   Composting

Composting proves to be the most crucial strategy to manage waste wherein the organic waste produced in landfills is converted to a material that can enhance agricultural productivity. This biological process reduces the volume of waste considerably and converts heterogeneous organic waste into substances with high organic matter. The biodegradation is done by microbial population under controlled conditions of moisture, temperature, and oxygen. The organic substances such as vegetables, leaves, fruits, manure, sludge, and other waste are converted by micro-organisms into humus-like materials which are used as an organic fertilizer in the fields. The microorganisms involved in composting are fungi, bacteria, and actino-mycetes. These decomposers release enzymes that convert complex materials into simpler ones. Aerobic composting is the composting done in presence of oxygen whereas anaerobic composting lacks oxygen involvement. Many other compost-ing systems are employed to convert waste to soil fertility enhancers, namely, bio-mineralization, vermicomposting, controlled microbial composting, aerated static pile composting, etc.

6.3.2 Alternative Disposal Technologies

6.3.2.1 Gasification

Gasification is a thermally driven reaction in presence of insufficient oxygen to con-vert hydrocarbons (compounds of carbon, hydrogen, and oxygen molecules) to CO_2 and H_2O. Fundamentally it's a process of chemical conversion of carbon compounds to composite gaseous mixture (hydrogen (H_2) and carbon monoxide (CO)) by par-tial oxidation using a gasification agent like air, oxygen, steam, etc. at 700–1600°C (Christensen, 2011; Golomeova et al., 2013). Temperatures are in the range of 800–1100°C with air as an oxidation agent and up to about 1500°C with oxygen. The overall gasification process is mainly exothermic involving some endothermic reactions, the heat for which may be supplied by the gasification agent. The products from gasification are in general gas (higher fractions of CO_2), liquid (tar and oil), and solid (ash). The percentage of output product of gasification mainly depends on the reaction temperature, the quality of waste as input, and process configurations (Christensen, 2011). Fixed bed, fluidized bed, and high temperature gasification are the major three categories of gasification technologies. High temperature gasification is a widely used technology. The coarsely shredded waste enters a gasifier wherein the carbonaceous fraction of waste reacts with the gasifying agent to produce the gaseous products. Sometimes the gasifier is fed with pyrolysed waste and even with fixed carbon (Chandrappa and Das, 2012).

The main reactions involved during gasification are as follows (Golomeova et al., 2013):

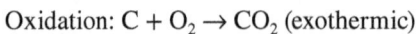

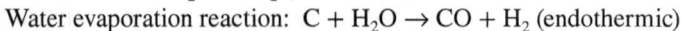

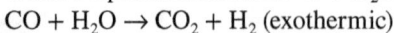

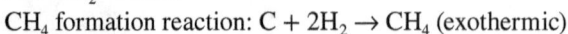

Oxidation: $C + O_2 \rightarrow CO_2$ (exothermic)
Water evaporation reaction: $C + H_2O \rightarrow CO + H_2$ (endothermic)
$CO + H_2O \rightarrow CO_2 + H_2$ (exothermic)
$C + CO_2 \rightarrow 2CO$
CH_4 formation reaction: $C + 2H_2 \rightarrow CH_4$ (exothermic)

Thus, CO, H_2, and CH_4 are the basic output of gasification. The resulting gas mixture (CO + H_2) is called syngas or synthesis gas. Heating values of syngas are generally around 4–10 MJ/m^3 (Golomeova et al., 2013). The carry-over particulate matter, chlorides/acid gases, and sulphur are removed from the raw syngas coming out from the reactor and then sent to the power generation plant to produce energy. The main advantage of gasification is it can be applied to convert organic waste to low calorific value gas.

6.3.2.2 Bio-drying

A decentralized system involving biological-mechanical conversion of solid waste is termed bio-drying. In this system, the solid waste is treated with the heat generated when the aerobic decomposition process of organic compounds is combined with excess air. The disadvantage of the bio-drying process is the emission of VOCs and other gases that potentially cause global warming through carbon dioxide (CO_2), methane (CH_4), and nitrous oxide (N_2O). Zaman et al. (2018) refer to bio-drying as an automatic and natural heating process wherein the drying process is reinforced by biological heat released at the on-site decomposition of organic matter used to treat moisture present in the mixed sludge. This alternative waste management technology is affected by moisture content in sludge, process temperature, aeration efficiency, and reactor design. Low moisture containing and enhanced calorific value waste after a bio-drying process, which is biologically stable, can be used as a source of refuse derived fuel (RDF) which is to be used immediately and not stored. High calorific fuel can be produced provided the bio-drying process is combined with the conventional mechanical biological treatment that removes moisture.

6.3.2.3 Mechanical Biological Treatment

A mechanical biological treatment (MBT) system is a combination of biological technologies like composting, AD, etc. with mechanical treatment technologies like sieves, screens, etc. (screens, sieves, magnets, etc.) which have the capacity to process mixed household waste as well as commercial and industrial wastes. This technology aims to reduce landfills and their post-closure inspection duration, biogas and leachate productions, odours during the waste deposit operations, etc. MBT can also be considered as a pre-treatment to improve the quality and quantity of biogas production. MBT of residual MSW involves sorting of recyclable materials such as paper, plastics, and metals by mechanical pre-processing followed by stabilizing biodegradable organic matter under controlled anaerobic and/or aerobic conditions. MBT also helps in reduction of greenhouse gas emissions.

6.3.2.3.1 Mechanical sorting

It's an automatic mechanical sorting stage applied before biodegradation to separate RDF/SRF and recyclables from the remaining waste. The equipment/machinery that may be involved in mechanical sorting involves eddy current separators, factory style conveyors, industrial magnets, etc. or it can be done through manual hand-picking (Christensen, 2011).

6.3.2.3.2 Biological processing

Biological processing can be either aerobic (composting) or anaerobic (digestion). Anaerobic digestion harnesses anaerobic microorganisms to break down the biodegradable component of the waste to produce biogas and soil improver. Biogas can be utilized to generate electricity and heat. Composting is also a type of biological processing that does not produce green energy but has the ability to produce green manure for soil enrichment.

6.3.2.4 Hydrothermal Carbonization

Hydrothermal carbonization (HTC) is a thermal conversion process for managing solid waste streams along with minimizing greenhouse gas production and producing residual material with good calorific value. It's a wet process to thermally convert biomass to a carbonaceous residue referred to as hydrochar at a relatively low temperature (180–350°C). During HTC, the feedstock is heated in subcritical water and at moderate pressures which results in decomposition by a series of simultaneous reactions, including hydrolysis, dehydration, decarboxylation, aromatization, and re-condensation (Lu et al., 2012). HTC treatment generates carbon-rich solid products, Hydrochar is a liquid product having furan derivatives and phenolic compounds and a minimum quantity of gas mostly CO_2. HTC of solid waste is also one of the potential alternative strategies to produce a solid fuel source and is advantageous in comparison to dry carbonization processes like pyrolysis, particularly for moisture containing feedstock. The basic requirement for this process is that the solid feedstock should be completely in the liquid phase during carbonization, requiring the process to occur in a closed system under saturated pressures. The presence of sufficient water is critical because on increase in temperature, the physical and chemical properties of water change significantly, ultimately mimicking that of organic solvents. Temperature, feedstock composition, water/solid ratio, and time of the reaction are important reaction parameters for the fate of the rate of conversion processes (Lu et al., 2012).

HTC as compared to other process like gasification, combustion, etc. is far better and simpler as it requires lower temperature and a wet feedstock which nullifies the need for additional water in the reaction. Moreover, it can be applied to heterogeneous wet organic residue as compared to the dry conversion process without preliminary separating and drying.

6.3.2.5 Hydrothermal Liquefaction

Hydrothermal liquefaction (HTL) is the solid waste treatment technology that can be used to efficiently process wet and dry biomass generating the product called bio-crude, which is the renewably equivalent to oil. Bio-crude is an energy intensive intermediate and can be upgraded to a variety of liquid fuels. HTL generates bio-crude from organic matter requiring the presence of water, that is, requires hydrothermal conditions, with temperatures ranging from 250°C to 450°C and pressures between 100 and 300 bar (Grande et al., 2021). The technique uses specific characteristics of compressed hot water and during the reaction, water remains in a supercritical state. During the entire liquefaction process, biomass is subjected to a series of depolymerization reactions involving hydrolysis, dehydration, and decarboxylation. The entire process results in insoluble products such as bio-carbon or bio-crude oil. Along with this, gases (CO_2, CO, H_2, or CH_4) or soluble organic substances (mainly acids or phenols) are also generated. The O/C and H/C ratio of the bio-crude decreases during the reaction along with an increase in temperature, and the increase in calorific value is also observed. Commonly used biomass that are subjected to HTL include oils (lipids), wet biomass from algae (proteins), and dry lignocellulosic biomass (consisting of cellulose, hemicellulose, and lignin) (Grande et al., 2021).

6.3.2.6 Pyrolysis

The thermo-chemical degradation of carbon-based materials through indirect, external sources of heat resulting in elevated temperatures (450–750°C or 300–800°C) and in the absence of oxidizing agents such as O_2, CO_2, or steam to produce carbonaceous char, oils, and combustible gases is termed as pyrolysis (Chandrappa and Das, 2012; Golomeova et al., 2013). As a result of elevated temperature, the volatile portions of the organic materials are driven off which results in syngas and useful complex liquid hydrocarbons. These products can then be added to fuel or solvent product or returned to a refinery where it is added to the feedstocks; the entire process is endothermic in nature. The composition and energy contents of the products from pyrolysis are highly dependent on the waste input and may vary significantly (Christensen, 2011). The products obtained on pyrolysis are as follows:

- Gas: Composite mixture of hydrogen, methane, carbon monoxide, carbon dioxide, and other volatile constituents yielding 20–50% by weight of the input quantity of waste and heating value around 3–12 MJ/Nm³.
- Liquid: A mixture of tar, oil, and water containing a complex range of hydrocarbons basically organic acids, phenols, PAHs, and alcohols. The amount of liquid may be around 30–50% by weight with heating values around 5–15 MJ/kg.
- Solid: A char-like material containing the remaining solid products (metals, glass, sand, etc.) in the order of 20–50% by weight and having a considerable ash content of 10–50%. The heating value of the char may be up to 10–35 MJ/kg.

Heating values and the yield of the products vary significantly for every process and are dependent on the waste input composition. In the pyrolysis process, waste is shredded and fed into a reactor and operated in the absence of oxygen under atmospheric pressure at 500–700°C for 0.5–1 h. After drying the waste, the moisture is released by heating to about 100–120°C (Christensen, 2011) followed by a series of complex reactions converting more complex compounds to simpler ones. With increasing temperatures from about 200°C to 800°C, gaseous outputs are obtained as a result of bond breaking between oxygen, hydrogen, and nitrogen. Followed by these primary reactions, the tar/oil is converted to gases and char which is referred to as the secondary reaction.

The initial reactions involve the decomposition of low volatile organic components to more volatile ones (Golomeova et al., 2013).

$$C_x H_y \rightarrow C_c H_d + C_m H_n$$

Moreover, at the early stages of the pyrolysis process, there occurs the formation of solid residue from a low volatile organic substance by condensation, hydrogen removal, and ring formation reactions.

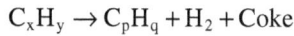

$$C_xH_y \rightarrow C_pH_q + H_2 + Coke$$

In the presence of oxygen, CO and CO_2 are produced or the interaction with water is possible. The produced coke can be vaporized into O_2 and CO_2. The products obtained from the pyrolysis process are solid residues (coke) (10–25%) and synthetic gas 'syngas' (75–90%). Syngas is used in the power generation plant to produce energy, such as steam and electricity. Synthetic gas typically has an energy value between 10 and 20 MJ/Nm^3 (Golomeova et al., 2013). However, the ash from the pyrolysis process is usually disposed of in a landfill. Pyrolysis is a method for the treatment in order to decrease leaching and emissions to the environment (Chandrappa and Das, 2012).

6.3.2.7 Molten Salt Oxidation

Molten salt oxidation (MSO) is a thermal treatment used to efficiently destroy the organic constituents of mixed and hazardous wastes, and energy-rich materials through oxidation. The technology involves flameless oxidation within the salt bath converting the organic components of the waste into CO_2, N_2, and water. The integrated MSO system consists of a reaction vessel, an off-gas treatment system, a salt recycle system, feed preparation equipment and immobilization system consisting of ceramic final waste. MSO process involves the following:

- Injection of organic-based wastes beneath a bed of molten carbonate salt at $900 \pm 950°C$
- Catalytic oxidation of organic constituents to inorganic products (H_2O, CO_2, etc.)
- Neutralization of acid gases (HCl) in the bed
- Periodic discharge of the salt for disposal or for processing and recycle

The molten salt, usually sodium carbonate, has varied applications, namely, as a dispersion medium, as a catalyst, as a stable heat transfer medium that resists thermal surges, helps in retaining soot, char and most of the ash in the melt itself. It also helps in retaining radionuclides, and other non-combustible material associated with the waste in the salt bed (Hsu et al., 2000). The top-feed injection system is designed to feed waste and air and throughputs up to 7 kg/h for chlorinated solvents. The off-gas systems remove entrained salt particulates, water vapour, and traces of gas species such as CO and NO_x.

6.3.2.8 Waste Autoclave

Waste autoclave uses heat, steam, and pressure for the processing of the waste. The purpose of autoclaving MSW is to physically treat it before processing for easier manipulation and further processing. Waste that is shredded is transferred to steam

TABLE 6.2
Heat Treatment Options

Type of Heat Treatment Process	Description
Autoclaving	• A batch process
	• Steam procession in a vessel under pressure
	• Waste subjected to steam under pressure followed by mechanical sorting
	• Sterilized waste separated at the end
Continuous	• Continuous heat treatment in absence of pressure in a vessel
	• Waste dried using externally applied heat
	• Mechanical sorting and separation of sanitized waste

autoclave router machines. Waste autoclaves basically pre-process the waste under conditions of saturated steam with the pressure reaching 5 bars or 0.5 MPa and temperatures up to 160°C.

6.3.2.9 Mechanical Heat Treatment

Mechanical heat treatment (MHT) is an alternative waste treatment technology that includes mechanical sorting or pre-processing stage followed by a form of thermal treatment. There are two types of heat treatment (Table 6.2).

6.3.2.9.1 Mechanical Preparation

Some initial screening processes are carried out to remove any large items from the waste stream unsuitable for further processing in the system.

6.3.2.9.2 Heat Treatment

The treatment vessel is loaded with waste and either an autoclave or continuous process is used for further treatment. Autoclave is a batch process and carried in a sealed vessel wherein steam is injected at a pressure of 5–7 bar and the vessel is rotated. The total reaction time is 1 h. Alternatively, other continuous heat treatment processes accept shredded waste and the process is carried at predetermined moisture content under atmospheric pressure. The vessel is rotated and a hot air stream passes through the vessel. The process completes in 45 min to remove the waste by mechanical separation.

6.3.3 FUTURE TECHNOLOGIES

6.3.3.1 Smart Waste Bins

Mukherjee et al. (2021) studied the smart waste bin concept incorporating the Internet of Things (IoT) and wireless sensor network technologies. Smart bins involve the integration of IoT devices with the latest technological equipment like solar cells, LEDs, RFID readers, etc. The red and green colour of indicators makes

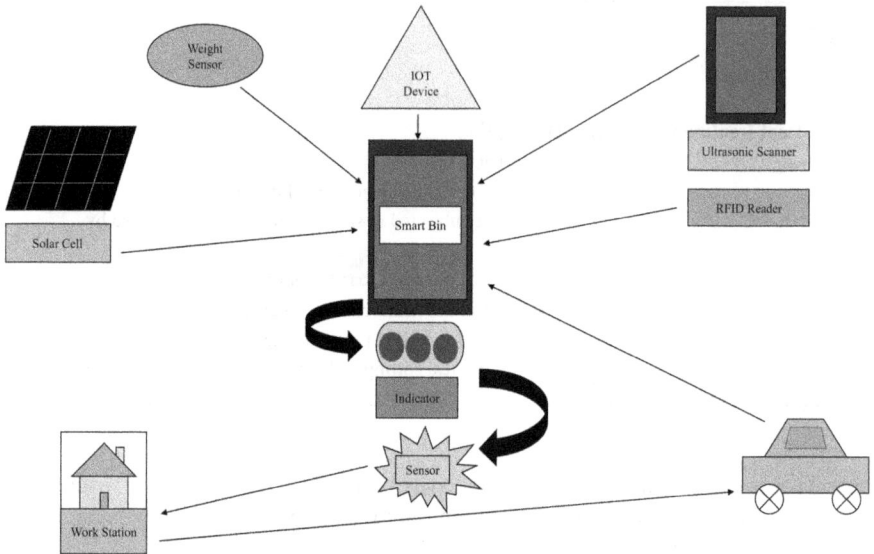

FIGURE 6.4   Schematic flow of smart waste bin management (Mukherjee et al., 2021).

detection easier and small batteries are used to charge the various sensor nodes in wireless sensor networks. The volume of garbage is detected by these sensors The sensors are engineered on the bins to sense the waste and the nodes help receive the sensor data, and then the data is forwarded to the backend server through different internet sources like Ethernet, Wi-Fi, or the 4G connections. The analytics module analyses the data collected by the bin and sends a signal to the nearby base station. This information is then communicated to the concerned authorities, who direct the garbage collection truck to the collect waste bin. Finally, the garbage is sent to the reuse and recycling unit (Figure 6.4).

The garbage is compressed using a piston associated with the cylindrical structure (Gupta et al., 2019). The trash plate is attached to the cylinder and the leaf switch is suspended upside down through the side hole at a point lower than the maximum level. In case of fault from the cleaning team, the above arrangement helps as precautionary measures of the garbage overflow. The compressing plate can reach down to press the switch. Once the threshold level is reached, the garbage is prevented from being dumped inside the trash can in order to avoid overflow.

Advantages of smart bins are as follows (Karthik et al., 2021):

- Reaching streets for the trucks becomes easy and saves fuel.
- Avoids overflow including noise reduction and lesser traffic flow.
- More efficient use of dustbins resulting in less pollution.
- Creates a clean, sanitary, and healthy environment.
- Sensors can reveal real time data on the level of the waste in the garbage bin.
- Process is accurate and speedy.

6.3.3.2 Pneumatic Waste Collection

Pneumatic waste collection helps with the negative impact of air pollution and also economic viability in terms of manual labour costs. One of the best techniques used is the pneumatic C&T technique which is practised in some urban areas (New York City and Helsinki (Finland)) (Yadav and Karmakar, 2020). In this technique, MSW is collected and transported through pipelines by using pressurized air and requires initial very high capital investment, skilled labour, and well-planned urban conglomeration. Proper design of this technique reduces overall GHG emissions significantly. The disadvantage of this technique is that it uses high energy and causes problems such as pipe blockages due to bulky waste which can be overcome using computational fluid dynamics (CFD) based simulation tools.

6.3.3.3 Fleet Management Systems

Fleet management systems aim at emptying waste bins for waste collection and transportation based on a predefined schedule. This system being conventional requires utmost care as it may involve overflowing, partial filling, to complete emptying, which would lead to unnecessary resource consumption. To overcome this problem, wireless sensor networks (WSN) have been deployed in MSW to achieve remote monitoring of filling levels of waste bins. The waste collection trucks can also communicate with waste bins through IoT to acquire the status of waste (Wu et al., 2020). The transportation system involving trucks, drivers, commercial and local customers, dispatch and back-office operations add up to waste and recycling fleet management (Martin, 2014).

The conventional method for waste collection and transportation involves trucks, waste bins, and standard routes and spots of collection involving fuel, labour, and maintenance costs making SWM unviable and less accounting. These problems call for extensive research and technological upgradation involving sensing of filled waste bins and digitization for proper solid waste management. The waste collection model considers priorities with two objectives, the shortest distance and the best service (Wu et al., 2020). This technology transfer shall shorten the distance, reduce GHG emissions, reduce costs, and lessen time and effort. Waste bins located in specific areas (e.g., hospitals, fuel stations, gas stations) are characterized as high priority bins which are collected on priority. The vehicles are located at the disposal centre and start their trips towards the allocated waste bins, collect the waste, and on completion (fully loaded), returns to the disposal centre to upload the collected waste (Wu et al., 2020).

The FMS includes an information system for garbage truck location, forecasting of the scheduled time of arrival, web services, cloud computing, and a GIS module (Chen et al., 2016). As a result of this, the FMS can exactly record the location of every garbage truck and their arrival and expected time cycle for collection of waste from all the bins. The arrival time forecasting method can analyse the historical data of travel time between each pair of collection points generating weighted multiple linear regression models. Map Reduce Models in the Hadoop platform are implemented using the associative laws of addition and multiplication in the weighted multiple linear regression model for big data processing. The current travel time can then be adopted in those weighted multiple linear regression models to obtain the forecasted arrival time via the GTA (Chen et al., 2016).

6.3.3.4 AI-Based Sorting of Waste

A significant contribution to the sorting of waste to generate high-quality secondary products can be achieved using artificial intelligence (AI) and robotics, in the sense of a recycling economy.

Such arrangements for waste sorting improvise the value chain of waste management by precisely sorting high value fractions and diverting them to local or global markets ultimately reducing the overall cost of waste disposal (Wilts et al., 2021). Several technologies can be implemented for sorting waste including air separation (suction hood, zigzag, rotary classifiers, etc.), film grippers, waste screening (trommel screen), ballistic separation, magnetic sorting, film grippers, sensor technology, and manual sorting. The combination of the above technologies makes the entire process more effective. Sorting performance can be highly enhanced using robotic technology and result in the segregation of high value streams without human intervention. One of the major advantages of robotic technology is the collection of multiple wastes at the same time in different categories and containers making the disposal process easier. ZenRobotics Recycler (ZRR) system, a recycling robot that is considered to be the world's first robot-driven waste sorting device with AI and is equipped with computer vision and deep learning algorithms, separate precisely selected waste fractions from solid construction and demolition (C&D) waste. Four different fractions with a purity of up to 98% can be handled using a single ZRR robot arm (Wilts et al., 2021). However, the efficiency and reduction of cost for selective separation of construction materials are achieved by the ZRR robot. Thus, the sorting efficiency can be achieved using sensor-based sorting systems exempting manual sorting options resulting in low operation cost, high recovery rate, and high reorganization capability. The value added product obtained from useless garbage thus can reduce carbon footprint and emissions (Wajeeha et al., 2016).

6.4 ENVIRONMENTAL IMPACT OF SOLID WASTE DISPOSALS

Environment and waste management are impacted by almost all anthropogenic activities resulted due to unrestricted and thoughtless human activities (Chandrappa and Das, 2012).

Solid waste production is inevitable for any process or activity of human beings and every segment of our society generates diverse waste with varied toxicity and quality (Davis et al., 1994; Vergara and George, 2012) resulting in a complex potential threat to the environment. The generation of this waste has potential health hazards and its management directly impacts the health of people, flora, fauna, and the environment (Davis et al., 1994; Vergara and George, 2012). Uncontrolled population growth has resulted in urbanization which in turn has contributed to high solid waste quantities imposing serious concerns and hazards for future generations. Solid waste management and handling is associated with both positive and negative impacts on the environment. Even though proper waste management does reduce the magnitude of impact, it will not eliminate the impact totally. The impact on the environment can occur at any stage of waste management (Chandrappa and Das, 2012).

The production and management of solid waste emit pollutants contributing to climate change and seriously impacting the entire ecosystem. Solid waste affects the environment through air, land, and water degradation which is invariably a part of its production and management (Vergara and George, 2012). The technologies designed to minimize the environmental impact of waste also impact the environment. Figure 6.5 shows the different associated risks of solid waste and their treatment technologies (Farooqi et al., 2021). Sustainable development has to combat associated environmental concerns like leachates from landfill, gaseous emissions from processes like incineration and pyrolysis, global warming, imbalanced nutrient cycle, severe odour from solid waste, and what not (Fatima et al., 2019).

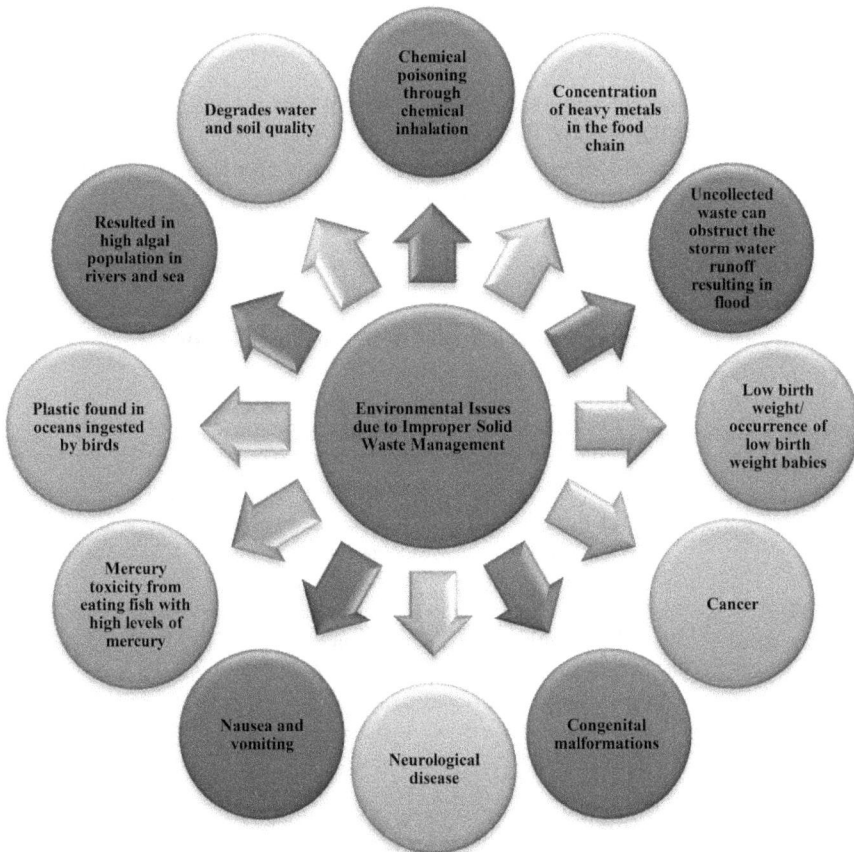

FIGURE 6.5 Environmental issues due to improper solid waste management. (Alam and Ahmade, 2013.)

Several different types of impacts can result from solid waste practices, including the following (Davis et al., 1994):

- Direct impacts resulting from the amounts of wastes produced, managed, and disposed (e.g., land-use impacts, transportation impacts)
- Direct impacts resulting from the constituents contained in the wastes (e.g., mercury in batteries, lead in paint residues)
- Indirect impacts (e.g., waste of materials or other resources) relating to solid waste practices

6.4.1 Aesthetic Degradation

The discarded polythene bags from illegal open roadsides and heavily sized solid waste storage containers often block drains and sewers. Ultimately these blockages are creating traffic blockage, flooding, and unhygienic conditions including rotten odour in the city leading to an aesthetic nuisance (Ejaz et al., 2010).

6.4.2 Pollution

Improper MSW disposal and management causes all types of pollution: air, soil, and water (Alam and Ahmade, 2013) (Table 6.3).

The most critical challenge to the society today is climate change which has threatened biodiversity and human security and has created extreme weather

TABLE 6.3
Summary of the Environmental Impacts of Solid Waste Disposal Technologies (Vergara and George, 2012)

Pollutants Emitted from Various Technologies	Environmental Sink		
	Air	Water	Soil
Dumping/landfill	CO_2, CH_4, odour, noise, VOCs, GHGs (CO_2, CH_4, N_2O)	Leachate, heavy metals, organic compounds	Heavy metals, organic compounds
Incineration	SO_2, HCl, CO, CO_2, N_2O, dioxins, furans, PAHs, VOCs, GHGs, Hg	Fallout of atmospheric pollutants	Fly ash, slag
Composting	Odour, GHGs (minor)	Leachate	Minor impact
Land Application	Bio-aerosols, odour, GHGs (minor)	Bacteria, viruses, heavy metals	Bacteria, viruses, heavy metals, PAH, PCBs
Recycling	GHGs (minor)	Wastewater from processing	Landfilling of residues
Transport	SO_2, CO_2, NO_x, odour	Fallout of atmospheric pollutants (e.g., nitrate)	

conditions, rise in sea level and melting glaciers, elevated temperatures, etc. The waste management systems currently in practice have tried to cope with these challenges and have the potential to be either a net source or sink of gases (Vergara and George, 2012). The decomposition of organic waste in landfills generates greenhouse gases and a complex soup called leachate (Alam and Ahmade, 2013).

Toxic chemicals like trichloroethane, toluene, tetrachloroethane, ethane, mercury, etc. severely affect human beings through air pollution (Fatima et al., 2019). Moreover, drastic pollution is created by open burning of the dumped solid waste collected in common bins (Ejaz et al., 2010). Open burning of MSW and improper incineration emit oxides of carbon, oxides of sulphur, ash, silicates, VOCs, hydrocarbons, dioxins, polychlorinated biphenyls, etc. which contribute significantly to urban pollution (Alam and Ahmade, 2013; Chandrappa and Das, 2012). Handling of waste and further combustion lead to issues of bottom ash, noise, fly ash, dust, and spores (Chandrappa and Das, 2012). Landfills are one of the major sources for emission of methane which has a serious impact if left uncontrolled. The build-up of landfill gas in buildings located near landfills can cause explosion or fire as it is heavier than air and thus is collected in sewers and manholes. This landfill gas generation (LFG) can cause asphyxiation (Chandrappa and Das, 2012). Chlorofluorocarbons (CFCs) are emitted due to dumping of old refrigerators, air conditioners, and old damaged compressors which also add to air pollution (Fatima et al., 2019). Vehicular movement and associated collected waste result in the generation of air pollutants.

Solid waste disposal also results in soil contamination mainly through molybdenum, toluene, nickel, tetrachloroethylene, etc. (Chandrappa and Das, 2012). These toxic chemicals travel through food chain and affect all the successive consumers as they bio-accumulate. The harmful effect of leachate (toxic soup of recalcitrant organics) generated due to indiscriminate dumping of waste on surface and groundwater supplies cannot be ruled out (Alam and Ahmade, 2013). Especially in monsoon, the leachate from open dump sites causes serious environmental issues as leachate generated from solid waste or industrial waste contains heavy metals and PAH (Benzo {a} pyrene) (Ejaz et al., 2010).

6.4.3 VECTORS

Improper waste management accumulates waste and stagnant water which provide burrows and breeding sites for biological disease vectors (disease transmitters). Solid waste in urban cities creates urban zoonosis, especially canine and rodent disease reservoirs typically when the garbage is accumulated for a longer time. The clogged drains create habitats, food, and nutrition for bees, flies, and mosquitos and even for dogs, cats, and rodents which may lead to severe diseases like flea-borne fever (Farooqi et al., 2021; Vergara and George, 2012). Not only the health effects but also the rodents breeding on open dumps and waste drains damage the physical assets including electrical cables. Landfills even host several birds, particularly seagulls, which create noise pollution and their sudden takeoff from the landfill causes severe accidents, especially to aircraft at near airports (Christensen, 2011).

6.4.4   HEALTH IMPACTS

Public health is highly impacted by unmanaged waste which is locally variable and at random sites.

Waste impacts public health and these wastes are locally specific and variable. Its mismanagement affects the normal human being. Farooqi et al. (2021) have studied the health impacts associated with solid wastes and have concluded that gas emissions from this waste can cause skin and eye infections and the dust can cause respiratory issues. Not only this but the waste sorters and collectors also suffer from parasitic, intestinal, respiratory, and skin diseases due to their direct exposure without any safety precautions. These risks multiply into severe infectious diseases when they are involved in biomedical wastes generated from hospitals and clinics. In general, human beings irrespective of whether poor or wealthy can face severe health issues by direct or indirect exposure to this unmanaged solid waste. However, poor people are more likely to be affected as they dwell nearer to these randomly dumped sites and are also involved in their management creating an occupational hazard for them.

Workers at the solid waste disposal sites tend to have higher infection and injury rates than the overall baseline population. The unregulated conditions, lack of protective clothing, and insufficient PPE (personal protective equipment) spikes congenital birth defects, exposure to chronic diseases like cancer (especially among workers at incinerator plants), and respiratory illness (Vergara and George, 2012). The food and water used by these dwellers for routine purposes or even drinking are polluted and cause severe health issues. As per the U.S. Public Health Service, there are 22 human diseases linked to improper municipal solid waste management. Dusting from disposal practices, exhaust fumes from waste collection vehicles, and open and uncontrolled burning emit a number of toxic substances harming the people and their overall health (Alam and Ahmade, 2013).

6.4.5   IMPACT ON FLORA AND FAUNA

The waste disposal and its management results in land-use change and alters the habitat of species (flora and fauna) which directly share the planet with human beings. The emissions of toxic gases and chemicals from these mismanaged sites have an acute impact on both flora and fauna (Vergara and George, 2012).

Animals in search of food often are attracted towards open solid waste, especially municipal solid waste having plastics, which is one of the main reasons for choking of the digestive tract leading to the death of stray animals. Not only animals but several birds also feed on these sites resulting in several health hazards. The solid waste that enters the food chain could be detrimental if the food is contaminated with toxic or infectious material.

Due to deforestation, urban developments have more or less encroached on the sanctuaries and forests and even the wild animals are now affected by feeding on the solid wastes. The water bodies are also fed with wastes which is affecting the aquatic ecosystem.

Bio-aerosols containing fungal or bacterial spores are released during compost-ing and not only this but also the buried deposits of organic waste generate toxic liquids and fumes causing severe community hazards (Ejaz et al., 2010).

Vergara and George (2012) in their studies on ocean pollution have revealed that millions of metric tons of plastics generated through MSW enter the oceans leading up to a 6:1 ratio of plastics to other marine debris in some places which has severe impact of aquatic health. Increased morbidity and mortality have been found in at least 267 species by ingestion or entangled by plastics. Plastics become poison vec-tors by absorbing persistent organic pollutants as they concentrate toxins and this bioaccumulation can propagate through the food chain. Plastic waste can also inhibit gas exchange from the sediments and act as a carrier for invasive species when it settles down.

6.5 CONCLUSION

Waste management, especially solid waste management, is a complex issue today and has to be addressed in a more scientific and professional manner. The exponen-tially increasing population is directly linked to a significant increase in municipal solid waste especially through urbanization and economic development in our coun-try. A spike in global population demands for pronounced need for food and essen-tials leading to the amount of waste generated. The environment, human beings, flora, and fauna are severely affected by poor solid waste management. This chapter thoroughly discussed the latest trends of generation, collection, recycling, and reuse of solid waste and also the waste to energy technologies like pyrolysis. These tech-nologies especially for plastic waste can generate hydrocarbons in the form of gas and oil which can be utilized for energy generation with permissible or no pollution. Thus, WTE technologies for solid waste management are futuristic technologies that are to be well accepted and adopted following all the technical and scientific condi-tions. After reuse, the best and sustainable practice can be recycling which is envi-ronmentally viable and beneficial, especially for solid waste management.

Co-processing is also today a well-adopted concept wherein wastes generated from one industry can be used as a by-product or in blending with other fuels for processing in another industry. However, when using any technologies or recycling methodologies, human health and exposure to toxins are to be taken care of, espe-cially when it involves excreta and other liquid and solid waste from households and the community, which may lead to serious infectious diseases leading to even death.

In this chapter, we have discussed different kinds of conventional and alternative technologies for waste management using which we can reduce the problems related to solid waste. Different types of future technologies also have been discussed to get clarity related to the future scope of solid waste management techniques There are different kinds of environmental impacts that have been studied in order to under-stand the importance of waste management and threats related to poor waste man-agement. At the broader and sustainable aspect, the primary solution to SWM is the minimization of waste and wherein the waste is unavoidable, the second option cer-tainly shall be the recovery of materials and energy from waste as well as recycling of waste. To reduce the environmental impact to its minimal level, the future needs a paradigm shift towards developing indigenous technologies to convert solid waste to clean and economically viable energy, preferably using greener and sustainable technologies.

REFERENCES

Alam, P., Ahmade, K., 2013. Impact of Solid Waste on Health and the Environment. *International Journal of Sustainable Development and Green Economics*, 2(1), 165–168.

Chandrappa, R., Das, D., 2012. Solid Waste Management Principles and Practice. Springer. https://doi.org/10.1007/978-3-642-28681-0

Chen, C., Yang, Y., Chang, C., Hsieh, C., Kuan, T., Lo, K., 2016. The Design and Implementation of a Garbage Truck Fleet Management System. *South African Journal of Industrial Engineering*, 27(01), 32–46. http://dx.doi.org/10.7166/27-1-982

Chen, Z., Zhang, H., 2013. Introduction to Solid Waste Pollution Control and Improvement. *Advanced Materials Research*, 664, 236–239. https://doi.org/10.4028/www.scientific.net/AMR.664.236.

Christensen, T., 2011. Solid Waste Technology and Management (I & II). John Wiley and Sons, Ltd., Publication. https://doi.org/10.1002/9780470666883

Davis, M., Holter, G., Deforest, T., Stapp, D., Dibari, J., 1994. Possible Global Environmental Impacts of Solid Waste Practices, PNL-10149, UC-249, Pacific Northwest Laboratory, Richland, Washington.

Ejaz, N., Akhtar, N., Nisar, H., Naeem, A., 2010. Environmental Impacts of Improper Solid Waste Management in Developing Countries: A Case Study of Rawalpindi City. *WIT Transactions on Ecology and the Environment: The Sustainable World*, 143, 379–387. https://doi.org/10.2495/SW100351

Farooqi, Z., Kareem, A., Faizan, R., Shujahat, A., 2021. Solid Waste, Treatment Technologies, and Environmental Sustainability: Solid Wastes and Their Sustainable Management Practices, Handbook of Research on Waste Diversion and Minimization Technologies for the Industrial Sector, 35–37. IGI Global. https://doi.org/10.4018/978-1-7998-4921-6.ch003

Fatima, S., Chaudhry, M., Batool, S., 2019. Environmental Impacts of the Existing Solid Waste Management System of Northern Lahore. *Chinese Journal of Urban and Environmental Studies*, 7(3), 1950013. https://doi.org/10.1142/S2345748119500131

Golomeova, S., Srebrenkoska, V., Krsteva, S., Spasova, S., 2013. Solid Waste Treatment Technologies. *Machines.Technologies.Matter*, 9, 59–61.

Grande, L., Pedroarena, I., Korili, S., Gill, A., 2021. Hydrothermal Liquefaction of Biomass as One of the Most Promising Alternatives for the Synthesis of Advanced Liquid Biofuels: A Review. Materials (Basel), 14(18), 5286. https://doi.org/10.3390/ma14185286

Gupta, P., Shree, V., Lingayya, H., Rajendran, S., 2019. The Use of Modern Technology in Smart Waste Management and Recycling: Artificial Intelligence and Machine Learning. 173–188, Springer. https://doi.org/10.1007/978-3-030-12500-4_11

Hsu, P., Foster, K., Ford, T., Wallman, P., Watkins, B., Pruneda, C., Adamson, M., 2000. Treatment of Solid Wastes with Molten Salt Oxidation. *Waste Management*, 20(5-6), 363–368.

Jain, A., Singhal, M., 2014. Waste Minimization. Environmental Sustainability: Concepts, Principles, Evidences and Innovations. 11–18, Excellent Publishing House.

Karthik, M., Sreevidya, L., Nithya Devi, R., Thangaraj, M., Hemalatha, G., Yamini, R., 2021. An Efficient Waste Management Technique with Iot Based Smart Garbage System. *Materials Today: Proceedings*. https://doi.org/10.1016/j.matpr.2021.07.179

Lu, X., Jordan, B., Berge, N., 2012. Thermal Conversion of Municipal Solid Waste via Hydrothermal Carbonization: Comparison of Carbonization Products to Products from Current Waste Management Techniques. *Waste Management*, 32, 1353–1365. http://dx.doi.org/10.1016/j.wasman.2012.02.012

Martin, D., 2014. Best Practices in Waste Fleet Management. *Waste Advantage Magazine (The Advantage in Waste and Recycling Industry),* November 7, 35.

Mukherjee, A., Wanjari, U., Chakraborty, R., Kaviyarasi, R., Vellingirid, B., George, A., Sundara Rajan, C.R., Gopalakrishnan, A., 2021. A Review on Modern and Smart Technologies for Efficient Waste Disposal and Management. *Journal of Environmental Management*, 297, 113347. https://doi.org/10.1016/j.jenvman.2021.113347

Vergara, S., George, T., 2012. Municipal Solid Waste and the Environment: A Global Perspective. *Annual Review of Environment and Resources*, 37, 277–309. https://doi.org/10.1146/annurev-environ-050511-122532

Wajeeha, S., Zulfiqar, A., Tahir, M., Asif, F., Ghazala, Y., 2016. Latest Technologies of Municipal Solid Waste Management in Developed and Developing Countries: A Review. *International Journal of Advanced Science and Research*, 1(10), 22–29.

Wilts, H., Garcia, B., Garlito, R., Gómez, L., Prieto, E., 2021. Artificial Intelligence in the Sorting of Municipal Waste as an Enabler of the Circular Economy. *Resources*, 10, 28. https://doi.org/10.3390/resources10040028

Wong, J., Surampalli, R., Zhang, T., Tyagi, R., Selvam, A. 2016. Waste Management and Sustainability: An Introduction, Sustainable Solid Waste Management. American Society of Civil Engineers (ASCE). https://doi.org/10.1061/9780784414101.ch01

Wu, H., Tao, F., Yang, B., 2020. Optimization of Vehicle Routing for Waste Collection and Transportation. *International Journal of Environmental Research and Public Health*, 17(14), 4963. https://doi.org/10.3390/ijerph17144963

Yadav, V., Karmakar, S., 2020. Sustainable Collection and Transportation of Municipal Solid Waste in Urban Centers. *Sustainable Cities and Society*, 53, 101937. https://doi.org/10.1016/j.scs.2019.101937

Zaman, B., Oktiawan, W., Hadiwidodo, M., Sutrisno, E., Purwono 2018. Bio-Drying Technology of Solid Waste to Reduce Greenhouse Gas. *E3S Web of Conferences*, 73, 05019. https://doi.org/10.1051/e3sconf/20187305019

7 Circular Economy and Solid Waste

Darshan Salunke, Pratibha Gautam, and Alok Garg

7.1 INTRODUCTION

As laid down by legislature, wastes are considered as any product, object, or material whose initial use has been exploited, and then it remains of no use for its consumer, and whose final fate is to become discarded material in the environment. Due to the increase in human population along with an increase in the standard of living, day by day the consumption of goods, services, and energy has increased many folds. To meet this demand for products, services, and energy, advancement in technology has occurred through industrialization and automation but the upgradation of technology means that the consumer can possess and utilize many more personal devices and tries to update them more frequently. On the contrary, the higher consumption rate promotes a higher generation of waste and ultimately leads to the degradation of the environment and health.

Conventionally, after the use of any product, the solid waste is rejected and dumped into land, water, or atmospheric environment. Various adverse health and environmental impact consequences are observed due to the inappropriate disposal of solid wastes. Such a disastrous situation is a result of linear economical model which works on the philosophies of "take-make-dispose" and is to date orthodoxically supported and followed by many manufacturers and industrialists. Ultimately, as a shift from a responsive approach to managing waste, the treatment of solid waste has evolved.

By the year 2050, the waste generated worldwide is expected to increase by more than 3 billion tons which is at present nearly 2 billion tons (Kaza et al., 2018). The present scenario of global waste composition is shown in Figure 7.1. The figure clearly depicts that around 60% of waste generation is of a biodegradable nature and can be utilized for composting purposes. The figure also shows that around 25% of waste has great potential for recycling and recovery from the waste generated.

It is estimated that if no proper improvements are made in the solid waste management sector, then the emissions related to solid waste will increase by around 2.40 billion tons per year by the end of the year 2050 (Eurostat, 2022). The global trend of waste disposal and treatment in the present scenario is shown in Figure 7.2. From the figure, it can be depicted that more than 55% of waste treatment and disposal facilities around the world are non-engineered and can be harmful to health and the environment. Roughly 13% of waste has been recycled and hardly 11% of waste has been incinerated to recover energy from waste. Efforts and technologies need to be made in direction of the managing this 33% of global open dumping practices.

DOI: 10.1201/9781003352396-7

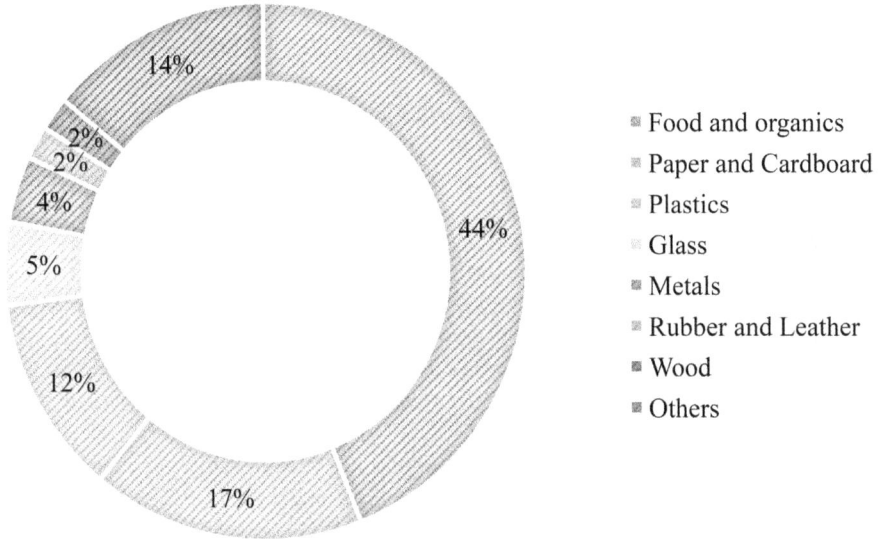

FIGURE 7.1   Global waste composition.

7.2   UNDERSTANDING CIRCULAR ECONOMY

The concept of circular economy (CE) emerged in the 1970s with the basic thought of reducing raw material consumption in industries, but this proved to be having greater potential in waste management. The acceptable definition of circular economy as per the Ellen MacArthur Foundation (EMF) is a production system that focuses on the aim of restoration and regeneration, shifting towards the use of green and renewable

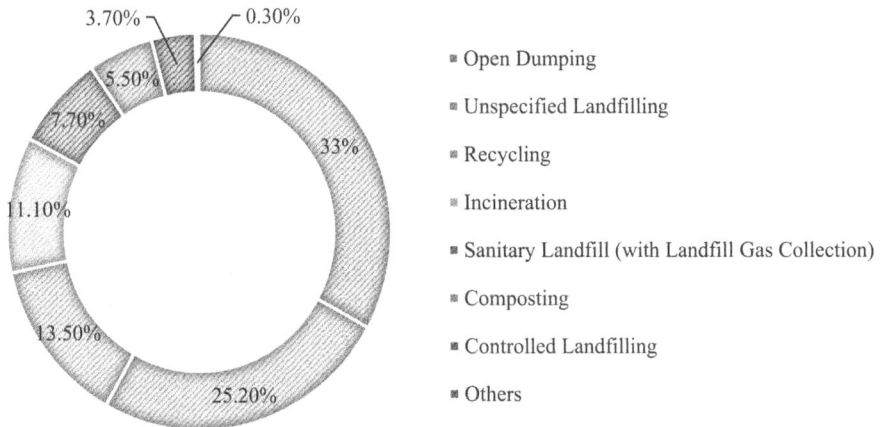

FIGURE 7.2   Global treatment and disposal of waste.

energy, rejects the use of toxic chemicals and raw materials, which supports reuse and focuses on eliminating generation of waste by higher quality design of products, materials, services, and system within the business model, and replaces the traditional concept of end-of-life (Ellen MacArthur Foundation 2016). The primary goals of the CE are to reduce resource wastage like energy and raw materials, to reduce environmental impacts, to dissociate the consumption of resources from the growth of GDP, and to increase human health and well-being by providing higher job opportunities (CIRAIG, 2015).

The foremost features of CE can be briefed as: it plays the role with communities or production plants that are emulating the behaviour of ecosystems; it depends on a closed circular loop system; it is a paradigm shift from "cradle-to-grave" to "cradle-to-cradle", from disposing off to restoration, from "take-make-dispose" to "take-make-recreate"; CE promotes on long usage rather than consumption; it is concentrated on eco-effectiveness rather than on eco-efficiency; it is based on systems thinking; and it requires resources which are biodegradable and are possible for permanent reuse (Bermejo, 2014; Ellen MacArthur Foundation 2016).

7.2.1 Barriers to Implementing CE

In the implementation of CE, a few huge barriers are creating numerous hindrances. These barriers as shown in Table 7.1 can be classified into three groups: (a) legislative barriers, (b) technological and skill barriers, and (c) lack of public participation (Geng and Doberstein, 2010).

TABLE 7.1
Barriers in Implementing CE

Legislative Barriers	Technological and Skill Barriers	Lack of Public Participation
Lack of a centralized platform to promote CE	Lack of superior environmental technologies to meet global demand	Insufficient knowledge of CE principles by governments and industries
Lack of corporate enthusiasm due to existing regulatory framework	Scarce technical competences	Lack of awareness campaigns
Low effluent discharge fees	Lack of financial support from the industries or government	Unaddressed and underdeveloped eco-industrial networks
Loopholes in taxation systems that demotivate recycling	Restrictions in competent training	Lack of willingness to accept the innovation
Lack of comprehensive policies to promote green technologies and products	Insufficient technology transfer from developed to developing countries	
Non uniform regional development	Lack of collaboration among various stakeholders	

7.3 CIRCULAR ECONOMY MODELS

As an economic domineering nature of CE, many circular business models are developed and applied at corporate levels. These models are described as follows (CIRAIG, 2015):

- **Circular Supplies Model:** It is based on the concept of cradle-to-cradle (C2C) which deals with the circular loop of decomposable resource raw materials as well as renewable and greener energy.
- **Resource Recovery Model:** It is based on the concept of converting wastes into useful raw materials for the self or other industries nearby.
- **Product Life Extension Model:** It is based on the main principle of executive economy and works on the aim of reuse, repair, upgradation, remaking of value-added products, and remanufacturing.
- **Sharing Platforms Model:** It operates on collaborative consumption along with a shared economy which accelerates the utilization rate of production and services by effective sharing and distribution among the users.
- **Product as a Service Model:** It provides users an economic arrangement as pay-for-use or on lease.
- **Ecological Transition Model:** This concept draws the system and process towards a more sustainable approach such as "low carbon economy transition", "transition based on socio-economic welfares", "sustainability transition", and many more.
- **Green Economy Model:** It focuses on tackling environmental problems by putting forward economic instruments. It results in better human health and socio equality, reducing risk to the environment and decreasing ecological scarcity.

7.4 MATERIAL RECOVERY FOR CE

The management of solid waste should not only be environmentally focused but should also be cost-efficient as well as globally and societally acceptable. There are various factors that affect the solid waste management process (Sorrell, 2015). As shown in Figure 7.3, the first factor that influences waste management is a political or legislative framework. Governmental schemes, regulations, and taxation patterns aid the new technologies and companies to look forward in managing their waste rather than just dumping it. Factors like environment, society, and the economy go hand in hand for a sustainable approach. Technical innovation and educational skills provide the backbone structure for solid waste management practices.

Proceeding further in the directions and concepts of CE, there is reduction in the utilization of virgin raw materials in the production process and instead preference for recovered materials. Material recovery is the eco-sustainable concept of CE implementation. In general, from various categories of solid wastes, different types of wastes can be recovered. From construction and demolition (C&D) waste, materials like metals, glass, wood, plastics, paper, mortar, cardboard, gypsum, stones, etc. can be recovered which can be utilized for construction of roads and pavements or

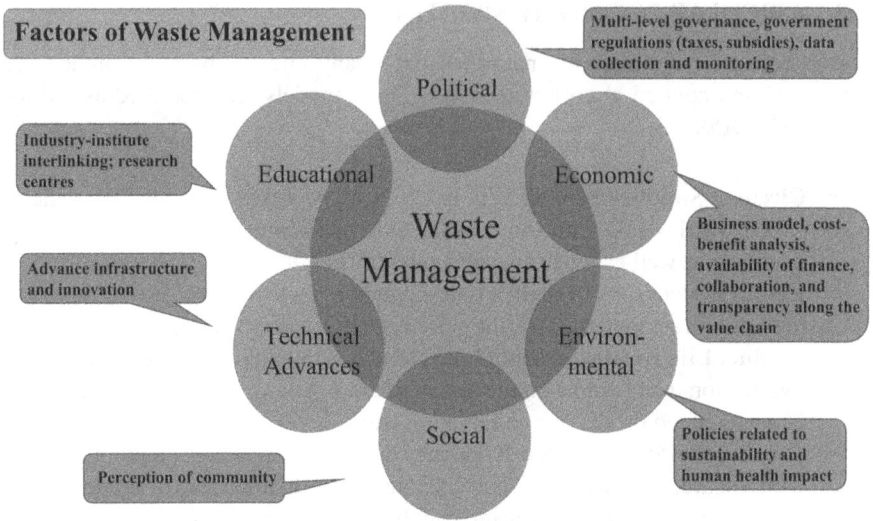

Factors of Waste Management

Political — Multi-level governance, government regulations (taxes, subsidies), data collection and monitoring

Educational — Industry-institute interlinking; research centres

Technical Advances — Advance infrastructure and innovation

Economic — Business model, cost-benefit analysis, availability of finance, collaboration, and transparency along the value chain

Environmental — Policies related to sustainability and human health impact

Social — Perception of community

Waste Management

FIGURE 7.3 Factors affecting waste management.

for low calorific fuels. From food waste including kitchen waste, materials like soil conditioners, animal and fish feeds, and specialty chemicals like pectin from citrus peel leftovers, as well as casein from milk by-products, phenols from grapes, etc. can be recovered, all of which have good market value. Rare earth elements and many metals like copper, mercury, and cerium can be recovered from batteries, motors, generators, and superconductors. From fluorescent lamps, recovery of phosphorus can be done, which is in high demand these days. Due to the increase in e-commerce online shopping, a variety of packaging waste is generated from which glass, plastic, paper, and cardboard can be recovered and recycled for further usage. The most common practices for the recovery of metals from e-waste are established and implemented at industrial levels. Even Japan has been in the media for a long time to recover gold, silver, and copper from e-waste and utilize these metals in preparation for Olympics medals for the Tokyo Olympics 2020 (Phys.org, 2022).

7.5 ENERGY RECOVERY FOR CE

The global population crossed 8 billion on 15th November 2022 (UN DE&SA, 2022) and with the tremendous rise in population, an increase in goods consumption and energy demand has been also witnessed. According to the bp Statistical Review of World Energy and reports of world final energy, global electricity consumption has risen by 6% in the last decade (bp Statistical Review of World Energy, 2022). Out of this electricity consumption, 61.4% of electricity has been derived from fossil fuels, of which 36% is from coal, 2.5% is from crude oil, and 22.9% is from gas (World Energy Data, 2022). On another side, the rise in consumption of goods leads to higher waste generation. Also, the increase in standard of living and wealth has a strong relation with greater consumption of energy. It is evident that alone municipal solid waste has an average heating value of roughly 10 MJ/kg, which does make sense for using waste as a source of energy (Malinauskaite et al., 2017).

Conventionally, only incineration has been associated with the concept of waste-to-energy (WtE). However, the concept has a much broader sense in the treatment of waste and energy generation from it. For example, producing heat or electricity from waste or making refused derived fuels from waste. Transforming waste into energy can be considered a significant element of the circular economy (Becidan et al. 2015). Similar to recovering materials from the waste stream, energy recovery is an attractive method of waste treatment and management. From food and processed food waste, biogas can be derived, and from cooking oils and fats, biofuels can be derived. In Poland, organic waste can be utilized to obtain high-quality solid fuels (Malinauskaite et al., 2017). In Slovenia, municipal waste treatment centres are equipped with landfill gas collection systems and have been reported to produce electricity approximately 16 GWh in the year 2015. In Norway, sewage sludge and food waste-based biogas plants are capable of producing annually 200 GWh of power (Malinauskaite et al., 2017).

7.6 INDUSTRIAL SYMBIOSIS

Industrial ecology (IE) is described as a theoretical outline, an execution method, and also referred to as industrial symbiosis where industries and companies go hand in hand such that one industry can benefit by taking its raw materials from the waste or by-product produced from another industry (CIRAIG 2015). As per the study, the main aim of industrial symbiosis can be summarized as: (a) limit the usage of products that are toxic in nature, (b) process and/or product should be designed in such a manner that it is environmentally friendly in nature, (c) replacement of materials that are non-renewable and/or dangerous, (d) to promote the concept of reuse, repair, and remanufacturing, (e) mining of waste and look for the valuable that can be derived from it, and (f) to develop the robust and advanced technologies that consume less resource (van Dijk et al., 2014).

7.7 CE INTEGRATED WITH SDGS

Management of solid waste is considered one of the global issues that is believed to affect the major dimensions of sustainability that are environmental, social, and economical (Rodic and Wilson, 2017). The advancement in solid waste management practices and the goals of sustainable development goals (SDGs) can be interlinked and can be dealt with a single intention. For example, SDG goal 3 of good well-being, goals 6 and 13 for solving environment-related issues, and goal 11 for looking for resources from waste or by-products that have values (Wilson, 2007).

While applying CE in the management of solid waste, many UN-SDGs can be achieved for a better environment, economy, and society. For instance, the concept of CE is based on reducing consumption of raw materials and optimizing production; by working on such ideology, SDG 12 can be easily achieved and the corporate can move towards sustainability. By focusing on the management of solid waste, particularly on plastic waste disposal, emerging wastes like micro-plastics can be handled and a robust solution to them can be achieved which will aid in the achievement of SDG 14 goal as this waste will be restricted from entering the marine ecosystem and life in marine ecosystem can be sustained. Sustainable goals 6 and 13 which are focused on human health and climate change can be achieved if the uncontrolled dumping of solid waste is taken care of. The finances involved in implementing CE in managing solid waste are much less than the actual budget that will be required to address these mentioned SDGs as such (Kaza et al., 2018). Solid waste management implemented with a circular economy can precisely be interlinked with 12 UN-SDGs out of 17 (Rodic and Wilson, 2017).

7.8   CONCLUSION

Regardless of many hindrances in implementing circular economy-based management of solid waste, it is evident that its practice is a sustainable approach and would aid in fetching direct economic benefits as well as indirect benefits by creating job opportunities; it would also be helpful environmentally by assuring the pollution free air, water, and land, and the approach would indeed be helping society by improved community health, better lifestyle, and better education and to sum up all of them, it would be able to achieve all major goals of SDGs.

Technological advancement as well as creating awareness in this regard needs to be the sowing element in the field of solid waste management and CE. Along with skill and technology, legislative and taxation policy should be made robust in this direction. This would encourage the stakeholders to divert their funds into it. Technological development should be focused more on the use of renewables and should look upon the scope of recoverable material or energy from waste. There is tremendous scope in the said direction in the near future to achieve SDGs by 2050!

REFERENCES

Becidan, M., Wang, L., Fossum, M., Midtbust, H., Stuen, J., Iver, J., et al., 2015. Norwegian waste-to-energy (WtE) in 2030: Challenges and opportunities. Chemical Engineering Transactions. 43, 2401–2406. https://www.aidic.it/cet/15/43/401.pdf

Bermejo, R., 2014. Handbook for a Sustainable Economy. Springer.

bp Statistical Review of World Energy, 2022; 71st edition. Retrieved from: https://www. bp.com/content/dam/bp/business-sites/en/global/corporate/pdfs/energy-economics/ statistical-review/bp-stats-review-2022-full-report.pdf

CIRAIG International Reference Centre for the Life Cycle of Products, Processes and Services, 2015. Circular economy: A critical review of concepts. https://ciraig.org/wp-content/uploads/2020/05/CIRAIG_Circular_Economy_Literature_Review_Oct2015. pdf

Ellen MacArthur Foundation, 2016. Circular Economy – UK, Europe, Asia, South America & USA. Ellen MacArthur Foundation. http://www.ellenmacarthurfoundation.org/ circular-economy

Eurostat, 2022. Greenhouse gas emissions from waste. https://ec.europa.eu/eurostat/web/ products-eurostat-news/-/DDN-20200123-1

Geng, Y., Doberstein, B., 2010. Developing the circular economy in China: Challenges and opportunities for achieving "leapfrog development". The International Journal of Sustainable Development & World Ecology. 15, 231–239. https://doi.org/10.3843/SusDev.15.3:6

Kaza, S., Yao, L.C., Bhada-Tata, P., Van Woerden, F., 2018, Washington, DC. What a Waste 2.0: A Global Snapshot of Solid Waste Management to 2050. Urban Development. World Bank Publications. https://elibrary.worldbank.org/doi/abs/10.1596/978-1-4648-1329-0.

Malinauskaite, J., Jouhara, H., Czajczyńska, D., Stanchev, P., Katsou, E., Rostkowsk, P., Thorne, R.J., Colón, J., Ponsá, S., Al-Mansour, F., Anguilano, L., Krzyżyńska, R., López, I.C., Vlasopoulos, A., Spencer, N., 2017. Municipal solid waste management and waste-to-energy in the context of a circular economy and energy recycling in Europe. Energy 2017. 10.1016/j.energy.2017.11.128

Phys.org, 2022. Retrieved from: https://phys.org/news/2019-02-tokyo-gold-silver-bronze-e-waste.html

Rodic, L., Wilson, D.C., 2017. Resolving governance issues to achieve priority sustainable development goals related to solid waste management in developing countries. Sustainability. 9(3), 404. https://doi.org/10.3390/su9030404

Sorrell, S., 2015. Reducing energy demand: A review of issues, challenges and approaches. Renewable and Sustainable Energy Reviews. 47, 74–82. http://dx.doi.org/10.1016/j.rser.2015.03.002

UN DE&SA, 2022. Retrieved from: https://www.un.org/en/desa/world-population-reach-8-billion-15-november-2022#:~:text=The%20global%20population%20is%20projected,today%20on%20World%20Population%20Day

van Dijk, S., Tenpierik, M., van den Dobbelsteen, A., 2014. Continuing the building's cycles: A literature review and analysis of current systems theories in comparison with the theory of cradle-to-cradle. Resources, Conservation and Recycling. 82, 21–34.

Wilson, D.C., 2007. Development drivers for waste management. Waste Management & Research. 25, 198–207. https://doi.org/10.1177/0734242X07079149

World Energy Data, 2022. Retrieved from: https://www.worldenergydata.org/world-final-energy/

8 Microplastics as Emerging Soil Pollutants
Ecological Impact and Management Strategies

Swati Sharma and Anita Saini

8.1 INTRODUCTION

The ubiquity of microplastics in the environment has become a major concern in the past few years. Microplastics refer to the plastic fragments less than 5 mm in size (Anik et al., 2021). Microplastics are used abundantly in a wide variety of materials including pharmaceutical, cosmetic, and cleaning products, from which they are discharged directly into the environment during the product's use. Also, the problem of microplastic pollution is rising with the rise in the use of larger plastics. This is because plastics remain persistent in nature for a very long time and are often fragmented naturally into microplastics of various types. No matter what their origin is, microplastics, being smaller and lighter, are transported to larger distances causing pollution of diverse ecosystems. Microplastics from various sources may reach the soil ecosystems through municipal solid wastes, sludge, floods, rain, surface runoff, etc. The agricultural soil may also get polluted during irrigation, mulching, and soil amendment processes (Zhao et al., 2017). The microplastic pollution of soil may influence the soil characteristics depending on their shape and chemical composition, which in turn causes serious ecological effects. Microplastics have been reported to affect the health of soil, plants, and humans. Their translocation within plant systems causes toxicity in plants (Zang et al., 2020; Sajjad et al., 2022). Microplastics even act as carriers for pathogenic microbes, harmful chemicals, and toxins leading to their distribution in different components of the environment (Miri et al., 2022; Sajjad et al., 2022). The soil biota is at risk of immediate and direct exposure. The microplastics are often transferred to food webs through their ingestion by soil organisms (Guo et al., 2020). Their accumulation in tissues of an organism, including human beings, has deleterious effects (Anik et al., 2021). Therefore, in order to protect the health of soil and organisms, effective management strategies are needed along with more efficient detection methods for the monitoring and mitigation of microplastics at the required points. This chapter gives a detailed account of the sources and transport of microplastics in soil along with their ecological effects, analysis, and mitigation approaches.

DOI: 10.1201/9781003352396-8

8.2 MICROPLASTICS: TYPES, SOURCES, AND TRANSPORT TO SOIL

Microplastics may be classified into primary and secondary microplastics based on their origin. The microplastics, when manufactured in a size range of 10 to 500μm, are called primary microplastics. They are produced purposefully for various applications (Guo et al., 2020; Anik et al., 2021; Woo et al., 2021). The primary microplastics may enter the soil ecosystems from either the manufactured products containing them or from the products during their utilization (Figure 8.1). The common sources include cosmetic products, cleaning agents, medical devices, air blasting, paints, coatings, ink, food supplements, nurdles used in plastic products, etc. (Guo et al., 2020; Chia et al., 2021). Drug delivery systems in pharmaceuticals and dental polishes are considerable sources of microplastics in ecosystems (Anik et al., 2021). The solid formulations (tablets and capsules) make use of microplastic additives that may be used as binders or lubricants. They are also used for microencapsulation purposes in pharmaceutical products. Additionally, microfibers found in the textile materials are released into the environment during manufacturing processes and laundry. The nets and ropes used in the agricultural systems are a significant contributor to microplastics in agricultural soils. Road transport is another source that generates large amounts of microplastics that are emitted as a result of tire abrasion during tires' wear and tear (Anik et al., 2021). The road marking paints and coatings consist of thermoplastic elastomers that are also worn out as abrasions and cause microplastic pollution. The microplastics from the tires and markings are mostly released in the air from where they are added to the soil through atmospheric transport and

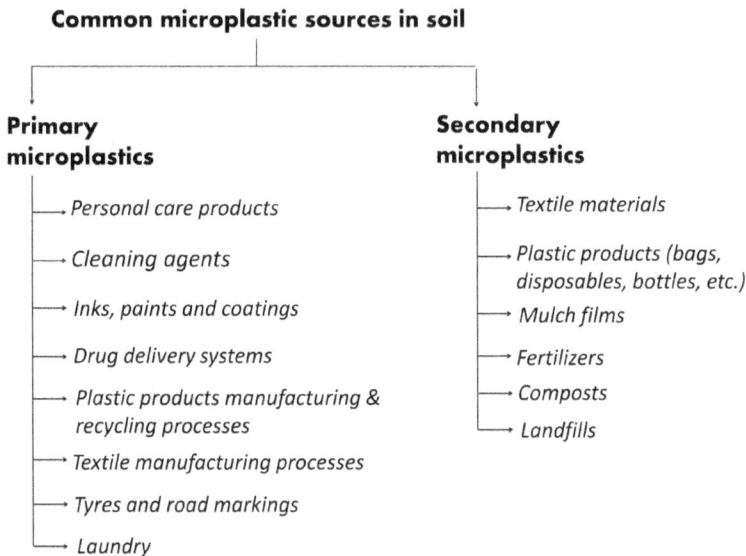

FIGURE 8.1 Common sources of microplastics in soil ecosystems.

precipitation. The runoff from roads also brings these microplastics to soil ecosystems. Microplastics are also released into the environment from plastic production and recycling sites during production, recycling, and transport.

Secondary microplastics are generated from larger plastics as a result of their fragmentation in the environment due to photo-oxidation (by UV), mechanical breakdown, and biological activities (Yang et al., 2021; Woo et al., 2021). Thus, most of the plastic refuse from different sources gives rise to large amounts of secondary microplastics (Figure 8.1). During photo-oxidation, the UV radiations cause plastic weakening and reduce the chain length of plastic polymers that ultimately causes fragmentation to smaller sizes. The UV-mediated oxidation even increases the susceptibility of plastics to further enhanced rates of photo-oxidation. The rate of fragmentation process, however, depends on many factors including plastic composition. The stabilizers and other additives used in the manufacturing of plastic products make them resistant to faster photodegradation. In soil, diverse microorganisms are actively involved in different types of biodegradation processes. Some microbes, though slowly, degrade plastic materials, especially weakened fragments, and release micro- and nano-microplastics in the soil (Dey et al., 2023). A significant amount of secondary microplastics is also generated from landfills and composts containing plastic wastes (de Souza Machado et al., 2018; Sajjad et al., 2022). This is because of high temperature conditions and microbial activity in these systems. Other soil organisms such as earthworms also play an important role in the generation of secondary microplastics when they ingest, partially degrade, and then defecate plastic wastes.

In agricultural soil, plastic mulch films used to conserve soil moisture and improve soil fertility are often a common source of secondary microplastics (Anik et al., 2021; Chia et al., 2021; Yang et al., 2021). The microplastics are released from the mulch films due to their breakdown as a result of exposure to solar UV and tillage activities (Zang et al., 2020). The composts are widely used in agricultural systems to improve soil fertility. The compost may consist of primary microplastics from varied sources or secondary microplastics may be generated due to natural degradation processes in the compost such as thermophilic conditions and microbial degradation (Zang et al., 2020). Other common sources of microplastics in agricultural soil include sewage sludge and wastewater (Chia et al., 2021; Yang et al., 2021). This is because a known treatment process for the sludge or wastewater does not ensure the complete removal of microplastics. The retained microplastics enter the agricultural soils when treated sludge and wastewater are used for fertilizer and irrigation purposes, respectively (Yang et al., 2021; Perez-Reveron et al., 2022).

The microplastics are also classified into different types based on their shape. The predominant shapes include microbeads, pellets, microfibers, particles, films, fragments and powders, etc. (Yang et al., 2021; Lamichhane et al., 2022). The microplastics vary in their types based on their composition as well (Chia et al., 2021; Lamichhane et al., 2022). Common microplastics, varying in their composition, that are found in the soil are given in Table 8.1. Microplastics may be transparent, black, white, and variously colored (Yang et al., 2021).

Microplastics may enter the soil from different routes, including air, water, wastewater, sludge, landfill dumping, etc. (Chia et al., 2021; Yang et al., 2021) (Figure 8.2). Once microplastics enter the soil systems, they are distributed heterogeneously within

TABLE 8.1

Common Microplastic Types in Soil

Polymer Type	Structure	Examples
Polyethylene (PE)		Plastic bags, containers & wraps
Polypropylene (PP)		Plastic bottle lids, tapes, carpets, ropes
Polystyrene (PS)		Disposable cutlery & utensils, food packaging materials, insulation boxes, paints
Polyethylene terephthalate (PET)		Construction materials, water bottles, jars
Polyamide (PA)/ Nylon		Textile, carpets, fishing ropes & gears
Polyvinyl chloride (PVC)		Pipes, cables, window frames, roofing, flooring, wall & electrical panels

the soil ecosystem. They are transported both vertically and horizontally through various natural processes and anthropogenic activities (Lamichhane et al., 2022; Perez-Reveron et al., 2022) (Figure 8.2). The horizontal migration of microplastics is governed by surface runoff, wind, erosion, and tillage activities. Wind erosion causes introduction of microplastics in the air that causes their transport to larger distances on the land. The runoff from the domestic, industrial, or agricultural sites may bring various types of microplastics into the terrestrial systems. Microplastics move vertically downwards under the effect of gravity through soil pores and cracks (Guo et al., 2020). To move rapidly through these channels, microplastics should have smaller sizes. Some of the soil animals produce continuous pores in the soil that provide channels for faster migration of microplastics. Smaller microplastics also tend to move rapidly along with infiltrating water in a process of leaching (Perez-Reveron

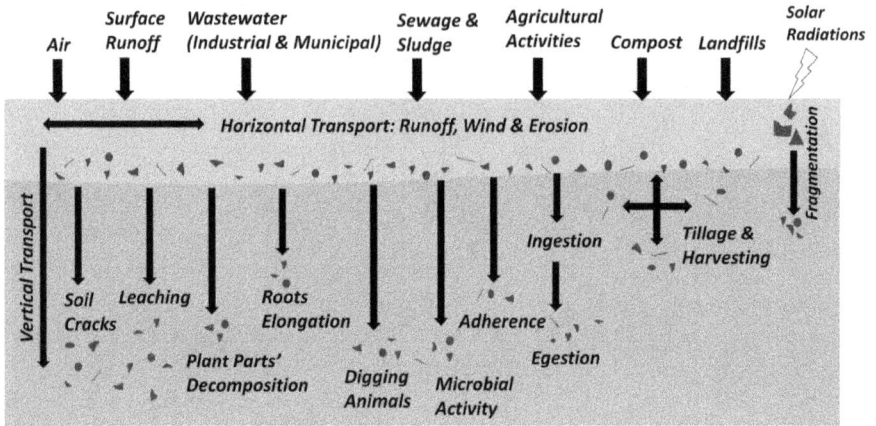

FIGURE 8.2 Transport of microplastics to and within soil.

et al., 2022). Their transport in the soil is influenced by characteristics of both soil and microplastics. For instance, high-density polymers tend to migrate to deeper soils, whereas low-density polymers may remain buoyant in saturated soils and may get diverted to different horizontal migration routes. Also, more coarse soils tend to accumulate more microplastics.

The tillage and ploughing activities also cause the vertical transport of microplastics but it is limited to different depths in only the upper horizon of the soil (Lamichhane et al., 2022; Perez-Reveron et al., 2022). The elongation of plant roots is another factor responsible for vertical transport of microplastics in soil (Guo et al., 2020). More the branching in the root system, the greater the number of pores created and hence microplastics migrate to different depths. When pores are filled with water, they provide buoyancy to the microplastic particles that tend to float at different locations in the soil layers. Some other biological factors also cause migration of microplastics in the soil. Presently, limited information is available on the interaction of microplastics with the microorganisms in soil. Some studies advocate the migration of microplastic through the surface of fungal mycelia and microbial pathogens (Perez-Reveron et al., 2022). Bacterial translocation through soil pores may contribute to the migration of microplastics swallowed by them. Contrarily, several bacterial secretions enhance the electrostatic interactions between microplastics and soil particles resulting in reduced migration rate of microplastics. Microbial biofilms also play an important role in microplastic transport within the soil matrix. The soil contains a large number of smaller invertebrates. Among them, earthworms perform activities such as excreting, burrowing, and feeding. The earthworms play a role in the vertical transport of microplastics through digging of burrows, ingestion-excretion, and also through attachment of microplastics to their body surface (cutaneous transport) (Guo et al., 2020; Lamichhane et al., 2022). The earthworms also retain microplastics in their casts that are transported to different routes (Guo et al., 2020). Micro-arthropods present in the soil that feed on these casts further distribute microplastics (Sajjad et al., 2022). The digging mammals (e.g., moles and gophers), mites, and collembolan are other biotic players in the transport of microplastics within the soil (Guo et al., 2020). Several studies have revealed presence of microplastics in excreta of sheep and chickens (Yang et al., 2021; Perez-Reveron et al., 2022). The microplastics have also been found in feces of deer, terrestrial birds' nests, etc. When microplastics are aged, they exhibit higher electrostatic forces over their surfaces that tend to promote their aggregation. The invertebrates move the aggregated microplastics easily. The factors that influent the biotic transfer of microplastics in soil include type, size, and shape of the plastic particles, type of animal, exposure time, organisms' defense system against microplastics, and their bioaccumulation in the organism's body.

8.3   ECOLOGICAL IMPACTS OF MICROPLASTICS

The migration of MPs in the soil is influenced by the nature of the soil and the presence of MPs in turn influences the properties of soil that may translate into biotic consequences that can potentially affect the quality and safety of food and thereby human health. The ecological and health impacts caused by microplastic pollution in soil are a matter of utmost concern. In soil, microplastics act as composite pollutants due to incorporation of additional chemicals such as additives like flame retardants, heat stabilizers, antioxidants, heavy metals, and persistent organic pollutants. All these substances as such have a number of ecological risks. Adding to it, the interactions between plastic particles and soil can make the behavior of these pollutants unpredictable (Guo et al., 2020). Using simulation experiments, Boots et al. (2019) showed that MPs can have an impact on soil ecosystem both above and under the ground. Further studies show transfer of microplastics from a lower trophic level to a higher trophic level in the food chain and their abundance increases as this transfer occurs (Guo et al., 2020).

8.3.1   Effect of Microplastics on Soil

MPs prevalence in soil can change important biophysical properties of soil such as its structure, nutrient composition, fertility, pH, and microbial communities. It can also integrate into soil and change the soil aggregation which plays a very important role in determining the porosity of soil and consequently movement of air and water through it, thus shaping soil habitat. They may loosely incorporate into soil into fragment types or more tightly in linear types (Yang et al., 2021). Wan et al. (2019) reported that MPs create channels for water movement in soil that can accelerate soil water evaporation in a concentration-dependent manner. This can result in soil cracking and desiccation. Additionally, reports also have shown changes in soil pH due to MPs. Exposure to HDPE in soil, planted with *Lolium perenne*, resulted in a drop in pH by 0.62 unit after 30 days (Boots et al., 2019). Furthermore, the type, shape, and size of MPs are important features that determine the type of effect they will have on soil. The addition of polyethylene high density (PEHD), polyester (PES), PET, PP, and PS resulted in decreased soil bulk density and increased soil density in the rhizosphere. Conversely, microplastics also promote the accumulation of high-molecular-weight humic-like materials that can improve soil quality by promoting soil stability, water holding capacity, nutrient availability, etc. However, most of the studies point toward effects such as altered water cycle in soil worsening soil water shortage and promoting translocation of pollutants in underlying layers of soil through cracks (Wan et al., 2019). MPs can also cause short-term changes in soil quality by affecting the activity of enzymes such as urease, catalase, fluorescein diacetate hydrolase (FDAse), and phenol oxidase that are closely associated with soil biochemical processes (Guo et al., 2020). Liu et al. (2017) showed that a 28% w/w level of MPs can significantly increase accumulation of dissolved organic matter (DOM) and result in release of nutrients into the soil like dissolved organic carbon (DOC), dissolved organic nitrogen (DON), and dissolved organic phosphorous (DOP), while with 7% w/w level of MPs, the effects were negligible, thus making it concentration-dependent effect. MPs in the soil also increase its adsorbing capacity as a result of their large surface area. In addition to organic pollutants that are present as additives in plastic, they also adsorb substances like polybrominated diphenyl ether (PBDE) and perfluorochemicals (PFCs), heavy metals like zinc, copper, lead, and antibiotics, thus increasing soil pollution levels (Guo et al., 2020; Yang et al., 2021).

8.3.2 EFFECT OF MICROPLASTICS ON SOIL ORGANISMS AND HUMANS

Studies show that transport of microplastics by different soil animals can in turn have an effect on their exposure to other organisms in soil, thereby changing the biophysical properties of soils. Chemicals derived from polyurethane foam microplastics were found to accumulate in earthworms that facilitate their transportation to new areas as well as to predators feeding on these earthworms (Gaylor et al., 2013). Microplastics have also been reported to increase bioavailability of metals such as zinc in soil ecosystems (Hodson et al., 2017). Polyethylene MPs in the size range of 250–1000 μm were found to cause histopathological damages and immune system changes in earthworms. Rosy-tipped earthworms (*Aporrectodea rosea*) exposed to high-density polyethylene (HDPE) showed a significant reduction in biomass (Boots et al., 2019). Apart from earthworms, other small soil invertebrates, nematodes, snails, and mice have also been reported to show presence of microplastics. Soil nematodes such as *Caenorhabditis elegans* showed a reduction in the level of intestinal calcium levels, oxidative damages, decreased rate of reproduction, survival, and body length on exposure to MPs (Lei et al., 2018a). Additionally, damages in cholinergic and GABAergic neurons have also been shown to result from exposure to polystyrene MPs (Lei et al., 2018b). Different species of collembolan have been shown to transport MPs depending on their type, size, and soil ecosystem. PVC microplastics after 28–56 days of exposure have been reported to reduce growth and reproduction in *Folsomia candida*. This report also showed an increase in elemental values (^{15}N and ^{13}C) in the collembolan tissues (Zhu et al., 2018). Casts, burrows, egestion, adherence, or close proximity to hosts are some of the ways through which these organisms further incorporate MPs in soil and increase the chances of their exposure to other organisms or their entry into groundwater. Additionally, as MPs get transferred along the food chain from soil to earthworm casts to chicken feces, there is an increase in their abundance at each level. In chicken feces, the concentration of MPs was around 129.8 ± 82.3 particles per gram, suggesting a potential threat to human health. Ingestion of MPs in poultry has been associated with toxicity, oxidative stress, inflammation, and tumors (Yang et al., 2021). Table 8.2 enlists several studies showing the effect of microplastics on soil organisms.

TABLE 8.2

Studies Showing Impact of Microplastics on Soil Biota

Organism	Species	Microplastic Type	Effects	References
Earthworm	*Eisenia fetida*	Polyurethane foam MPs and PBDE	Accumulation and translocation to predators and new areas	Gaylor et al., 2013
	Lumbricus terrestris	Zinc and high-density polyethylene	MPs serve a pathway for the metals in soil ecosystem, no adverse effects on survival and body weight of the earthworm	Hodson et al., 2017
	L. terrestris	Low-density polyethylene (LDPE) microplastics (<150 μm)	On 60-day exposure, microplastics can exert toxicity to earthworms; mortality was higher at 28%, 45%, and 60% of MPs than at 7% w/w	Huerta Lwanga et al., 2016
	L. terrestris		Movement of MPs from surface into burrows via the earthworms could result in pollution of groundwater and terrestrial food web	Huerta Lwanga et al., 2017
	Eisenia andrei		Histopathological and immune system damages	Rodriguez-Seijo et al., 2017
Snails	*Achatina fulica*	Microfibers	Intake and excretion reduced after 28-day exposure, villous injury of gastrointestinal wall	Song et al., 2019
Nematodes	*Caenorhabditis elegans*		Intestinal and oxidative damage	Lei et al., 2018a
		Polystyrene MPs	Size-dependent excitatory toxicity on locomotor behaviors	Lei et al., 2018b
		Polystyrene particles	Transgenerational toxicity at the concentrations higher than 100 μg/L	Zhao et al., 2017
Collembolan	*Folsomia candida*	PVC (80–250 μm)	Changes in gut microbiota, growth, reproduction, and isotope composition in the soil ecosystem	Zhu et al., 2018
Isopods	*Procellio scaber*	Polyethylene microplastics (mean size 183 ± 93 μm)	No impacts on rate of food ingestion, defecation, food assimilation. No change in body mass, survival rates, and energy reserves in digestive gland	Kokalj et al., 2018
Mite	*Hypoaspis aculeifer*	PVC	Improved transport of microplastics when predator and prey coexist in soil than when either is alone, transport leads to the exposure of MPs to other soil biota and changes in biophysical property of soils	Zhu et al., 2018

(Continued)

TABLE 8.2 *(Continued)*
Studies Showing Impact of Microplastics on Soil Biota

Organism	Species	Microplastic Type	Effects	References
Fungi	*Aspergillus oryzae* and *A. nidulans*	Polystyrene latex (PSL)	Particles were readily taken up by *A. oryzae* through its relatively softer cell wall in comparison to *A. nidulans*	Nomura et al., 2016
Plants	*Vicia faba*	Polystyrene MPs (100 mg/)	Inhibition of growth and potentially blocked intercellular connections and cell wall pores	Jiang et al., 2019
	Lolium perenne (*rye grass*)	Biodegradable PLA	Reduced germination rate and shoot height	Boots et al., 2019
	–		Suppressed activity of enzymes such as β-glucosidase	Zang et al., 2020
Animals	Mice		Hepatic lipid metabolism disorder, decreased gut mucin secretion, decreased expression of gene involved in lipogenesis genes and hepatic triglyceride synthesis	Lu et al., 2018

A study exploring transport of microplastics in soil involving three soil micro arthropod species – *Folsomia candida, Hypoaspis aculeifer,* and *Damaeus exspinosus* – revealed that soil micro arthropods were able to transport polyvinyl chloride particles (80–250 μm) up to 9 cm. Micro arthropods commonly found on soil surfaces can enter pores in soil due to their small size, and thus this transport of microplastics by them can have a significant effect on physical properties of soil and exposure of underneath soil biota to MPs. Additionally, this study also showed that predator and prey, when present together in soil, improved transport of microplastics significantly when compared to either predator or prey alone in soil (Zhu et al., 2018). Conversely, terrestrial isopod *Procellio scaber* when exposed for 14 days to polyethylene microplastics showed no significant impact on parameters like body mass change, survival rates, and food ingestion or assimilation rate. However, studies involving prolonged exposure time and other MPs are required to fully understand the effect of MPs on isopods (Kokalj et al., 2018).

Studying the effects of MPs on soil microbiota has not been easy due to the fact that microbes have a short life span and relatively small size, making a study of how microplastics interact with them in soil environment difficult. Some studies highlight how microplastics disturb the soil and water relationship that consequently results in altered soil structure and microbial functions in that respective soil environment. However, there are studies that show an accumulation of microplastics in yeasts and filamentous fungi. Further fungi like *Zalerion maritimum* have also been recently reported to be able to utilize polyethylene. Studies done with polystyrene latex nanoparticles (NPs) on filamentous fungi *Aspergillus oryzae* and *Aspergillus nidulans* showed that NPs were readily taken up by *A. oryzae* through its relatively softer cell wall in comparison to

A. nidulans. This indicates that different microbial species have different sensitivity to different plastic particles depending on properties of both (Nomura et al., 2016).

The changed physical and chemical parameters of soil due to presence of MPs can significantly affect the root system and vegetative stage of plants, thereby affecting their growth. Ramos et al. (2015) found that while pesticide concentration in soil was 13–32 µg pesticide/g soil, its concentration in PE film was found to be far more, in the range of 584–2284 µg pesticide/g plastic. In a study that investigated effects of PS MPs on *Vicia faba*, it was found that at a concentration of 100 mg/L, the PS MPs inhibited the growth and potentially blocked intercellular connections and cell wall pores (Jiang et al., 2019). Small particles of polyethylene have been reported to have an adverse effect on wheat growth. Further exposure to MPs obtained from biodegradable polylactic acid (PLA) and microplastic clothing fibers resulted in reduced germination rate and shoot height in ryegrass (*Lolium perenne*) (Boots et al., 2019). Another report by de Souza Machado et al. (2018), however, also shows root biomass gain in presence of HDPE. This probably occurs due to less number of organisms dependent on the resources available, thus providing the plants an advantage to attain higher biomass levels. MPs have also been reported to suppress the activity of enzymes such as β-glucosidase that are involved in carbon cycling. Additionally, the type and amount of MPs can also greatly alter carbon flow through the plant-soil system, thus significantly impacting the key pools and fluxes within the terrestrial carbon cycle (Zang et al., 2020). There have been reports on the impact of MPs on aquatic plants that show MPs within the size range of 1–5 µm and concentration up to 41.5 mg/L not having an effect on the growth rate of marine microalgae while that between 0.9 and 2.1 mg/L significantly reducing the development of chlorophyll. However, very few studies report specific impacts of microplastics on terrestrial plants, and further research is required in this area. Figure 8.3 depicts the

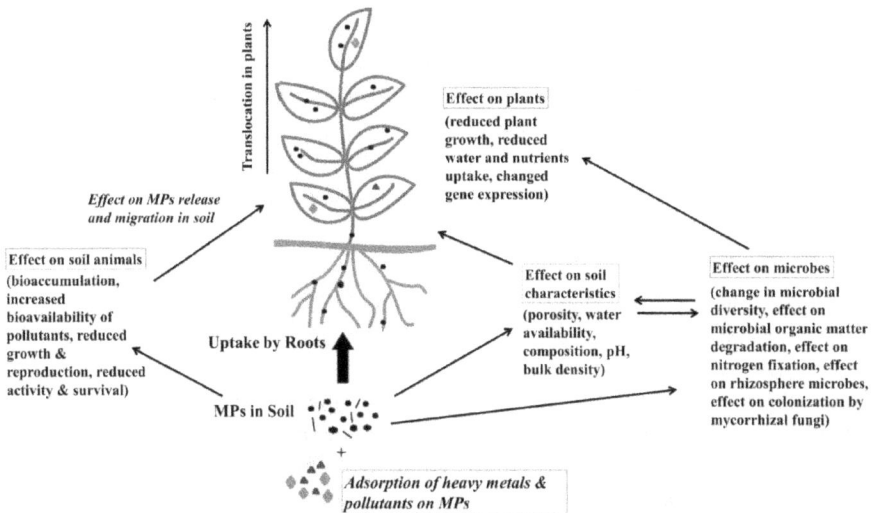

FIGURE 8.3 Direct and indirect impact of microplastics soil pollution on plants.

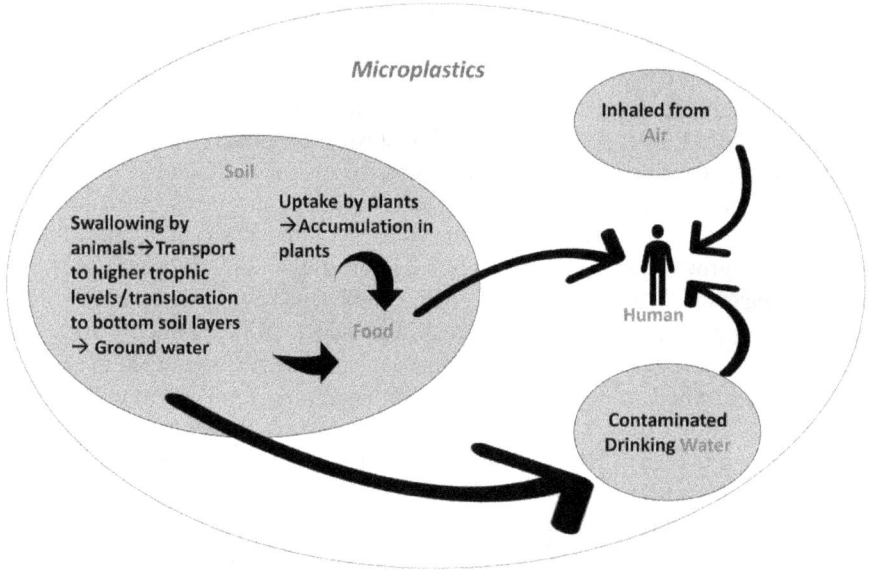

FIGURE 8.4   Transport of microplastics to human beings.

impact of microplastic pollution of soil on plants. Because all components of the soil ecosystem are interconnected, plant growth is affected by different factors directly or indirectly.

Studies examining human stool samples and colectomy specimens have confirmed the presence of MPs in the human gut as well (Ibrahim et al., 2021). Sources through which humans are likely to obtain these particles are through conventional food chain, consumption of salt, water, or food items contaminated by MPs, and inhalation of $PM_{2.5}$ (Yu et al., 2022) (Figure 8.4). The microplastics and additives in them cause serious health issues in human beings such as endocrine problems, cardiovascular diseases, asthma, diabetes, intestinal obstruction, and increased risk of cancer (Anik et al., 2021; Yu et al., 2022).

8.4   DETECTION AND ANALYSIS OF MICROPLASTICS IN SOIL

8.4.1   SAMPLING

Sampling is a very important step in the analysis. To get nearly accurate results during analysis, sampling methods need to be followed strategically to obtain all soil microplastics in the collected sample. Sampling may be done using different approaches such as random (equal chances of getting different microplastics), systematic (ensures uniformity), stratified (all layers of soil are explored), or composite (composite of different subsamples) sampling (Junhao et al., 2021; Perez et al., 2022). The selection of the approach depends on multiple factors. It has been suggested that more accuracy is ensured if more than one sampling method is combined. Also, collection of the minimum required number of samples should be high enough for accurate analysis. Another important point to be considered is sampling depth (Junhao et al., 2021; Perez et al., 2022). The selection of depth depends on the site as well as the purpose of the experiment. For example, if studying the leaching of microplastics, then they need to be analyzed up to greater depths.

8.4.2 SIEVING

After the collection of samples, sieving is an essential step to remove gravel, soil aggregates, or organic matter which can have a negative impact on the extraction process. Sieves of different mesh sizes are employed (Chia et al., 2021). Commonly, particles larger than 5 mm are removed. It is better to use small mesh-sized screens (1 mm) to screen out larger particles affecting the subsequent steps during analysis (Junhao et al., 2021; Perez et al., 2022). A sieving gradient may also be made in which sieves of decreasing mesh size are arranged in a stack (Junhao et al., 2021).

8.4.3 SAMPLE PROCESSING

During processing, microplastics are separated from the soil in a way that they are not damaged. To achieve separation, the density separation method is employed most commonly, which makes use of salt solutions of high density for extraction purposes (Moller et al., 2020; Chia et al., 2021). Before separation, the sample is generally subjected to ultrasonic treatment which enhances the recovery rate of particles of different densities. Suspension solutions of different compositions (sodium chloride, zinc chloride, sodium iodide, sodium bromide, calcium chloride, and zinc bromide) can be used (Yang et al., 2021). Munich Plastic Sediment Separator (MPSS) and its modified separators have been made which consist of parts such as a stirring sediment container, standpipe, and dividing chamber with valve as well as a holder for filter ((Junhao et al., 2021)). In such separators, separated microplastics are directly subjected to the filtration process. To improve the recovery rate, even multi-stage processes and separators have been designed.

The separation can also be achieved using other methods such as electrostatic separation, froth flotation, and magnetic extraction (Moller et al., 2020). In electrostatic separation, an electric field is applied that separates charged soil particles from uncharged microplastics. In the froth floatation, froth selectively attaches to hydrophobic plastics causing their flotation. In the magnetic extraction method, magnetized hydrophobic iron nanoparticles are used that bind the microplastics and are then extracted by applying a magnetic field.

8.4.4 FILTRATION AND DIGESTION

The separated microplastics are filtered through filters that can be made from paper, glass, polycarbonate membrane, or cellulose acetate materials (Chia et al., 2021; Perez et al., 2022). The pore size of the filter may vary. Also, stacked filter systems may be employed (filtration gradient). Because separated particles are often mixed with organic materials in soil, they need to be digested before analysis of microplastics. The organic matter can be digested using hydrogen peroxide,

Sampling & Storage

Sieving

Sample Processing

Filtration

Digestion

Analysis

Visual Identification
- Optical microscopy
- Scanning electron microscopy (SEM)
- Transmission electron microscopy (TEM)
- Polarized light microscopy
- Nile red staining

Spectroscopy
- Raman spectroscopy
- Fourier transform infrared (FTIR)
- Micro-FTIR
- Attenuated total reflectance (ATR)
- Energy-dispersive X-ray spectroscopy (EDS)
- Surface-enhanced Raman spectroscopy (SERS)
- Near IR (NIR) spectroscopy
- Vis-NIR spectroscopy

Thermal Analyses
- Thermogravimetric analysis (TGA)
- Pyrolysis-gas chromatography mass spectrometry (Py-GC-MS)
- Differential scanning calorimetry (DSC)
- TGA-FTIR
- TGA-FTIR-GC-MS
- Nanothermal analysis (nano-TA)

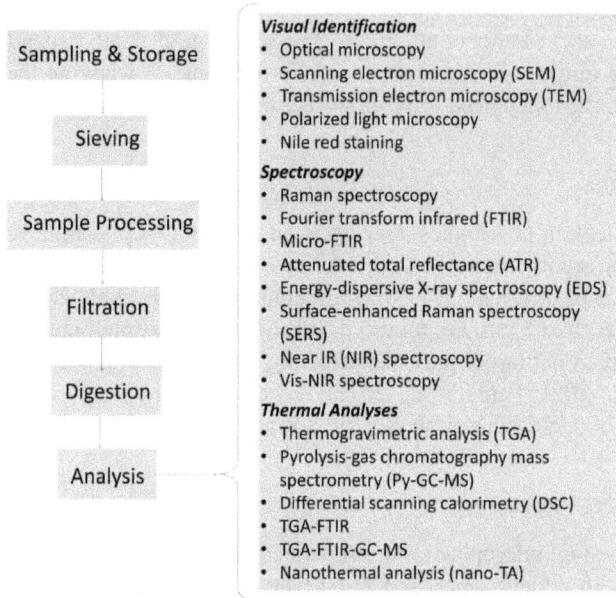

FIGURE 8.5   Process and methods for analysis of microplastics in soil.

alkali (KOH and NaOH), acid (nitric or hydrochloric acid), or enzymes (Moller et al., 2020). There are not as many studies done on organic matter digestion in soil as in marine water. Suitable methods are, therefore, needed to be designed to standardize the process.

8.4.5   ANALYSIS OF MICROPLASTICS

The analysis involves identification, characterization, and quantification. The purified and digested samples can be analyzed using different methods (Figure 8.5).

8.4.5.1   Visual Identification and Microscopy

Visual characterization, done by the naked eye or microscopic methods, is used for preliminary identification and is subject to errors (Chia et al., 2021). Several studies have used hot needle test which involves microscopic examination before and after heating treatment (Moller et al., 2020). Microscopic observation may be done using light or electron microscope. The optical microscope, owing to its resolution limitations, can be used to identify plastics that are several hundred micrometers in size. Often, poor separation and digestion of samples may leave some biological and other materials adhered to microplastics that interfere with their microscopic study. Smaller microplastics may be visualized and analyzed through scanning electron microscopy (SEM) producing images with higher magnification and resolution that help in distinguishing microplastics from organic particles (Shim et al., 2017). Transmission electron microscopy (TEM) is employed for visual identification of plastics at the nanoscale (Huang et al., 2022). Polarized light microscopy may be used to identify different microplastic crystals that vary in their degree of crystallinity (Shim et al., 2017).

The small and transparent particles can be visualized by staining with Nile red, a fluorescent dye that is selective for hydrophobic microplastics (Moller et al., 2020). It gives results in only 10–30 min and can aid in preliminary analysis of hidden microplastics. Researchers have even suggested that Nile red staining followed by fluorescence microscopy and FTIR confirmation can be effective in minimizing errors in identification. The organic matter removal from the sample, however, is essential before Nile red staining because it stains natural lipids and organic matter.

8.4.5.2 Spectroscopy

Raman spectroscopy and Fourier transform infrared (FTIR) are the most widely used techniques to characterize and identify plastics (Moller et al., 2020). FTIR spectroscopy reveals chemical bonding in different microplastic types. A spectrum depicting the bonding configurations is created that is helpful in distinguishing plastic from non-plastic particles and identification of transparent (colorless) microplastics. An advanced technique of micro-FTIR coupled to a microscope provides information about both the microscopic and spectroscopic features of the particles that aid in better identification of microplastics (Perez et al., 2022). Attenuated total reflectance (ATR) is useful for characterization of very small (10 μm), irregular, thick, and opaque microplastic particles without requiring any sample preparation (Shim et al., 2017). A focal plane array (FPA) detector can analyze multiple samples simultaneously. These techniques are based on mass and cannot be used to quantify numbers. Interpretation of results, however, requires skilled analyses because microplastics get modified in the soil and may even be a part of composites. Raman spectroscopy is based on the principle of development of a spectrum unique to different polymers (varying in their molecular structure and atoms) upon irradiation with a laser beam. It gives information about polymer composition (Junhao et al., 2021). The energy-dispersive X-ray spectroscopy (EDS) may also be used for the analysis of compositional elements (Shim et al., 2017). The additives may also be detected as markers of microplastics in this type of spectroscopy. The particles lesser than 1 μm can be analyzed using surface-enhanced Raman spectroscopy (SERS). Raman spectroscopy is considered better because it gives better resolution than FTIR (Junhao et al., 2021).

Atomic force microscopy (AFM) combined with IR or Raman spectroscopy can reveal the chemical structure of even nano-sized microplastics (Shim et al., 2017). Near IR (NIR) spectroscopy (spectrum range over 4000–15,000 cm^{-1}) can classify multiple samples in less time (Woo et al., 2021). Vis-NIR spectroscopy is based on light reflected from the sample surface including visible wavelength range that may help in quantification of microplastics as it corresponds to the chemical composition.

8.4.5.3 Thermal Analyses

The thermal analysis is done on the products formed from degradation of the sample. Thermogravimetric analysis (TGA) involves determination of reduction in weight of the sample after heating at a specific temperature and pressure. The method is suitable for quantitative analysis. Different plastic products vary in their physical and thermal properties which can be analyzed by differential scanning calorimetry (DSC) that can be used to identify microplastics (Moller et al., 2020; Junhao et al., 2021). One more method, i.e., pyrolysis-gas chromatography mass spectrometry (Py-GC-MS), is used in which the sample is pyrolyzed at high temperatures and then products are analyzed (Junhao et al., 2021). The results obtained are compared with the standard. The major limitation of this method is that it is not useful for all microplastic types. For the analysis of bulk samples, combined thermal analysis and GC-MS can give information about total microplastic content (weight). TGA-FTIR can be used to perform real-time qualitative analysis and TGA-FTIR-GC-MS can be used to perform real-time quantitative analysis (Woo et al., 2021). The TGA analysis followed by TGA thermal extraction and desorption (TED)-GC-MS can determine levels of even small-sized microplastics in large samples (Moller et al., 2020; Woo et al., 2021). The thermal analysis technique can be used to study the surface characteristics of the sample in combination with high spatial resolution in nanothermal analysis (nano-TA).

8.5 MITIGATION STRATEGIES FOR MICROPLASTICS POLLUTION

Owing to the negative effects of microplastics on the soil ecosystem, mitigation of microplastic pollution is very critical to save the environment. The mitigation can be done by working at various levels. Broadly, mitigation strategies are based on reduction in use and manufacture of plastics as well as microplastics and management of already generated plastic waste (Figure 8.6). The management of generated plastic waste can be done through various means. Wastewater treatment is an approach to reduce microplastics in industrial or municipal wastewater. Though some of the existing treatment methods have been found to have limited capability

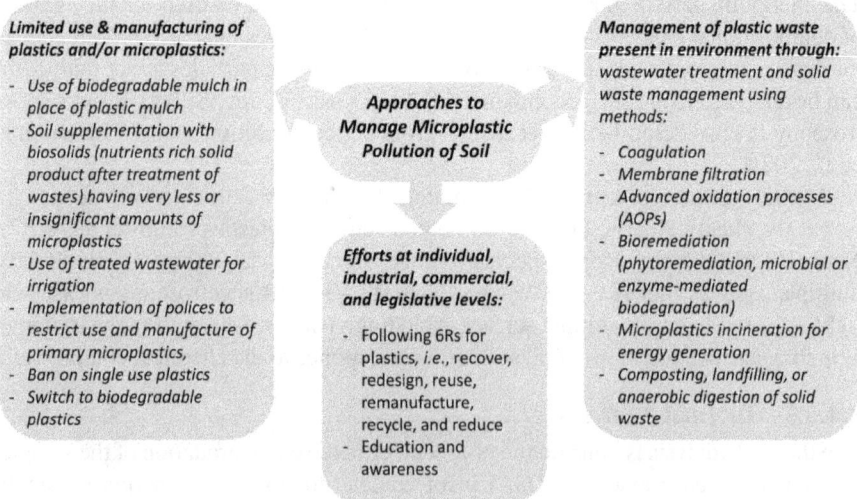

Limited use & manufacturing of plastics and/or microplastics:

- Use of biodegradable mulch in place of plastic mulch
- Soil supplementation with biosolids (nutrients rich solid product after treatment of wastes) having very less or insignificant amounts of microplastics
- Use of treated wastewater for irrigation
- Implementation of polices to restrict use and manufacture of primary microplastics,
- Ban on single use plastics
- Switch to biodegradable plastics

Approaches to Manage Microplastic Pollution of Soil

Efforts at individual, industrial, commercial, and legislative levels:

- Following 6Rs for plastics, i.e., recover, redesign, reuse, remanufacture, recycle, and reduce
- Education and awareness

Management of plastic waste present in environment through: wastewater treatment and solid waste management using methods:

- Coagulation
- Membrane filtration
- Advanced oxidation processes (AOPs)
- Bioremediation (phytoremediation, microbial or enzyme-mediated biodegradation)
- Microplastics incineration for energy generation
- Composting, landfilling, or anaerobic digestion of solid waste

FIGURE 8.6 Methods for control of microplastics pollution of soil.

of reducing microplastics in treated water, improvements are being made in the known methods and new methods are being developed to increase the efficiency of treatment processes. The solid waste can be treated at point sources. The treatment, however, is not always feasible at non-point sources that need to be dealt with a different approach (Chen et al., 2021). Identification of microplastics is essential to remove them from their source for which more efficient extraction and analytical methods are needed. Standardization of these methods is required for samples from different origins such as air, water, sludge, or soil (Chen et al., 2021; Roy et al., 2022).

Several methods can be employed to remove or reduce microplastics from the wastewater or solid waste. The coagulation process (chemical or electrical) can remove microplastics from the wastewater (Anik et al., 2021; Roy et al., 2022). Chemical coagulation involves use of different iron and aluminum salts that adsorb dissolved solids in wastewater through flocculation, which are sedimented subsequently in the coagulation tank. Electrocoagulation involves use of metal electrodes that lead to production of coagulants that remove MPs. The membrane filtration process has also been used as a method for removal of MPs from wastewater (Anik et al., 2021; Roy et al., 2022). Research efforts are in progress to make this method more effective for the removal of even smaller microplastics. This method has a major limitation of membrane clogging. All of these methods are non-destructive techniques wherein separated microplastics are not discharged uncontrolled in the environment but still need to be managed. The degradation of separated or non-separated microplastics is a potent strategy. One such technique is advanced oxidation processes (AOPs) in which the material is transformed physicochemically such that modifications in chemical and molecular structure lead to its disintegration and ultimately mineralization of the material. An example includes photocatalysis (Roy et al., 2022), a redox process driven by light, in which nano-sized semiconductors are used that produce highly reactive species when excited with light. The reactive radicals (hydroxyl, superoxide, alkoxyl, sulfate, and chlorine radicals) oxidize the plastic polymers producing volatile organics, CO_2, and H_2O. Bioremediation is an environment-friendly approach that is being considered widely. It involves processes such as phytoremediation and microbial or enzyme-mediated biodegradation (Miri et al., 2022; Rozman et al., 2022; Saini and Sharma, 2022). Several microorganisms including fungi (e.g., *Aspergillus* sp., *Penicillium* sp., *Fusarium* sp.), bacteria (e.g., *Azotobacter* sp., *Bacillus* sp., *Pseudomonas* sp., *Rhodococcus ruber, Enterobacter* sp., *Thermomonospora* sp.), and actinomycetes (e.g., *Actinobacter* sp., *Amycolatopsis* sp., *Actinomadura* sp., *Streptomyces* sp., *Thermobifida fusca*) are capable of degrading natural or synthetic plastics (Guo et al., 2020; Anik et al., 2021; Miri et al., 2022; Roy et al., 2022). The degradation rate, however, is governed by several factors such type of ecosystem (terrestrial or marine), nature (synthetic or natural), and characteristics of plastics (e.g., hydrophobicity, molecular weight, crystallinity, chemical composition, etc.) (Roy et al., 2022). The bacterial species *Ideonella sakaiensis* has been found to have the capability of degradation of PET while using it as a sole source of carbon and energy (Lamichhane et al., 2022). Recently, research studies have been targeted toward development of effective bioaugmentation processes involving use of monoculture, microbial consortia, or genetically modified microbes to enhance the biodegradation of plastic polymers. Studies conducted in the past few decades have revealed that agricultural soil has become one of the major sinks for plastic materials of different sizes including microplastics. There is an urgent need to replace plastic mulches with biodegradable mulches and use treated bio-solids or wastewater for irrigation. Another approach can be aimed at generating energy through the incineration of microplastics in furnaces and ovens in industrial processes. The resulting ash can also be used in metal recovery or road construction (Prata et al., 2019).

The reduction of plastic waste is a long-term approach to mitigate microplastic pollution in all ecosystems. Many countries have taken the initiative to reduce plastic pollution through various waste management policies such as restrictions on use and manufacturing of primary microplastics. For example, China initiated the "Plastic Restriction Order" in 2008 and reconsidered the problem in 2020 through "Opinions on Further Strengthening the Control of Plastic Pollution." The United States and France made and enacted upon "Microbead-Free Waters Act" through which they banned the use of microbeads in the rinse-off cosmetics (Saini and Sharma, 2022). Microbeads are also banned in the UK and Canada. They can be replaced with other alternatives such as chito-beads or cellulose microbeads (Roy et al., 2022; Saini and Sharma, 2022). The single use plastics (water bottles, plastic bags, food packages, etc.) have been banned in different regions of the world (Prata et al., 2019; Chen et al., 2021). Other actions include implementation of manufacturing, recycling, and waste emission regulations. Another approach is to reduce production of non-degradable plastics for which alternatives such as recyclable and biodegradable materials can be used (Saini and Sharma, 2022). The life cycle assessment of plastics should be done extensively in order to eco-design the manufacturing process to reduce harmful wastes (Prata et al., 2019). This approach can be aimed at making the plastic degradable or recyclable through reduced use of additives and polymers in the final product. The bio-based plastics are mostly degradable as they are made from materials such as starch, cellulose, and lignin (Chen et al., 2021). The biodegradable bioplastics can be prepared from bacterial polyhydroxyalkanoates (PHA) and renewable sources such as plants, food wastes, and microalgae (Chen et al., 2021; Roy et al., 2022). Several microalgae can be employed for removal of microplastics through adsorption on their surface (Anik et al., 2021). Depending on the type and characteristics of plastics and the environment, biodegradable plastics may be subjected to recycling, incineration, or biological waste treatments, such as composting, landfilling, or anaerobic digestion. Landfilling, however, is not encouraged much as it leads to leaching of microplastics (Chen et al., 2021). Composting and anaerobic digestion resulting in complete degradation are more relevant methods. Nowadays, management of solid wastes is being encouraged through the 6Rs which include recovery, redesign, remanufacture, reuse, recycle, and reduce (Saini and Sharma, 2022). The regulations and policies emphasize on following these Rs at individual, industrial, commercial, and legislative levels. Education and awareness is another tool that can produce significant output in terms of reduced plastic waste emissions. The awareness can be done through various means such as activities in educational institutions, the internet, news, clean-ups, open online courses, media, apps, etc. (Prata et al., 2019;

Chen et al., 2021; Saini and Sharma, 2022). From the observation of the situation, this can be concluded that multiple control approaches are needed to manage soil microplastic pollution effectively and efforts are mandatory at regional, national, and international levels.

8.6 CONCLUSION AND FUTURE PROSPECTS

Presently, the ecological risks from soil microplastic pollution are likely to rise in the near future owing to increased production of plastic products and prolonged persistence of plastics in the environment. This chapter summarizes the current problems of microplastic pollution in the soil ecosystem, including types, sources, migration, and ecological impacts of soil microplastics along with the detection methods and remedial measures for them. However, more studies are required to fill the knowledge gaps in relation to different aspects of microplastic pollution. Advancements are needed to assess the pollution level at different sources in the environment, for which more efficient techniques are to be developed from sampling to characterization. The methods are to be standardized for different samples and microplastic types. Transport routes need to be elucidated for applying effective control measures at required locations. Furthermore, control strategies need to be implemented worldwide to reduce plastics/microplastic emissions and accumulation in the environment. Legislative and other regulatory bodies should encourage redesign and recycling of plastic waste. The remedial measures should be followed rigorously in order to remediate soil and other environmental components from the existing plastic wastes. Also, public awareness with the efforts of government and non-government organizations can prove beneficial in the long-term control of microplastic pollution.

REFERENCES

Anik, A. H., Hossain, S., Alam, M., Sultan, M. B., Hasnine, M. T., and Rahman, M. M. 2021. Microplastics pollution: A comprehensive review on the sources, fates, effects, and potential remediation. *Environmental Nanotechnology, Monitoring & Management* 16:100530.

Boots, B., Russell, C. W., and Green, D. S. 2019. Effects of microplastics in soil ecosystems: Above and below ground. *Environmental Science & Technology* 53(19):11496–11506.

Chen, X., Huang, G., and Dionysiou, D. D. 2021. Editorial overview: Emissions of microplastics and their control in the environment. *Journal of Environmental Engineering* 147(9):01821002.

Chia, R. W., Lee, J. Y., Kim, H., and Jang, J. 2021. Microplastic pollution in soil and groundwater: A review. *Environmental Chemistry Letters* 19(6):4211–4224.

de Souza Machado, A. A., Kloas, W., Zarfl, C., Hempel, S., and Rillig, M. C. 2018. Microplastics as an emerging threat to terrestrial ecosystems. *Global Change Biology* 24(4):1405–1416.

Dey, S., Anand, U., Kumar, V., Kumar, S., Ghoraj, M., Ghosh, A., Kant, N., Suresh, S., Bhattacharya, S., Bontempi, E., Bhat, S. A., and Dey, A., 2023. Microbial strategies for degradation of microplastic generated from COVID-19 healthcare waste. *Environmental Research* 216, 114438. https://doi.org/10.1016/j.envres.2022.114438

Gaylor, M. O., Harvey, E., and Hale, R. C. 2013. Polybrominated diphenyl ether (PBDE) accumulation by earthworms (*Eisenia fetida*) exposed to biosolids-, polyurethane foam microparticle-, and penta-BDE-amended soils. *Environmental Science & Technology* 47(23):13831–13839.

Guo, J. J., Huang, X. P., Xiang, L., Wang, Y. Z., Li, Y. W., Li, Hui, L., Cai, Q. Y., Mo, C. H., and Wong, M. H. 2020. Source, migration and toxicology of microplastics in soil. *Environment International* 137:105263.

Hodson, M. E., Duffus-Hodson, C. A., Clark, A., Prendergast-Miller, M. T., and Thorpe, K. L. 2017. Plastic bag derived-microplastics as a vector for metal exposure in terrestrial invertebrates. *Environmental Science & Technology* 51(8):4714–4721.

Huang, Z., Hu, B., and Wang, H. 2022. Analytical methods for microplastics in the environment: A review. *Environmental Chemistry Letters* 21(1)1–19.

Huerta Lwanga, E., Gertsen, H., Gooren, H., Peters, P., Salánki, T., Van Der Ploeg, Besseling, E., Koelmans, A. A., and Geissen, V. 2016. Microplastics in the terrestrial ecosystem: Implications for *Lumbricus terrestris* (Oligochaeta, Lumbricidae). *Environmental Science & Technology* 50(5):2685–2691.

Huerta Lwanga, E., Mendoza Vega, J., Ku Quej, V., Chi, J. D. L. A., Sanchez del Cid, L., Chi, C., Escalona Segura, G., Gertsen, H., Salánki, T., van der Ploeg, M., and Koelmans, A. A. 2017. Field evidence for transfer of plastic debris along a terrestrial food chain. *Scientific Reports* 7(1):1–7.

Ibrahim, Y. S., Tuan Anuar, S., Azmi, A. A., Wan Mohd Khalik, W. M. A., Lehata, S., Hamzah, S. R., and Dzulkifleem, I. et al. 2021. Detection of microplastics in human colectomy specimens. *JGH Open* 5(1):116–121.

Jiang, X., Chen, H., Liao, Y., Ye, Z., Li, M., and Klobučar, G. 2019. Ecotoxicity and genotoxicity of polystyrene microplastics on higher plant *vicia faba*. *Environmental Pollution* 250:831–838.

Junhao, C., Xining, Z., Xiaodong, G., Li, Z., Qi, H., and Siddique, K. H. 2021. Extraction and identification methods of microplastics and nanoplastics in agricultural soil: A review. *Journal of Environmental Management* 294:112997.

Kokalj, A. J., Horvat, P., Skalar, T., and Kržan, A., 2018. Plastic bag and facial cleanser derived microplastic do not affect feeding behaviour and energy reserves of terrestrial isopods. *Science of the Total Environment* 615: 761–766.

Lamichhane, G., Acharya, A., Marahatha, R., Modi, B., Paudel, R., Adhikari, A., Raut, B. K., Aryal, R. S., and Parajuli, N. 2022. Microplastics in environment: Global concern, challenges, and controlling measures. *International Journal of Environmental Science and Technology* 20(4) 1–22.

Lei, L., Wu, S., Lu, S., Liu, M., Song, Y., Fu, Z., Shi, H., Raley-Susman, K. M., and He, D. 2018a. Microplastic particles cause intestinal damage and other adverse effects in zebrafish *danio rerio* and nematode *Caenorhabditis elegans*. *Science of Total Environment* 619:1–8.

Lei, L., Liu, M., Song, Y., Lu, S., Hu, J., Cao, C., Xie, B., Shi, H., and He, D. 2018b. Polystyrene (nano) microplastics cause size-dependent neurotoxicity, oxidative damage and other adverse effects in *caenorhabditis elegans*. *Environmental Science: Nano* 5(8):2009–2020.

Liu, H., Yang, X., Liu, G., Liang, C., Xue, S., Chen, H., Ritsema, C. J., and Geissen, V. 2017. Response of soil dissolved organic matter to microplastic addition in Chinese loess soil. *Chemosphere* 185:907–917.

Lu, L., Wan, Z., Luo, T., Fu, Z., and Jin, Y. 2018. Polystyrene microplastics induce gut microbiota dysbiosis and hepatic lipid metabolism disorder in mice. *Science of Total Environment* 631:449–458.

Miri, S., Saini, R., Davoodi, S. M., Pulicharla, R., Brar, S. K., and Magdouli, S. 2022. Biodegradation of microplastics: Better late than never. *Chemosphere* 286:131670.

Möller, J. N., Löder, M. G., and Laforsch, C. 2020. Finding microplastics in soils: A review of analytical methods. *Environmental Science & Technology* 54(4):2078–2090.

Nomura, T., Tani, S., Yamamoto, M., Nakagawa, T., Toyoda, S., Fujisawa, E., Yasui, A., and Konishi, Y. 2016. Cytotoxicity and colloidal behavior of polystyrene latex nanoparticles toward filamentous fungi in isotonic solutions. *Chemosphere* 149:84–90.

Perez, C. N., Carré, F., Hoarau-Belkhiri, A., Joris, A., Leonards, P. E., and Lamoree, M. H. 2022. Innovations in analytical methods to assess the occurrence of microplastics in soil. *Journal of Environmental Chemical Engineering* 10(3) 107421.

Pérez-Reverón, R., Álvarez-Méndez, S. J., Kropp, R. M., Perdomo-González, A., Hernández-Borges, J., and Díaz-Peña, F. J. 2022. Microplastics in agricultural systems: Analytical methodologies and effects on soil quality and crop yield. *Agriculture* 12(8):1162.

Prata, J. C., Silva, A. L. P., Da Costa, J. P., Mouneyrac, C., Walker, T. R., Duarte, A. C., and Rocha-Santos, T. 2019. Solutions and integrated strategies for the control and mitigation of plastic and microplastic pollution. *International Journal of Environmental Research and Public Health* 16(13):2411.

Ramos, L., Berenstein, G., Hughes, E.A., Zalts, A., and Montserrat, J. M., 2015. Polyethylene film incorporation into the horticultural soil of small periurban production units in Argentina. *Science of the Total Environment* 523: 74–81.

Rodriguez-Seijo, A., Lourenço, J., Rocha-Santos, T. A. P., Da Costa, J., Duarte, A. C., Vala, H., and Pereira, R. 2017. Histopathological and molecular effects of microplastics in *Eisenia andrei* Bouché. *Environmental Pollution* 220:495–503.

Roy, P., Mohanty, A. K., and Misra, M. 2022. Microplastics in ecosystems: Their implications and mitigation pathways. *Environmental Science: Advances* 1(1):9–29.

Rozman, U., Kokalj, A. J., Dolar, A., Drobne, D., and Kalčíková, G. 2022. Long-term interactions between microplastics and floating macrophyte *Lemna minor*: The potential for phytoremediation of microplastics in the aquatic environment. *Science of Total Environment* 831:c154866.

Saini, A., and Sharma, J. G. 2022. Emerging microplastic contamination in ecosystem: An urge for environmental sustainability. *Journal of Applied Biology and Biotechnology* 10(5):66–75.

Sajjad, M., Huang, Q., Khan, S., Khan, M. A., Yin, L., Wang, J., Lian, F., Wang, Q., and Guo, G. 2022. Microplastics in the soil environment: A critical review. *Environmental Technology & Innovation* 27: 102408.

Shim, W. J., Hong, S. H., and Eo, S. E. 2017. Identification methods in microplastic analysis: A review. *Analytical Methods* 9(9):1384–1391.

Song, Y., Cao, C., Qiu, R., Hu, J., Liu, M., Lu, S., Shi, H., Raley-Susman, K. M., and He, D. 2019. Uptake and adverse effects of polyethylene terephthalate microplastics fibers on terrestrial snails (*Achatina fulica*) after soil exposure. *Environmental Pollution* 250:447–455.

Wan, Y., Wu, C., Xue, Q., and Hui, X. 2019. Effects of plastic contamination on water evaporation and desiccation cracking in soil. *Science of the Total Environment* 654:576–582.

Woo, H., Seo, K., Choi, Y., Kim, J., Tanaka, M., Lee, K., and Choi, J. 2021. Methods of analyzing microsized plastics in the environment. *Applied Sciences* 11(22):10640.

Yang, L., Zhang, Y., Kang, S., Wang, Z., and Wu, C. 2021. Microplastics in soil: A review on methods, occurrence, sources, and potential risk. *Science of the Total Environment* 780:146546.

Yu, H., Zhang, Y., Tan, W., and Zhang, Z. 2022. Microplastics as an emerging environmental pollutant in agricultural soils: Effects on ecosystems and human health. *Frontiers in Environmental Science* 10: 217.

Zang, H., Zhou, J., Marshall, M. R., Chadwick, D. R., Wen, Y., and Jones, D. L. 2020. Microplastics in the agroecosystem: Are they an emerging threat to the plant-soil system? *Soil Biology and Biochemistry* 148:107926.

Zhao, L., Qu, M., Wong, G., and Wang, D. 2017. Transgenerational toxicity of nanopolystyrene particles in the range of μg L^{-1} in the nematode *Caenorhabditis elegans*. *Environmental Science: Nano* 4(12):2356–2366.

Zhu, D., Bi, Q. F., Xiang, Q., Chen, Q. L., Christie, P., Ke, X., Wu, L. H., and Zhu, Y. G. 2018. Trophic predator-prey relationships promote transport of microplastics compared with the single *Hypoaspis aculeifer* and *Folsomia candida*. *Environmental Pollution* 235:150–154.

9 Fate and Behavior of Micro- and Nanoplastics in Wastes

Ana Emília M. de Freitas,
Wyvirlany Valente Lobo,
Airi dos Santos Sousa,
Antônio José de Andrade Junior,
Silma de Sá Barros, Orlando A. R. L. Paes,
and Flávio A. de Freitas

9.1 INTRODUCTION

Polymeric materials such as thermoplastics, thermosets, and elastomers allowed technological advancement and the production of various products present in our daily lives. However, there are growing concerns about the destination of these materials and the impacts they generate. According to Geyer, Jambeck, and Law (2017), between 1950 and 2015, about 8.3 billion tons of plastics were produced for application and commerce. In this same period, approximately 6.3 million tons of plastic waste were generated. Of these, approximately 79% were destined for landfills, and in the natural environment, 12% were incinerated and only about 9% were recycled.

Plastic waste presents a high risk of pollution, as a result of its inadequate disposal and exposure to the environment. In addition, another impact lies in its tendency to form microplastics and nanoplastics. It is known that plastic waste is ingested by various animals such as fish, birds, mammals, and various marine species, increasing the mortality of these species (Thiel et al. 2018; Thushari and Senevirathna 2020; Roman et al. 2021). Macroplastics (>5 mm) are easier to identify and study and, depending on the state of the waste, it is possible to determine the source. On the other hand, micro- and nanoplastics (MNPs) are difficult to detect and analyze. Another worrying factor is the fact that some plastic materials can adsorb hydrophobic pollutants, increasing the risk of having high toxicity (Moore 2008; Dey et al. 2023).

The main polymers produced are polyethylene (PE), polypropylene (PP), elastomers, polystyrene (PS), polyvinyl chloride (PVC), poly(ethylene terephthalate) (PET), polyurethane (PU), resins, polyester, and polyamide (Geyer, Jambeck, and Law 2017). Depending on the material, the resulting MNPs may have different properties, such as differences in density, crystallinity, oxidative degradation behavior, additives, or surface properties (Andrady 2017).

DOI: 10.1201/9781003352396-9

MNPs are currently ubiquitous in the oceans, generating impacts associated with their presence and encouraging research to discover their origin, behavior, and the extent of their environmental and health effects. Due to their reduced dimensions, they are difficult to detect, making their collection and recovery unfeasible. In addition, they are present across the globe and are widely distributed, mainly by sea currents (Andrady 2017).

Recently, the COVID-19 pandemic led to the generation of microplastic waste from items such as gloves, masks, and disposable personal protective equipment (PPE) (Dey et al., 2023). These items are made of different materials, such as polyethylene, polyacrylonitrile, polystyrene, polypropylene, and polyurethane, consequently resulting in a variety of microplastics produced (Dey et al. 2023). In 2020, between 2.4 and 2.5 billion masks were produced, of which 1–10% are destined for oceans, producing 72 to 31,200 tons of microplastics (Saliu et al. 2021).

MNPs can be primary or secondary: the primers are industrially manufactured and are released directly into the environment in the produced form. They are present in several products such as cleaning agents, personal hygiene, cosmetics, and exfoliants. They can also be produced indirectly during the manufacture or maintenance of objects, such as tire erosion and the washing of synthetic fabrics; secondary products are produced by the degradation of larger products, such as bags, tires, and packaging. These MNPs are the result of poor waste management, leading to material fragmentation by abrasive processes and photo, thermal, and chemical degradation (Boucher and Friot 2017; Saliu et al. 2021).

Thermoplastic waste has high chemical stability and, consequently, resistance to aging and biological degradation, which prevents its rapid degradation and makes it remain for decades irregularly in nature (Moore 2008; Andrady 2017). However, some factors are responsible for the fragmentation of these materials, such as exposure to ultraviolet solar radiation, oxidative properties of the atmosphere, hydrolytic properties of water, thermal decomposition, or physical fragmentation due to mechanical stress and friction, and from these mechanisms particles are formed that reach the micro and nanometer scale (Moore 2008; Velis, Lerpiniere, and Tsakona 2017; Saliu et al. 2021).

A worrying aspect about the presence of plastic particles in nature is that the smaller their dimensions, the greater their bioavailability, allowing their assimilation by smaller and smaller organisms which can lead to the filling of the digestive tract of these organisms (Ugwu, Herrera, and Gómez 2021). This assimilation of plastic particles can occur directly or through the food chain, from the consumption of animals that have microplastics or nanoplastics in their bodies. A worrying factor is that ingestion of MNPs already occurs in animals at low trophic levels in the food chain, such as zooplankton (Desforges, Galbraith, and Ross 2015). Several marine species have already presented microplastics, such as anemones (Romanó de Orte, Clowez, and Caldeira 2019), turtles, sea birds (Ugwu, Herrera, and Gómez 2021), zooplankton, and fishes (Neves et al. 2015; Desforges, Galbraith, and Ross 2015).

The concentration of these pollutants is higher in regions with higher population density, as it is directly related to habits, products used, and waste produced by people in urban centers (Browne et al. 2011). This contamination also affects humans, as studies have already identified microplastics in human feces (Yan et al. 2022), lungs (Amato-Lourenço et al. 2021), and blood samples (Leslie et al. 2022).

Human ingestion of these particles is inevitable, mainly due to their presence in various foods and water, in addition to being inhaled through the air (Amato-Lourenço et al. 2021). There are already studies that have proven the presence of MNPs in fruits and vegetables (Oliveri Conti et al. 2020), fishes (Neves et al. 2015), and bottled beverages, such as mineral water (Zuccarello et al. 2019). This presence of MNPs in the human body may be related to health problems, as shown in the work of Yan et al. (2022), where patients with intestinal inflammatory problems had a higher concentration of plastic fragments in their feces.

The reduced size of MNPs results in difficulties in their identification, separation, and quantification, but there are already ways to identify them, such as using Fourier transform infrared spectroscopy (FT-IR), Raman spectroscopy (Ugwu, Herrera, and Gómez 2021); Phan, Padilla-Gamiño, and Luscombe 2022), and pyrolysis gas chromatography (Albignac et al. 2022). The choice of method depends on the environment in which the residue is found, the types of contaminants involved, and the type of information to be obtained, such as the characteristic of the particle or the weathering suffered (Phan et al. 2022).

Furthermore, there are recent studies that aim to mitigate the impacts caused by microplastics, such as the use of microorganisms to promote the degradation of these residues, using fungi, bacteria, and microalgae (Othman et al. 2021; Dey et al. 2023), through the action of enzymes (Othman et al. 2021), membrane bioreactors, ultrafiltration, coagulation-flocculation (Shaik, Mohammed, and Rao 2022), and studies aimed at removal in effluent treatment plants (Iyare, Ouki, and Bond 2020).

Thus, microplastics and nanoplastics are present in water, soil, and air, contaminating the environment and human beings. These factors demonstrate the need for further studies to facilitate the detection, quantification, and taking of measures (management) to contain the formation of these residues and remedy their effects.

9.2 MICROPLASTICS AND NANOPLASTICS

In recent years, there has been an increase in environmental concern regarding plastics that are released into the environment in different ways, which can be degraded by abiotic factors (temperature, humidity, salinity, UV radiation from sunlight, and mechanical abrasion) or biologically (Caixeta et al., 2018; Ranjan and Goel, 2021). These weather conditions allow the "erosion" of plastic, transforming it into nano- or microparticles that reach aquatic environments and are ingested by animals, being passed on throughout the food chain (Ranjan and Goel, 2021; Wang et al. 2022). Thus, they can accumulate in the environment, generating a series of hostile factors to the flora and fauna (Caixeta et al., 2018; Ranjan and Goel, 2021).

About 18% of plastic waste is recycled and 24% is incinerated. The remaining 58% are deposited in landfills or enter the environment, taking into account global data.

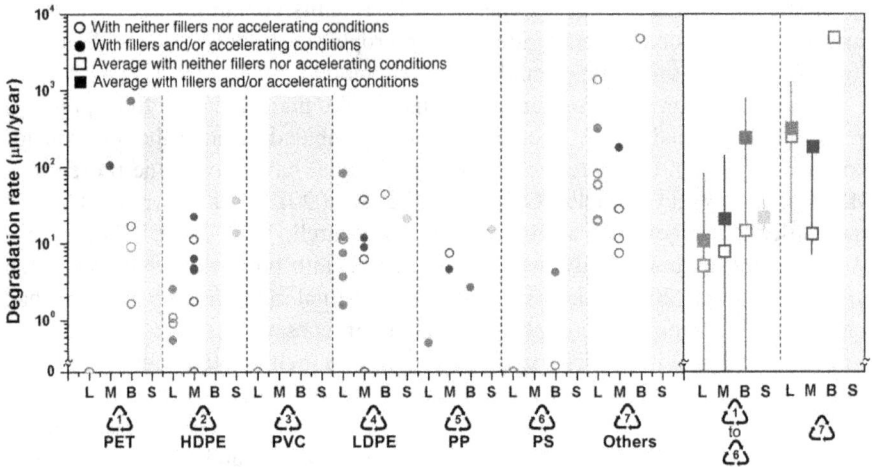

FIGURE 9.1 Degradation rates of major plastics under different environmental conditions (L, landfill/compost/soil; M, marine; B, biological; S, sunlight). The type 7 corresponds to several plastics that are nominally biodegradable. The range and mean values are shown on the right as lines and squares. Data points representing extremely slow degradation rates are shown on the X-axis. Columns in gray denote that no data was found (Chamas et al. 2020).

These plastics build up and stay for a long time (Chamas et al. 2020). In Figure 9.1 we can observe the degradation rates of the plastics most used by the industry based on the polymers used for their production. Thus, in the graph, they are organized by type of plastic and degradation environment.

When degradation occurs, the particles decrease in scale. When dimensions are less than 5 mm, they are classified as microplastics, and when these particles are in the nanometric scale (10^{-9} m), they are classified as nanoplastics. They normally occur in mixtures of these sizes, being recognized by the acronym MNPs (micro- and nanoplastics). Depending on the polymeric material that forms this plastic, the environment in which it is found, and the presence or absence of degradation accelerators, these materials present a greater or lesser facility to degrade (Chamas et al. 2020).

The presence of these pollutants in the environment represents a threat to the biota. Due to their very small size, they can reach even remote areas and become available to a wide variety of organisms – from lower trophic levels – and also because of their vast surface area, they can adsorb highly toxic compounds, such as hydrocarbons and heavy metals (Olivatto et al. 2018; Caixeta et al., 2018).

One of the main sources of microplastics is washing clothes made of synthetic textile fibers, which release microfibers of varying sizes into sewers, being mainly polyester and presenting mostly 360–660 µm in length and 12–16 µm in average diameter. The release occurs mainly by mechanical stresses and chemicals that displace the microfibers of the fabric threads, and this event is responsible for about 35% of the global release (Boucher and Friot 2017; Falco et al. 2019).

Another example of contamination occurs through the air, through the tires. These release microplastics and nanoplastics in suspension from abrasion with the ground during vehicle use. In addition, these MNPs can also be carried by rain, contaminating water bodies and soils (Boucher and Friot 2017).

These tiny plastic particles have become a looming problem. In this current decade, several studies began to be concerned with degraded plastic, since they have already been found in highly consumed animals – mainly fish (Neves et al. 2015), drinking water (Zhang et al. 2020a), in the food we eat (Oliveri Conti et al. 2020), and even in the human body, that is, in placentas of pregnant women and human bloodstream of 80% of blood donors (Leslie et al. 2022) (Table 9.1).

The effects of the presence of these plastics on the human body are still being revealed. What is known so far is that they can accumulate in some organs (Leslie et al. 2022) and undergo several biotransformations in the gastrointestinal tract and colon, in addition to reducing the bacterial cultures present in the gut flora, which can cause several other consequences (Tamargo et al. 2022). In addition to the plastic

TABLE 9.1
Plastic Particles Found in Water, Foods, and Human Body

Sample	Polymer	Particle size	Reference
Drinking water	Polyethylene and polystyrene	10–20 µm	Zhang et al. 2020b
Fishes	–	>500 µm	Chen et al. 2022
Copepods (crustaceans)	Polystyrene	–	Kim et al. 2022
Plants	Polystyrene	–	Liu et al. 2022
Fruits and vegetables	–	<10 µm	Oliveri Conti et al. 2020
Cow milk	Polyethersulfone and polysulfone	0.1 µm to 5 mm	Kutralam-Muniasamy et al. 2020
Table salt	Polyethylene and polypropylene	Between 20 µm and 5 mm in size (average size of 2.32 mm)	Gündoğdu 2018
Breast milk	Polyethylene, polyvinyl chloride, and polypropylene	2 to 12 µm	Ragusa et al. 2022
Human placenta	Polyvinyl chloride (43.27%), polypropylene (14.55%), and polybutylene succinate (10.90%)	20.34–307.29 µm (80.29% of particles are < 100 µm)	Zhu et al. 2023
Human blood	Polyethylene terephthalate, polyethylene, and polymers of styrene and poly(methyl methacrylate)	–	Leslie et al. 2022
Human feces	Poly(ethylene terephthalate) and polyamide	–	Yan et al. 2022
Human lungs	Polyethylene and polypropylene	<5.5 µm	Amato-Lourenço et al. 2021

itself, various additives present in them can also cause serious problems, for example, altering fetal programming at an epigenetic level (changes in DNA that do not change its sequence but affect the activity of one or more genes). This effect can be passed down through generations and may play a role in the development of various chronic disorders, such as metabolic, reproductive, and degenerative diseases, as well as some forms of cancer (Gruber et al. 2022).

9.3   MNPs QUANTIFICATION/QUALIFICATION TECHNIQUES

Several methods can be used in the analysis and identification of MNPs that can vary in efficiency and cost to be performed. The most important characteristics in the analysis of these pollutants are the physical (size, shape, and color) and chemical (type of polymer) characteristics.

Research related to MNPs has been ongoing for years. However, there is still no consensus on the collection, treatment, and characterization procedures (Figure 9.2). In this way, numerous methods are used, resulting in a wide variation of results in the research carried out, making it difficult to compare and standardize the techniques (Chia et al. 2022).

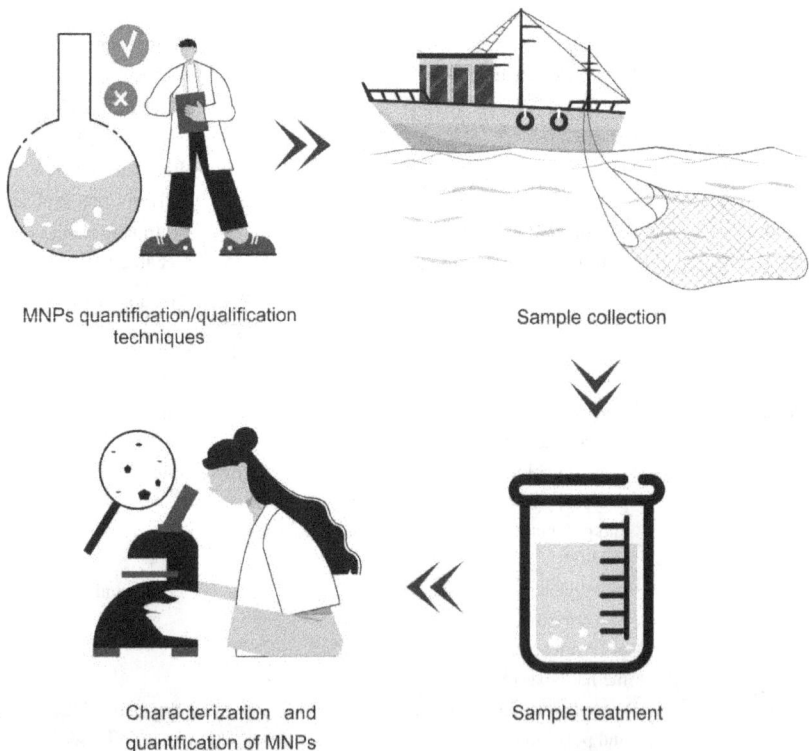

MNPs quantification/qualification
techniques

Sample collection

Characterization and
quantification of MNPs

Sample treatment

FIGURE 9.2   Scheme showing the main steps to qualify and quantify plastic particles, in this case in water samples.

9.3.1 SAMPLE COLLECTION

One of the most used sampling techniques is low-volume sampling. This technique consists of a mesh net being dragged by a boat and having a volume meter to calculate the amount of volume filtered (Cha, Lee, and Chia 2022). According to the recommendations of the NOAA (National Oceanic and Atmospheric Administration), the mesh to be used is 333 µm. However, smaller mesh sizes can collect hundreds of times more plastic (Wang et al. 2020).

There are also several other collection techniques such as the surface microlayer method, manual net collection, and bulk water sampling. These techniques are rarely used due to the large volume of samples required (Cha, Lee, and Chia 2022). When comparing the techniques used in the sampling, they found that the method used in the collection influenced the number of plastics found.

9.3.2 SAMPLE TREATMENT

Before the identification of MNPs, the samples need to go through the purification process due to the impurities present in the samples. These impurities are particles that were present in the environment from which the samples were collected, such as organic matter, inorganic sediments, etc. The presence of these impurities makes the identification process imprecise, and this treatment is necessary to facilitate and improve the identification of MNPs (Chia et al. 2022).

Density separation is one of the most used methods for separating plastic particles from impurities. In this method, the sample of MNPs is saturated with a salt solution. After that, the sample is shaken, keeping the remaining at rest for the extraction of MNPs. Thus, the method is based on the difference in density between the particles, while the density (ρ) of MNPs varies between 0.8 (silicone) and 1.4 g/cm³ (polyethylene terephthalate – PET), while the density of inorganic sediments is around 2.65 g/cm³.

Among the solutions used in density separation, the saturation of water with sodium chloride (NaCl; ρ = 1.2 g/cm³) is the most used because of its low cost and non-toxicity. However, denser microplastics are not captured in this system. Different substances can also be used, depending on the cost and dangerousness, such as sodium polytungstate (ρ = 1.4–1.5 g/cm³), calcium chloride (CaCl$_2$; ρ = 1.3 g/cm³), sodium iodide (NaI; ρ = 1.8 g/cm³), and zinc chloride (ZnCl$_2$; ρ = 1.6 g/cm³) (Priya et al. 2022).

Another way of purification, more modern and precise, is Munich Plastic Sediment Separator (MPSS), which was developed by Imhof et al. (2012). Unlike the conventional method, MPSS can separate plastic particles smaller than 309 µm, in addition to extracting about 96% of the particles present in the samples in just one separation step. In different methods, several steps are required to achieve this efficiency (Jin et al. 2022).

Two methods are used to separate organic matter: chemical and enzymatic digestion. In chemical digestion, samples of MNPs are mixed with a chemical that degrades the organic matter without compromising the structure of the fragments. The great care that must be taken in this methodology is related to the substance used. On the one hand, oxidizing acids such as sulfuric and nitric acids can degrade organic matter, but compromise and damage MNPs. On the other hand, non-oxidizing acids, such as hydrochloric acid, cannot completely degrade organic matter (Löder et al. 2017). Thus, depending on the substance used, the quantification of MNPs can be compromised, which is a major disadvantage. The most suitable and most commonly used substance is hydrogen peroxide (H_2O_2) with volumes ranging from 10% to 30% due to the low risk of deteriorating MNPs (Tsering et al. 2022; Chia et al. 2022). In addition, it is possible to use a catalyst to make the digestion process faster and more efficient. According to the NOAA, a solution of Fe(II) can be mixed with hydrogen peroxide as a catalyst in a method called "wet peroxide oxidation (WPO)" (Tsering et al. 2022).

In enzymatic digestion, the sample is incubated with a combination of several enzymes, such as lipase, amylase, proteinase, chitinase, and cellulose, to degrade organic matter. Its advantage over the chemical approach is that MNPs are not damaged. However, the digestion time is very long, reaching up to 15 days (Löder et al. 2017; Tirkey and Upadhyay 2021).

This treatment step must be carried out very carefully since one of the problems that can occur is to generate an overestimation of the number of MNPs due to brittle particles (Löder et al. 2017).

9.3.3 CHARACTERIZATION AND QUANTIFICATION OF MNPs

Visual identification was one of the first ways of analyzing the identification of plastic fragments. After the sample treatment, the microplastics can be identified both with the naked eye and with the aid of a microscope. Generally, naked eye identification is used for particles with a size of 1–5 mm, normally found on the beach. To identify the MNPs from 1 mm to the scale of nanometers (nm), the microscope is used – the instrument most used in works related to MNPs (Tirkey and Upadhyay 2021). The ease of the process, low cost, and fast counting are the main advantages of this technique. However, the disadvantages are the difficulty in identifying the plastics, requiring another identification technique, and the lack of precision in the quantification and recognition of MNPs (Priya et al. 2022). When compared to more accurate techniques, this method presents an identification error between 20 and 70% (Cha, Lee, and Chia 2022).

Faced with these factors that can lead to errors in the identification of MNPs, Nóren (2007) established some criteria to improve identification accuracy:

1. The cellular or organic structure is not visible in microplastics
2. The length of the fibers must be equally thick throughout their length
3. The colors of microplastics must be clear and homogeneous
4. Clear or white, they should be examined under a microscope at high magnification and under a fluorescence microscope to exclude an organic origin

Another way also used to differentiate MNPs from non-plastic particles is the hot needle method (Syafina et al. 2022). According to "The guide to microplastic identification," created by the Marine & Environmental Research Institute (MERI), this test proved to be very efficient in distinguishing plastic particles and organic matter. The method consists of using a very hot needle on the particles, where the plastic particles will melt or curl, while organic matter or other non-plastic particles will not show the same behavior (MARI 2012).

The most accurate techniques for identifying MNPs are Fourier transform infrared spectroscopy (FT-IR), Raman spectroscopy, gas chromatography coupled to mass spectrometry (GC-MS), thermal analysis (TGA-M pyrolysis), liquid chromatography, and a staining method using the red Nile dye (Chia et al. 2022).

FTIR spectroscopy is based on the irradiation of infrared light with a predetermined wavelength, where the sample absorbs this radiation which is collected by the equipment and provides data on the structure of the sample (Chia et al. 2022). Raman spectroscopy is based on laser scattering through the sample resulting in different frequencies of diffuse light according to the molecular structure and atoms that form the polymer (Priya et al. 2022).

The advantages of using these two techniques are the identification of MNPs with high precision, the determination of the types of plastic polymers, and the keeping of the structure of the MNPs intact without any type of wear or damage. FTIR spectrometry also provides data on the physical or chemical wear of plastics and can be used on microplastics up to 500 µm and on particles up to 20 µm with the "accentuated total reflectance" (ATR) mode (Phan et al. 2022). However, spectroscopy techniques also have some limitations, such as high cost, long analysis time, complex, and trained and experienced personnel required in handling the equipment, in addition to knowledge of the characteristics of the various types of polymers. There is also the possibility of error in the counting of MNPs due to overlapping particles. In addition, Raman spectrometry is sensitive to chemical additives and pigments, which can lead to errors in the identification of some polymers (Jin et al. 2022).

The GC-pyrolysis and TGA-MS techniques thermally degrade the sample to be analyzed in a mass spectrometer and by gas chromatography. After this process, the results are compared with data from common plastics. In this procedure, it is possible to obtain identification and concentration data. However, this method precludes further analysis, since the sample is degraded, in addition to being expensive, complex, and limited in its ability to analyze certain sizes and masses of MNPs (Funck et al. 2020).

Liquid chromatography is based on the solubility of MNPs to perform the analysis. In this method, a specific solvent is required for each type of polymer, and the sample is solubilized with this solvent and then analyzed in high-performance liquid chromatography (HPLC). The main advantage of this technique is the high recovery of the analyzed polymers. And its main limitations are the need for a relatively high mass to perform the analysis and the impossibility of determining the physical characteristics of the polymers (Priya et al. 2022).

The staining technique with the red Nile dye consists of identifying the MNPs through the fluorescence emitted when the particles are irradiated with blue light. This method is one of the easiest and simplest techniques to use. However, it can be somewhat imprecise if the sample does not have adequate pre-treatment to eliminate impurities, as it can also be adsorbed by organic matter (Meyers et al. 2022).

9.4   PROBLEMS CAUSED BY MNPs

Currently, about 45 different types of plastic are used by the population and end up being discarded, for the most part, incorrectly (Lin et al. 2018). The long period of degradation results in fragments in micro- and nanometric scales, which end up dispersed in terrestrial and aquatic environments and can also, due to low density, be carried by the wind and dispersed in the atmosphere (Wright et al. 2020). The toxic effects of MNPs are associated, in part, with their chemical and physical characteristics, considering their composition and degradation time (Yang et al. 2020). The longer the time for total degradation, the longer the plastic will exist as a pollutant in different ecosystems, affecting the environment and human health.

9.4.1   ENVIRONMENTAL PROBLEMS

MNPs are considered ubiquitous in the oceans, being transported by runoff, dumping of domestic, industrial, and wastewater treatment plants, aquaculture, fisheries, and atmospheric emissions (Talvitie et al. 2017; Simon, van Alst, and Vollertsen 2018; Wright et al. 2020). In addition to existing in marine waters, MNPs are also present in sediments and end up being ingested by various marine organisms (Gall and Thompson 2015). Trophic bioaccumulation and biomagnification are consequences of the accumulation of MNPs in the environment and also the cause of different ecological problems (Phuong et al. 2018).

All animals and plants need water to maintain their vital functions, and therefore all, at some point, have or will have contact with MNPs via air, ingestion, or dermal contact. Due to the high rate of MNPs dispersed in the waters, all animals, especially aquatic ones, can ingest these contaminated plastic fragments and develop health problems (Auta, Emenike, and Fauziah 2017). According to Ugwu, Herrera, and Gómez (2021), microplastics can accumulate in the gastrointestinal system of aquatic animals such as mollusks. This indicates that not only large and medium-sized aquatic animals are susceptible to the ingestion of macroplastics, but also small-sized animals can ingest MNPs.

A study carried out by Roch, Friedrich, and Brinker (2020) with microplastics collected from the epidermis of marine animals shows that these particulate materials serve as transport channels for bacteria and also adsorb chemical compounds on their surface. Shen et al. (2019) also showed that MNPs can be vehicles responsible for increasing the speed of diffusion of organisms in environments since such microorganisms are adsorbed in colonies on plastic fragments, causing an increase in biological invasions and, therefore, an increase in the existing genetic exchange between thesecommunities, especially in the transfer of genes resistant to pathogens and antibiotics. These data suggest that the increased flow of energy, materials, and information between ecosystems could be an indicator of environmental imbalance.

MNPs are capable of altering the chemical and physical characteristics of soils, altering their functions, the communities of existing organisms, nutrient cycling, and microbial activities, which directly affect the health of the soil and of all the animals and plants that use it (de Souza Machado et al. 2019; Yu et al. 2021). Plants, like animals, end up suffering interference in their metabolism due to the absorption of water-containing MNPs.

Yin et al. (2021) observed that MNPs taken up by vascular plants can induce several phytotoxic effects, affecting growth and photosynthesis, besides causing oxidative stress.

In addition to the already mentioned harmful effects, MNPs are capable of interfering with the reproduction, growth, and metabolism of animals, causing histopathological damage, intestinal microflora disorders, DNA damage, genotoxicity, reproductive toxicity, neurotoxicity, and metabolic disorders. These problems are directly linked to the particle size, type, shape, and dosage of MNPs assimilated by living beings (de Souza Machado et al. 2019).

Given the information about the negative impacts of MNPs, it is common consensus and a trend that more studies are conducted to assess and measure all the damage that MNPs can cause in different biotas around the world. However, for future studies to be of better quality and not underestimate the impacts caused by MNPs on nature, researchers must develop more efficient methodologies (Ugwu, Herrera, and Gómez 2021).

9.4.2 Plastic and Human Health

It is well known that MNPs cause serious impacts on the health of the environment and public health through the contact of living beings with contaminated environments. One of the ways to better understand the possible effects caused by MNPs is by understanding the sources and exposure routes that humans are subjected to. The routes can be inhalation, ingestion, dermal contact, and the use of prostheses (Sangkham et al. 2022), as shown in Figure 9.3.

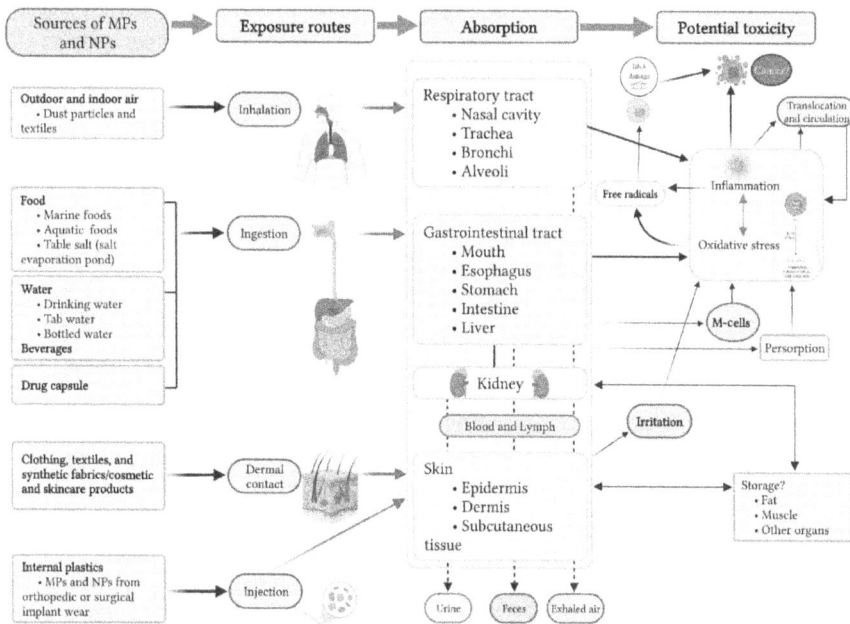

FIGURE 9.3 Potential pathways and routes of exposure to MNPs and potential toxic effects on humans (Sangkham et al. 2022).

Ingestion is considered the major route of exposure of humans to MNPs. It usually occurs through the ingestion of food, drinks, and drugs in capsules. Studies show that MNPs can be found in a range of products consumed daily by the population, such as fish and other marine animals, milk, water, and table salt (Sangkham et al. 2022). Consumption of these contaminated foods can lead to an accumulation of MNPs in the stomach, kidneys, and liver (Revel, Châtel, and Mouneyrac 2018). In addition, the process of endocytosis may allow MNPs to enter cells (Alimba et al. 2021). Although there are some studies associated with the sources and exposure routes, more studies should be carried out to prove the negative effects caused by the ingestion of MNPs in the human body (Sangkham et al., 2022).

Inhalation is the second major route of exposure of humans to MNPs and it can occur in two ways: through the nasal passages or aspirated through the mouth (Sangkham et al. 2022). Although most MNPs are blocked by mucociliary clearance in the nasal cavities (the main defense mechanism of the upper and lower airways), many particles end up entering the respiratory system and migrating to the lungs, accumulating in the tissues, bronchi, and pulmonary alveoli. This accumulation, associated with the chemical composition of the particles, can cause acute and even chronic respiratory inflammation, especially in people with compromised clearance systems (Gasperi et al. 2018; Chen, Feng, and Wang 2020).

Dermal contact is considered another source of interaction with MNPs, mainly through the use of cosmetics, personal care products, and clothing. Studies show that because of their particle size, microplastics are unlikely to be able to pass through the layers of the skin and be absorbed into the body. However, other studies suspect that nanoplastics can cross the stratum corneum of the epidermis (outer layer of the epidermis) and deposit in deeper layers. As there are not enough studies to prove the permeability of MNPs in the skin, the risks to human health by dermal contact are associated with the triggering of allergic processes in epithelial cells, caused by contact and deposition of MNPs on the skin surface, and, if they have adsorbed chemical compounds, may cause increased dermal irritability (Schneider et al. 2009; Revel et al. 2018; Sangkham et al. 2022).

Another source of MNPs is related to orthopedic prostheses and surgical implants produced with plastic resins and injected (inserted) into the human body. In this way, there is the possibility of particles detaching and being released by the body (Sangkham et al. 2022). Despite being considered a source of MNPs, studies proving its potential harm are still lacking.

According to Sangkham et al. (2022), MNPs can still affect the respiratory, digestive, circulatory, immune, nervous, embryonic, and placental systems and can also generate disturbances in metabolism, neurotoxicity, DNA damage, and tumors in human cells. Although there are already several studies on the damage to human health caused by MNPs, they are still insufficient to elucidate all the pathogenic processes that occur at the cellular level and the consequences associated with their accumulation for long periods in the human body. Only after these studies will human beings be able to understand the real harms of MNPs and seek to correct the mistakes of the past, to leave a healthier planet for future generations.

9.5 PLASTIC WASTE MANAGEMENT

Due to population growth and the development of some countries, a significant increase in urban solid waste (USW) is observed, among them are organic waste (food), paper waste (cardboard), metals (aluminum cans, steel objects), glass waste (broken glass, bottles), and mainly plastic waste (bottles, packaging materials, some plastic-based home appliances) (Tejaswini et al. 2022).

Most plastics will come from waste that goes to landfills without any kind of treatment and this lack of waste management causes numerous environmental impacts such as pollution, global climate change, and numerous health problems for living beings (Tejaswini et al. 2022).

Effective actions related to the management and recycling of plastic waste are applied in developed countries, such as Singapore and South Korea, where they have the highest recycling rate (around 59%), followed by Netherlands (50%), UK (43%), USA (35%), Japan (19%), and India (13%) (Lee et al. 2021; Tejaswini et al. 2022).

These solid waste management projects are a way to prevent plastic waste from being disposed of incorrectly. As a result, management policy arises, such as the implementation of various environmental laws and regulations.

The "polluter pays" principle (PPP) is an alternative (environmental policy) that exists in developed countries. According to the law, the polluter must bear the costs to reduce the pollution caused, and these costs recovered by the polluters are invested in recycling processes, reuse, and proper disposal of the generated waste (Tejaswini et al. 2022).

An example of this PPP model is present in the Netherlands, Sweden, and Singapore: if companies use more than 50 tons of single-use plastic packaging, they must pay taxes to the government, and these taxes are used for waste recycling. In the Dutch retail sector, it is forbidden to give plastic bags, and customers must carry their carrying bags (Tejaswini et al. 2022).

There is also recycling as an alternative to contain the increase in plastic waste in the environment, such as primary recycling, which consists of reprocessing scraps and non-conforming materials in plastic production processes; the secondary is the use of post-consumer materials, where these residues go through the process of separation, washing, drying, and grinding of the material to process them again; the tertiary consists of the addition of chemical reagents to recover the monomers; the quaternary is used for energy generation in the middle of the process. However, this recycling is extremely harmful because incineration generates toxic gases (Babaremu et al. 2022).

One of the most recent waste management alternatives is the circular economy (CE – Figure 9.4), used to close the cycle of plastic waste management and convert this waste into secondary products, but this model still needs improvement, as other processes are required for its application, such as the waste decontamination phase (Babaremu et al. 2022).

As they are rich in carbon atoms, plastic waste has great potential to be applied in various value-added materials, as a low-cost alternative, as mitigation of plastic pollution, and as an example of a circular economy. Thus, Several studies are currently underway to find new applications for these waste materials. From plastic

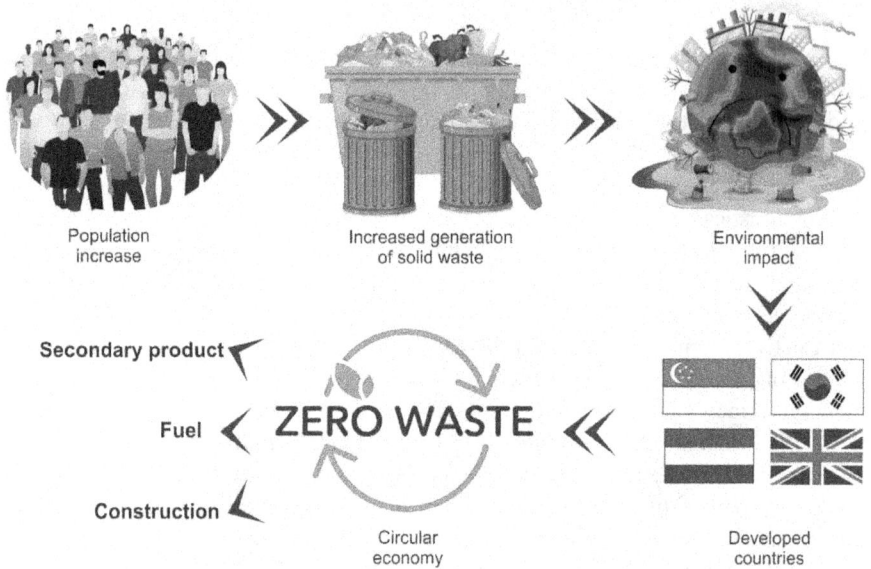

FIGURE 9.4 Consequences of population growth in relation to waste generation and environmental impacts and example of some countries that apply circular economy through public policies to produce products from waste.

waste, it was possible to produce graphene nanotubes that can be used for sustainable energy conservation through supercapacitors, with economic and ecological benefits (Karakoti et al. 2022). In addition, other carbon nanomaterials can be manufactured such as nanofibers, nanosheets, carbon spheres, and porous carbon (Ren et al. 2022).

These residues can also be used as an alternative raw material for the production of fuels due to the hydrocarbons present in their structures, being converted into alternative energy for boilers, industrial engines, marine engines, and locomotive diesel engines. Compared to diesel, fuel savings of up to 20% and a reduction in exhaust emissions of about 13% for CO and 16% for HC were found (Padmanabhan et al. 2022). The conversion of plastic waste to fuel is traditionally done by high-temperature pyrolysis, resulting in volatile products that pollute the environment. A more promising alternative for future applications would be advanced oxidation that can convert plastic into fuel at ambient temperature and pressure (Li et al. 2022).

In the field of civil construction, these residues can be used as plastic aggregates to reduce density to produce lightweight concrete, and improve impact resistance, abrasion, and insulation properties, increasing their durability, but mechanical properties, such as resistance, were reduced (Abbas and Hilal 2021; Nodehi and Taghvaee 2022).

Many alternatives in the management of plastic waste have been proposed over the last few years, some even consolidated, but the management of MNPs is still seen as a challenge, and some research has considered using filtration membranes in efflu-ent treatment plants for the removal of plastic particles from wastewater (Rout et al. 2022). However, more studies, more funding, and more guidance for people all over our planet are needed so that the exaggerated consumption of plastics is avoided, reducing the demand to produce this material so present in our lives, which we often do not even imagine.

9.6 CONCLUSIONS

Plastics have become a significant problem worldwide since they are produced in large quantities and most reach the environment, either in their original or frag-mented form (micro- and nanometric). In this context, MNPs are considered emerg-ing contaminants, as they are already commonly found in water, soil, and air, as well as in animals, plants, and even humans. For the identification and quantifica-tion of MNPs, the collection and preparation of samples are of paramount impor-tance. There are several techniques for identification and quantification, some being more expensive than others and some more precise. However, much effort still needs to be put into studies related to quantifying and identifying these particles so that any research group can collaborate in the fight against this great evil. Through consumption of food and drink, as well as inhalation of particles or use of surgical prostheses, human beings have been contaminated. The effect of the presence of plastic on the environment and human bodies is still not well known. But what we have observed is frightening (damage to cells and organs, bioaccumulation, trig-gering of various diseases). Therefore, it is crucial that global measures be imple-mented to manage waste and inform the population to mitigate the proliferation of these damgerous particles.

ACKNOWLEDGMENTS

The authors would like to thank FAPEAM – Fundação de Amparo a Pesquisa do Estado do Amazonas – for the scholarships granted to Ana Emília M. de Freitas, Airi dos S. Sousa, and Wyvirlany V. Lobo; to CAPES – Coordenação de Aperfeiçoamento de Pessoal de Nível Superior – for the scholarships granted to Silma de Sá Barros and Antônio J. de Andrade Júnior; and to Pronametro Program (Suframa/Inmetro) for the research fellowships granted to Orlando A. da R. L. Paes and Flávio A. de Freitas.

REFERENCES

Abbas, Aya M., and Ameer A. Hilal. 2021. "Prediction of Strength of Foamed Concrete Containing Waste Plastic as Coarse and Fine Aggregates Using Minitab Software." In *2021 14th International Conference on Developments in eSystems Engineering (DeSE)*, 313–17. IEEE. https://doi.org/10.1109/DeSE54285.2021.9719519
Albignac, Magali, Jean François Ghiglione, Céline Labrune, and Alexandra ter Halle. 2022. "Determination of the Microplastic Content in Mediterranean Benthic Macrofauna by Pyrolysis-Gas Chromatography-Tandem Mass Spectrometry." *Marine Pollution Bulletin* 181 (August). https://doi.org/10.1016/j.marpolbul.2022.113882

Alimba, Chibuisi G., Caterina Faggio, Saravanadevi Sivanesan, Adebayo L. Ogunkanmi, and Kannan Krishnamurthi. 2021. "Micro(nano)-plastics in the Environment and Risk of Carcinogenesis: Insight into Possible Mechanisms." *Journal of Hazardous Materials.* https://doi.org/10.1016/j.jhazmat.2021.126143

Amato-Lourenço, Luís Fernando, Regiani Carvalho-Oliveira, Gabriel Ribeiro Júnior, Luciana dos Santos Galvão, Rômulo Augusto Ando, and Thais Mauad. 2021a. "Presence of Airborne Microplastics in Human Lung Tissue." *Journal of Hazardous Materials* 416 (August): 126124. https://doi.org/10.1016/j.jhazmat.2021.126124

Andrady, Anthony L. 2017a. "The Plastic in Microplastics: A Review." *Marine Pollution Bulletin* 119 (1): 12–22. https://doi.org/10.1016/j.marpolbul.2017.01.082

Auta, H.S., C.U. Emenike, and S.H. Fauziah. 2017. "Distribution and Importance of Microplastics in the Marine Environment: A Review of the Sources, Fate, Effects, and Potential Solutions." *Environment International.* https://doi.org/10.1016/j.envint.2017.02.013

Babaremu, K.O., S.A. Okoya, E. Hughes, B. Tijani, D. Teidi, A. Akpan, J. Igwe, S. Karera, M. Oyinlola, and E.T. Akinlabi. 2022. "Sustainable Plastic Waste Management in a Circular Economy." *Heliyon* 8 (7): e09984. https://doi.org/10.1016/j.heliyon.2022.e09984

Boucher, J., and D Friot. 2017. *Primary Microplastics in the Oceans: A Global Evaluation of Sources.* IUCN, Gland, Switzerland. p. 43. https://doi.org/10.2305/IUCN.CH.2017.01.en

Boucher, Julien, and Damien Friot. 2017. *International Union for Conservation of Nature: A Global Evaluation of Sources Primary Microplastics in the Oceans.* Edited by João Matos de Sousa Marine. IUCN, Gland, Switzerland.

Browne, Mark Anthony, Phillip Crump, Stewart J. Niven, Emma Teuten, Andrew Tonkin, Tamara Galloway, and Richard Thompson. 2011. "Accumulation of Microplastic on Shorelines Worldwide: Sources and Sinks." *Environmental Science & Technology* 45 (21): 9175–79. https://doi.org/10.1021/es201811s

Caixeta, Danila, Frederico Caixeta, and Frederico Menezes Filho. 2018. "Nano e microplásticos nos ecossistemas: impactos ambientais e efeitos sobre os organismos." *Enciclopédia Biosfera* 15 (27): 19–34. https://doi.org/10.18677/EnciBio_2018A92

Cha, Jihye, Jin-Yong Lee, and Rogers Wainkwa Chia. 2022. "Comment on the Paper 'Microplastic Contamination of an Unconfined Groundwater Aquifer in Victoria, Australia.'" *Science of the Total Environment* 820 (May): 153121. https://doi.org/10.1016/j.scitotenv.2022.153121

Chamas, Ali, Hyunjin Moon, Jiajia Zheng, Yang Qiu, Tarnuma Tabassum, Jun Hee Jang, Mahdi Abu-Omar, Susannah L. Scott, and Sangwon Suh. 2020. "Degradation Rates of Plastics in the Environment." *ACS Sustainable Chemistry and Engineering* 8 (9): 3494–3511. https://doi.org/10.1021/acssuschemeng.9b06635

Chen, Guanglong, Qingyuan Feng, and Jun Wang. 2020. "Mini-Review of Microplastics in the Atmosphere and Their Risks to Humans." *Science of the Total Environment.* https://doi.org/10.1016/j.scitotenv.2019.135504

Chen, Yuling, Zhixin Shen, Gaojun Li, Kehuan Wang, Xingwei Cai, Xiong Xiong, and Chenxi Wu. 2022. "Factors Affecting Microplastic Accumulation by Wild Fish: A Case Study in the Nandu River, South China." *Science of the Total Environment* 847 (November): 157486. https://doi.org/10.1016/j.scitotenv.2022.157486

Chia, Rogers Wainkwa, Jin-Yong Lee, Jiwook Jang, and Jihye Cha. 2022. "Errors and Recommended Practices That Should Be Identified to Reduce Suspected Concentrations of Microplastics in Soil and Groundwater: A Review." *Environmental Technology & Innovation* 102933. https://doi.org/10.1016/j.eti.2022.102933

Desforges, Jean-Pierre W., Moira Galbraith, and Peter S. Ross. 2015a. "Ingestion of Microplastics by Zooplankton in the Northeast Pacific Ocean." *Archives of Environmental Contamination and Toxicology* 69 (3): 320–30. https://doi.org/10.1007/s00244-015-0172-5

Dey, S., U. Anand, V. Kumar, S. Kumar, M. Ghoraj, A. Ghosh, N. Kant, S. Suresh, S. Bhattacharya, E. Bontempi, S.A. Bhat, and A. Dey. 2023. "Microbial Strategies for Degradation of Microplastics Generated from COVID-19 Healthcare Waste." *Environmental Research* 114438. https://doi.org/10.1016/j.envres.2022.114438

Falco, Francesca De, Emilia Di Pace, Mariacristina Cocca, and Maurizio Avella. 2019. "The Contribution of Washing Processes of Synthetic Clothes to Microplastic Pollution." *Scientific Reports* 9 (1): 6633. https://doi.org/10.1038/s41598-019-43023-x

Funck, Matin, Aylin Yildirim, Carmen Nickel, Jürgen Schram, Torsten C. Schmidt, and Jochen Tuerk. 2020. "Identification of Microplastics in Wastewater after Cascade Filtration Using Pyrolysis-GC–MS." *MethodsX* 7: 100778. https://doi.org/10.1016/j.mex.2019.100778

Gall, S.C., and R.C. Thompson. 2015. "The Impact of Debris on Marine Life." *Marine Pollution Bulletin* 92 (1–2): 170–79. https://doi.org/10.1016/j.marpolbul.2014.12.041

Gasperi, Johnny, Stephanie L. Wright, Rachid Dris, France Collard, Corinne Mandin, Mohamed Guerrouache, Valérie Langlois, Frank J. Kelly, and Bruno Tassin. 2018. "Microplastics in Air: Are We Breathing It In?" *Current Opinion in Environmental Science and Health.* https://doi.org/10.1016/j.coesh.2017.10.002

Geyer, Roland, Jenna R. Jambeck, and Kara Lavender Law. 2017. "Production, Use, and Fate of All Plastics Ever Made." *Science Advances* 3 (7). https://doi.org/10.1126/sciadv.1700782

Gruber, Elisabeth S., Vanessa Stadlbauer, Verena Pichler, Katharina Resch-Fauster, Andrea Todorovic, Thomas C. Meisel, and Sibylle Trawoeger, et al. 2022. "To Waste or Not to Waste: Questioning Potential Health Risks of Micro- and Nanoplastics with a Focus on Their Ingestion and Potential Carcinogenicity." *Exposure and Health.* https://doi.org/10.1007/s12403-022-00470-8

Gündoğdu, Sedat. 2018. "Contamination of Table Salts from Turkey with Microplastics." *Food Additives & Contaminants: Part A* 35 (5): 1006–14. https://doi.org/10.1080/19440049.2018.1447694

Imhof, Hannes K., Johannes Schmid, Reinhard Niessner, Natalia P. Ivleva, and Christian Laforsch. 2012. "A Novel, Highly Efficient Method for the Separation and Quantification of Plastic Particles in Sediments of Aquatic Environments." *Limnology and Oceanography: Methods* 10 (7): 524–37. https://doi.org/10.4319/lom.2012.10.524

Iyare, Paul U., Sabeha K. Ouki, and Tom Bond. 2020. "Microplastics Removal in Wastewater Treatment Plants: A Critical Review." *Environmental Science: Water Research and Technology.* Royal Society of Chemistry. https://doi.org/10.1039/d0ew00397b

Jin, Naifu, Yizhi Song, Rui Ma, Junyi Li, Guanghe Li, and Dayi Zhang. 2022. "Characterization and Identification of Microplastics Using Raman Spectroscopy Coupled with Multivariate Analysis." *Analytica Chimica Acta* 1197 (March): 339519. https://doi.org/10.1016/j.aca.2022.339519

Karakoti, Manoj, Sandeep Pandey, Gaurav Tatrari, Pawan Singh Dhapola, Ritu Jangra, Sunil Dhali, Mayank Pathak, Suman Mahendia, and Nanda Gopal Sahoo. 2022. "A Waste to Energy Approach for the Effective Conversion of Solid Waste Plastics into Graphene Nanosheets Using Different Catalysts for High Performance Supercapacitors: A Comparative Study." *Materials Advances* 3 (4): 2146–57. https://doi.org/10.1039/d1ma01136g

Kim, Kanghee, Hakwon Yoon, Jin Soo Choi, Youn-Joo Jung, and June-Woo Park. 2022. "Chronic Effects of Nano and Microplastics on Reproduction and Development of Marine Copepod Tigriopus Japonicus." *Ecotoxicology and Environmental Safety* 243 (September): 113962. https://doi.org/10.1016/j.ecoenv.2022.113962

Kutralam-Muniasamy, Gurusamy, Fermín Pérez-Guevara, I. Elizalde-Martínez, and V.C. Shruti. 2020. "Branded Milks – Are They Immune from Microplastics Contamination?" *Science of the Total Environment* 714 (April): 136823. https://doi.org/10.1016/j.scitotenv.2020.136823

Lee, Min-Yong, Na-Hyeon Cho, Sun-Ju Lee, Namil Um, Tae-Wan Jeon, and Young-Yeul Kang. 2021. "Application of Material Flow Analysis for Plastic Waste Management in the Republic of Korea." *Journal of Environmental Management* 299 (December): 113625. https://doi.org/10.1016/j.jenvman.2021.113625

Leslie, Heather A., Martin J.M. van Velzen, Sicco H. Brandsma, A. Dick Vethaak, Juan J. Garcia-Vallejo, and Marja H. Lamoree. 2022a. "Discovery and Quantification of Plastic Particle Pollution in Human Blood." *Environment International* 163 (May): 107199. https://doi.org/10.1016/j.envint.2022.107199

Li, Ning, Hengxin Liu, Zhanjun Cheng, Beibei Yan, Guanyi Chen, and Shaobin Wang. 2022. "Conversion of Plastic Waste into Fuels: A Critical Review." *Journal of Hazardous Materials.* https://doi.org/10.1016/j.jhazmat.2021.127460

Lin, Lang, Lin Zi Zuo, Jin Ping Peng, Li Qi Cai, Lincoln Fok, Yan Yan, Heng Xiang Li, and Xiang Rong Xu. 2018. "Occurrence and Distribution of Microplastics in an Urban River: A Case Study in the Pearl River along Guangzhou City, China." *Science of the Total Environment* 644: 375–81. https://doi.org/10.1016/j.scitotenv.2018.06.327

Liu, Yingying, Rong Guo, Shuwu Zhang, Yuhuan Sun, and Fayuan Wang. 2022. "Uptake and Translocation of Nano/Microplastics by Rice Seedlings: Evidence from a Hydroponic Experiment." *Journal of Hazardous Materials* 421 (January): 126700. https://doi.org/10.1016/j.jhazmat.2021.126700

Löder, Martin G.J., Hannes K. Imhof, Maike Ladehoff, Lena A. Löschel, Claudia Lorenz, Svenja Mintenig, and Sarah Piehl, et al. 2017. "Enzymatic Purification of Microplastics in Environmental Samples." *Environmental Science & Technology* 51 (24): 14283–92. https://doi.org/10.1021/acs.est.7b03055

MARI. 2012. "*Guide to Microplastic Identification.*" Marine & Environmental Research Institute, Center for Environmental Studies, Blue Hill, Maine. https://ise.usj.edu.mo/wp-content/uploads/2019/05/MERI_Guide-to-Microplastic-Identification_s.pdf accessed 07/12/2023.

Meyers, Nelle, Ana I. Catarino, Annelies M. Declercq, Aisling Brenan, Lisa Devriese, Michiel Vandegehuchte, Bavo De Witte, Colin Janssen, and Gert Everaert. 2022. "Microplastic Detection and Identification by Nile Red Staining: Towards a Semi-Automated, Cost- and Time-Effective Technique." *Science of the Total Environment* 823 (June): 153441. https://doi.org/10.1016/j.scitotenv.2022.153441

Moore, Charles James. 2008a. "Synthetic Polymers in the Marine Environment: A Rapidly Increasing, Long-Term Threat." *Environmental Research* 108 (2): 131–39. https://doi.org/10.1016/j.envres.2008.07.025

Neves, Diogo, Paula Sobral, Joana Lia Ferreira, and Tânia Pereira. 2015. "Ingestion of Microplastics by Commercial Fish off the Portuguese Coast." *Marine Pollution Bulletin* 101 (1): 119–126. https://doi.org/10.1016/j.marpolbul.2015.11.008

Nodehi, Mehrab, and Vahid Mohammad Taghvaee. 2022. "Applying Circular Economy to Construction Industry through Use of Waste Materials: A Review of Supplementary Cementitious Materials, Plastics, and Ceramics." *Circular Economy and Sustainability* 2 (3): 987–1020. https://doi.org/10.1007/s43615-022-00149-x

Nóren, Fredrik. 2007. *Small Plastic Particles in Coastal Swedish Waters.* N-Research Report, Commissioned by KIMO, Sweden, pp. 1–11.

Olivatto, G.P., R. Carreira, V.L. Tornisielo, and C.C. Montagner. 2018. "Microplastics: Contaminants of Global Concern in the Anthropocene | Microplásticos: Contaminantes De Preocupação Global No Antropoceno." *Revista Virtual De Quimica* 10 (6): 1968–89.

Oliveri Conti, Gea, Margherita Ferrante, Mohamed Banni, Claudia Favara, Ilenia Nicolosi, Antonio Cristaldi, Maria Fiore, and Pietro Zuccarello. 2020. "Micro- and Nano-Plastics in Edible Fruit and Vegetables. The First Diet Risks Assessment for the General Population." *Environmental Research* 187 (August): 109677. https://doi.org/10.1016/j.envres.2020.109677

Othman, Ahmad Razi, Hassimi Abu Hasan, Mohd Hafizuddin Muhamad, Nur 'Izzati Ismail, and Siti Rozaimah Sheikh Abdullah. 2021. "Microbial Degradation of Microplastics by Enzymatic Processes: A Review." *Environmental Chemistry Letters*. Springer Science and Business Media Deutschland GmbH. https://doi.org/10.1007/s10311-021-01197-9

Padmanabhan, Sambandam, K. Giridharan, Balasubramaniam Stalin, Subramanian Kumaran, V. Kavimani, N. Nagaprasad, Leta Tesfaye Jule, and Ramaswamy Krishnaraj. 2022. "Energy Recovery of Waste Plastics into Diesel Fuel with Ethanol and Ethoxy Ethyl Acetate Additives on Circular Economy Strategy." *Scientific Reports* 12 (1). https://doi.org/10.1038/s41598-022-09148-2

Phan, Samantha, Jacqueline L. Padilla-Gamiño, and Christine K. Luscombe. 2022. "The Effect of Weathering Environments on Microplastic Chemical Identification with Raman and IR Spectroscopy: Part I. Polyethylene and Polypropylene." *Polymer Testing* 116 (December): 107752. https://doi.org/10.1016/j.polymertesting.2022.107752

Phuong, Nam Ngoc, Laurence Poirier, Quoc Tuan Pham, Fabienne Lagarde, and Aurore Zalouk-Vergnoux. 2018. "Factors Influencing the Microplastic Contamination of Bivalves from the French Atlantic Coast: Location, Season and/or Mode of Life?" *Marine Pollution Bulletin* 129 (2): 664–74. https://doi.org/10.1016/j.marpolbul.2017.10.054

Priya, A.K., A.A. Jalil, Kingshuk Dutta, Saravanan Rajendran, Yasser Vasseghian, Jiaqian Qin, and Matias Soto-Moscoso. 2022. "Microplastics in the Environment: Recent Developments in Characteristic, Occurrence, Identification and Ecological Risk." *Chemosphere* 298 (July): 134161. https://doi.org/10.1016/j.chemosphere.2022.134161

Ragusa, Antonio, Valentina Notarstefano, Alessandro Svelato, Alessia Belloni, Giorgia Gioacchini, Christine Blondeel, and Emma Zucchelli, et al. 2022. "Raman Microspectroscopy Detection and Characterisation of Microplastics in Human Breastmilk." *Polymers* 14 (13): 2700. https://doi.org/10.3390/polym14132700

Ranjan, Ved Prakash, and Sudha Goel. 2021. "Recyclability of Polypropylene after Exposure to Four Different Environmental Conditions." *Resources, Conservation and Recycling* 169 (June): 105494. https://doi.org/10.1016/j.resconrec.2021.105494

Ren, Shiying, Xin Xu, Kunsheng Hu, Wenjie Tian, Xiaoguang Duan, Jiabao Yi, and Shaobin Wang. 2022. "Structure-Oriented Conversions of Plastics to Carbon Nanomaterials." *Carbon Research* 1 (1). https://doi.org/10.1007/s44246-022-00016-2

Revel, Messika, Amélie Châtel, and Catherine Mouneyrac. 2018. "Micro(nano)plastics: A Threat to Human Health?" *Current Opinion in Environmental Science and Health*. https://doi.org/10.1016/j.coesh.2017.10.003

Roch, S., C. Friedrich, and A. Brinker. 2020. "Uptake Routes of Microplastics in Fishes: Practical and Theoretical Approaches to Test Existing Theories." *Scientific Reports* 10 (1). https://doi.org/10.1038/s41598-020-60630-1

Roman, Lauren, Qamar Schuyler, Chris Wilcox, and Britta Denise Hardesty. 2021. "Plastic Pollution Is Killing Marine Megafauna, But How Do We Prioritize Policies to Reduce Mortality?" *Conservation Letters* 14 (2). https://doi.org/10.1111/conl.12781

Romanó de Orte, Manoela, Sophie Clowez, and Ken Caldeira. 2019. "Response of Bleached and Symbiotic Sea Anemones to Plastic Microfiber Exposure." *Environmental Pollution* 249 (June): 512–17. https://doi.org/10.1016/j.envpol.2019.02.100

Rout, Prangya Ranjan, Mohanty, Anee, Sharma, Ana, Miglani, Mehak, Liu, Dezhao, and Varjani, Sunita. 2022. "Micro- and Nanoplastics Removal Mechanisms in Wastewater Treatment Plants: A Review." *Journal of Hazardous Materials Advances* 6 (May): 100070. https://doi.org/10.1016/j.hazadv.2022.100070

Saliu, Francesco, Maurizio Veronelli, Clarissa Raguso, Davide Barana, Paolo Galli, and Marina Lasagni. 2021. "The Release Process of Microfibers: From Surgical Face Masks into the Marine Environment." *Environmental Advances* 4 (July). https://doi.org/10.1016/j.envadv.2021.100042

Sangkham, Sarawut, Orasai Faikhaw, Narongsuk Munkong, Pornpun Sakunkoo, Chumlong Arunlertaree, Murthy Chavali, Milad Mousazadeh, and Ananda Tiwari. 2022. "A Review on Microplastics and Nanoplastics in the Environment: Their Occurrence, Exposure Routes, Toxic Studies, and Potential Effects on Human Health." *Marine Pollution Bulletin* 181 (August): 113832. https://doi.org/10.1016/j.marpolbul.2022.113832

Schneider, Marc, Frank Stracke, Steffi Hansen, and Ulrich F. Schaefer. 2009. "Nanoparticles and Their Interactions with the Dermal Barrier." *Dermato-Endocrinology* 1 (4): 197–206. https://doi.org/10.4161/derm.1.4.9501

Shaik, Feroz, Nayeemuddin Mohammed, and Lakkimsetty Nageswara Rao. 2022. "Treatment Technologies for the Removal of Micro Plastics from Aqueous Medium." In *AIP Conference Proceedings*, 030009. AIP Publishing LLC. https://doi.org/10.1063/5.0080232

Shen, Maocai, Yuan Zhu, Yaxin Zhang, Guangming Zeng, Xiaofeng Wen, Huan Yi, Shujing Ye, Xiaoya Ren, and Biao Song. 2019. "Micro(nano)plastics: Unignorable Vectors for Organisms." *Marine Pollution Bulletin* 139: 328–31. https://doi.org/10.1016/j.marpolbul.2019.01.004

Simon, Márta, Nikki van Alst, and Jes Vollertsen. 2018. "Quantification of Microplastic Mass and Removal Rates at Wastewater Treatment Plants Applying Focal Plane Array (FPA)-Based Fourier Transform Infrared (FT-IR) Imaging." *Water Research* 142: 1–9. https://doi.org/10.1016/j.watres.2018.05.019

de Souza Machado, Anderson Abel, Chung W. Lau, Werner Kloas, Joana Bergmann, Julien B. Bachelier, Erik Faltin, Roland Becker, Anna S. Görlich, and Matthias C. Rillig. 2019. "Microplastics Can Change Soil Properties and Affect Plant Performance." *Environmental Science and Technology* 53 (10): 6044–52. https://doi.org/10.1021/acs.est.9b01339

Syafina, Paramastri Rahmi, Adyati Pradini Yudison, Emenda Sembiring, Mohammad Irsyad, and Haryo Satriyo Tomo. 2022. "Identification of Fibrous Suspended Atmospheric Microplastics in Bandung Metropolitan Area, Indonesia." *Chemosphere* 308 (December): 136194. https://doi.org/10.1016/j.chemosphere.2022.136194

Talvitie, Julia, Anna Mikola, Outi Setälä, Mari Heinonen, and Arto Koistinen. 2017. "How Well Is Microlitter Purified from Wastewater? – A Detailed Study on the Stepwise Removal of Microlitter in a Tertiary Level Wastewater Treatment Plant." *Water Research* 109: 164–72. https://doi.org/10.1016/j.watres.2016.11.046

Tamargo, Alba, Natalia Molinero, Julián J. Reinosa, Victor Alcolea-Rodriguez, Raquel Portela, Miguel A. Bañares, Jose F. Fernández, and M. Victoria Moreno-Arribas. 2022. "PET Microplastics Affect Human Gut Microbiota Communities During Simulated Gastrointestinal Digestion, First Evidence of Plausible Polymer Biodegradation During Human Digestion." *Scientific Reports* 12 (1): 528. https://doi.org/10.1038/s41598-021-04489-w

Tejaswini, M.S.S.R., Pankaj Pathak, Seeram Ramkrishna, and P. Sankar Ganesh. 2022. "A Comprehensive Review on Integrative Approach for Sustainable Management of Plastic Waste and Its Associated Externalities." *Science of the Total Environment* 825 (June): 153973. https://doi.org/10.1016/j.scitotenv.2022.153973

Thiel, Martin, Guillermo Luna-Jorquera, Rocío Álvarez-Varas, Camila Gallardo, Iván A. Hinojosa, Nicolás Luna, and Diego Miranda-Urbina, et al. 2018. "Impacts of Marine Plastic Pollution from Continental Coasts to Subtropical Gyres—Fish, Seabirds, and Other Vertebrates in the SE Pacific." *Frontiers in Marine Science* 5 (July). https://doi.org/10.3389/fmars.2018.00238

Thushari, G.G.N., and J.D.M. Senevirathna. 2020. "Plastic Pollution in the Marine Environment." *Heliyon* 6 (8): e04709. https://doi.org/10.1016/j.heliyon.2020.e04709

Tirkey, Anita, and Lata Sheo Bachan Upadhyay. 2021. "Microplastics: An Overview on Separation, Identification and Characterization of Microplastics." *Marine Pollution Bulletin* 170 (September): 112604. https://doi.org/10.1016/j.marpolbul.2021.112604

Tsering, Tenzin, Mika Sillanpää, Mirka Viitala, and Satu-Pia Reinikainen. 2022. "Variation of Microplastics in the Shore Sediment of High-Altitude Lakes of the Indian Himalaya Using Different Pretreatment Methods." *Science of the Total Environment* 849 (November): 157870. https://doi.org/10.1016/j.scitotenv.2022.157870

Ugwu, Kevin, Alicia Herrera, and May Gómez. 2021. "Microplastics in Marine Biota: A Review." *Marine Pollution Bulletin.* https://doi.org/10.1016/j.marpolbul.2021.112540

Velis, C.A., D. Lerpiniere, and M. Tsakona. 2017. *Prevent Marine Plastic Litter-Now! An ISWA Facilitated Partnership to Prevent Marine Litter, with a Global Call to Action for Investing in Sustainable Waste and Resources Management Worldwide.* International Solid Waste Association (ISWA), Vienna.

Wang, Pengcheng, Zongxiong Huang, Shan Chen, Miaoyu Jing, Zhihao Ge, Junyan Chen, Shuangli Yang, Jinghu Chen, and Yimin Fang. 2022. "Sustainable Removal of Nano/ Microplastics in Water by Solar Energy." *Chemical Engineering Journal* 428 (January): 131196. https://doi.org/10.1016/j.cej.2021.131196

Wang, Sumin, Hongzhe Chen, Xiwu Zhou, Yongqing Tian, Cai Lin, Weili Wang, Kaiwen Zhou, Yuanbiao Zhang, and Hui Lin. 2020. "Microplastic Abundance, Distribution and Composition in the Mid-West Pacific Ocean." *Environmental Pollution* 264 (September): 114125. https://doi.org/10.1016/j.envpol.2020.114125

Wright, S.L., J. Ulke, A. Font, K.L.A. Chan, and F.J. Kelly. 2020. "Atmospheric Microplastic Deposition in an Urban Environment and an Evaluation of Transport." *Environment International* 136. https://doi.org/10.1016/j.envint.2019.105411

Yan, Zehua, Yafei Liu, Ting Zhang, Faming Zhang, Hongqiang Ren, and Yan Zhang. 2022. "Analysis of Microplastics in Human Feces Reveals a Correlation between Fecal Microplastics and Inflammatory Bowel Disease Status." *Environmental Science & Technology* 56 (1): 414–21. https://doi.org/10.1021/acs.est.1c03924

Yang, Hui, Haoran Xiong, Kaihang Mi, Wen Xue, Wenzhi Wei, and Yingying Zhang. 2020. "Toxicity Comparison of Nano-Sized and Micron-Sized Microplastics to Goldfish Carassius Auratus Larvae." *Journal of Hazardous Materials* 388. https://doi.org/10.1016/j.jhazmat.2020.122058

Yin, Lingshi, Xiaofeng Wen, Danlian Huang, Chunyan Du, Rui Deng, Zhenyu Zhou, and Jiaxi Tao, et al. 2021. "Interactions between Microplastics/Nanoplastics and Vascular Plants." *Environmental Pollution.* https://doi.org/10.1016/j.envpol.2021.117999

Yu, Zhe fu, Shuang Song, Xiao lu Xu, Qing Ma, and Yin Lu. 2021. "Sources, Migration, Accumulation and Influence of Microplastics in Terrestrial Plant Communities." *Environmental and Experimental Botany.* https://doi.org/10.1016/j.envexpbot.2021.104635

Zhang, Yongli, Allison Diehl, Ashton Lewandowski, Kishore Gopalakrishnan, and Tracie Baker. 2020a. "Removal Efficiency of Micro- and Nanoplastics (180 nm–125 µm) during Drinking Water Treatment." *Science of the Total Environment* 720 (June): 137383. https://doi.org/10.1016/j.scitotenv.2020.137383

Zhu, Long, Jingying Zhu, Rui Zuo, Qiujin Xu, Yanhua Qian, and Lihui AN. 2023. "Identification of Microplastics in Human Placenta Using Laser Direct Infrared Spectroscopy." *Science of the Total Environment* 856 (January): 159060. https://doi.org/10.1016/j.scitotenv.2022.159060

Zuccarello, P., M. Ferrante, A. Cristaldi, C. Copat, A. Grasso, D. Sangregorio, M. Fiore, and G. Oliveri Conti. 2019. "Exposure to Microplastics (<10 µm) Associated to Plastic Bottles Mineral Water Consumption: The First Quantitative Study." *Water Research* 157 (June): 365–71. https://doi.org/10.1016/j.watres.2019.03.091

10 Environmental Fate, Behavior, and Risk Management Approaches of Nanoplastics in the Environment
Current Scenario and Future Insights

Shikha Gulati, Anoushka Amar, and Sheetal Olihan

10.1 INTRODUCTION

Due to the widespread use of microplastics (MPs) or nanoplastics (NPs) in the manufacturing and packaging sectors today, it would be impossible to live without the use of plastics. Initially, balls, containers, and waterproof coatings were made from organic natural materials like rubbers with a latex foundation. The first synthetic polymers were created in the nineteenth century and were based on nitrocellulose (Gangadoo et al. 2020). Leo Baekeland created the first synthetic plastic in the world in 1907 (Kiran, Kopperi, and Mohan 2022). Plastics have been a huge success as a foundation material; thanks to their low cost of manufacturing and adaptability, they are now used in practically every aspect of daily life. Because of the gradual deterioration that comes with plastic's durability, it was once thought that plastics may remain in the surroundings for decades or even millennia. Globally, 335 million tonnes (MT) of polymers are used to make plastic. Asia produces over half of all plastics, while the NAFTA and EU countries each produce approximately 20% (Wayman and Niemann 2021). Plastics have been discovered in a diverse range of environmental media, including the atmosphere, groundwater, fresh surface water, sediment, marine surface water, and the seabed. Because plastics are chemically stable, persistent, and bioaccumulative, the issue of plastic pollution in the atmosphere is getting more and more attention. Plastics are less dense than soil minerals,

DOI: 10.1201/9781003352396-10

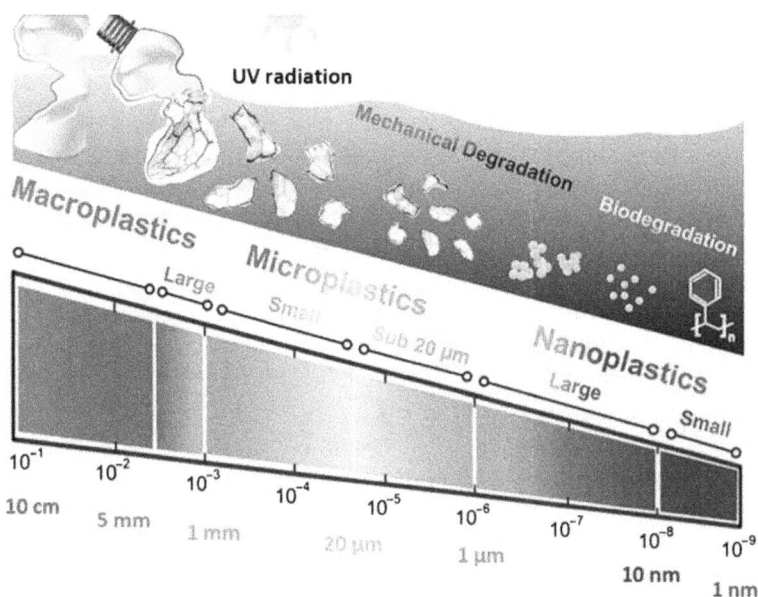

FIGURE 10.1 Fragmentation and degradation flow and size-based differentiation of plastics (reprinted with permission from Kundu et al. 2021).

and once they are incorporated into the soil, they create aggregates and change the physicochemical characteristics of the soil. It has an impact on the soil ecosystem's nutrient cycle process. According to size, plastics are divided into macroplastics, microplastics (MPs), and nanoplastics (NPs). Weathering, UV exposure, and biodegradation all have the potential to break down plastic trash into tiny fragments or particles (microplastics) (Figure 10.1) (Kundu et al. 2021).

MP is characterized as plastic particles ranging in size from 1 m to 5 mm. Due to their surface and structural characteristics, they are prone to harmful plasticizers' liquidation as well as the adhesion of organic pollutants carried by water. Therefore, consuming microplastics can increase the amount of toxins in the food chain, which could lead to bioaccumulation (Mofijur et al. 2021). Additionally, nanoplastics, which have a particle size between 1 and 1000 nm, can be produced from microplastics. Polyethylene (PE), polypropylene (PP), polystyrene (PS), polyvinyl chloride (PVC), nylon (PA), cellulose acetate (CA), and thermoplastic polyester are the most prevalent plastic polymers. Only 6–14% of plastics are recycled, leaving almost >80% in the air, in landfills, and in the natural world or aquatic bodies through countless paths, posing a huge threat to the macrocosm. Nanoplastics research is still in its early stages. NPs pose a greater risk than MPs because they can enter cells and tissues more easily. These nanoparticles are the principal cause of plastic trash toxicity since they build up and stay in the environment for many years, if not centuries. Nanoplastics are abundantly present in the aquatic environment and are easily swallowed by creatures because of their very small particle size. Ingesting and building up in the body of zebrafish, PS nano-plastics (70 nm) can lead to localized infection and lipid aggregation in the liver, which disrupts metabolic processes and energy cycling. More seriously, PS nano-plastics may be able to cross the blood-brain barrier, a highly selective barrier that works through an active transport mechanism mediated by P-glycoprotein to block the entry of potential neurotoxins, and enter the brain tissues, where they may have a profound impact on organisms. These tiny polymers enter the body mostly by food, inhalation, and skin contact (Mofijur et al. 2021). Due to their low den-sity and size, micro- and nanoplastics are easily disseminated by wind and water currents in the environment. Heavy metals, polycyclic aromatic hydrocarbons, polychlorinated biphenyls, and other chemicals can be adsorbed by nanoplastic particles from water, which may exacerbate the harmful effects of nanoplastics on living things. Biota may be adversely affected by micro- and nanoplastic (e.g., by causing inflammation, oxidative stress, and disruption of hormone signaling). According to reports, between 100 and 250 million tonnes (MT) would report-edly reach the ocean by 2025 (Peng et al. 2020). Plastic is produced, consumed, and dumped mostly in terrestrial environments, from which significant amounts of plastic are released into the ocean. As a result, the land and marine environments are common sources of micro- and nanoplastics. According to recent studies, the estimated 98% of MP/NP contamination in the sea comes from the land (Kiran et al. 2022). In response to the consequences of environmental plastic waste on the growth, development, and life of numerous living species, including humans, the scientific community has developed novel monitoring and purification methods. Plastics' environmental fate may now be examined using a plastic cycle, similar to the carbon or nitrogen cycle, as a result of this global contamination (Pradel 2022). There have been few experimental studies on the toxic effects of nanoplastics in mammals, compared to studies on model species like zebrafish and earthworms. Despite the fact that nanoplastics can enter creatures' cell membranes and blood circulation, lower trophic levels have metabolic systems that are distinct from those of humans. The same is true of their capacity for clearance and resistance to nanoplastics (Mofijur et al. 2021). As a result, it is urgently necessary to cre-ate new technologies to eliminate or drastically reduce the amount of micro- and nanoplastics in natural sources like the environment and water, both for human health and to maintain a healthy, sustainable environment. Governments, busi-nesses, and individuals are moving toward greater sustainability as a result, for example, by reducing the consumption of single-use plastics. An integrated DPSIR (driving forces, pressures, states, impacts, and reactions) paradigm is suggested in order to better comprehend how ecosystems, human society, and nanoplastics are related. This chapter focuses on the origins of MPs/NPs, ways that plastic waste enters the ocean, behaviors of nanoplastics in the environment, impacts of MPs/NPs on human and marine health, migration/dispersion mechanisms of micro- and nanoplastics, and their fate in the environment as a whole. Nanoplastics' potential toxicity studies and biological safety evaluation, potential identification methods used in microplastic analyses, and management strategies of nanoplastics and some future research needs are proposed.

10.2 FORMATION OF MICROPLASTICS AND NANOPLASTICS

Figure 10.2 shows the various methods of formation of microplastics and nanoplastics.

Naturally, abiotic and biotic degradation occurs simultaneously. Mechanical degradation of plastics caused by climatic changes is known as abiotic changes. The molecular bonds of plastics are unaffected, only analytical changes happen.

Thermal degradation is a detailed simplification of plastics at temperatures from 375°C to 500°C. Owing to the intensive heat, the central polymeric chains undergo bond scission which results in alterations in crystallinity, molecular weight, color, and tensile strength. This process is unrealistic in the environment as such high temperatures are not reached (Lambert, Sinclair, and Boxall 2014; Boyle and Örmeci 2020).

Photodegradation is a photo-induced process that institutes oxygen into the polymer matrix, resulting in the formation of carbonyl and hydroxyl groups, which then contribute to biotic degradation. Ozone oxidation in the environment aids in the

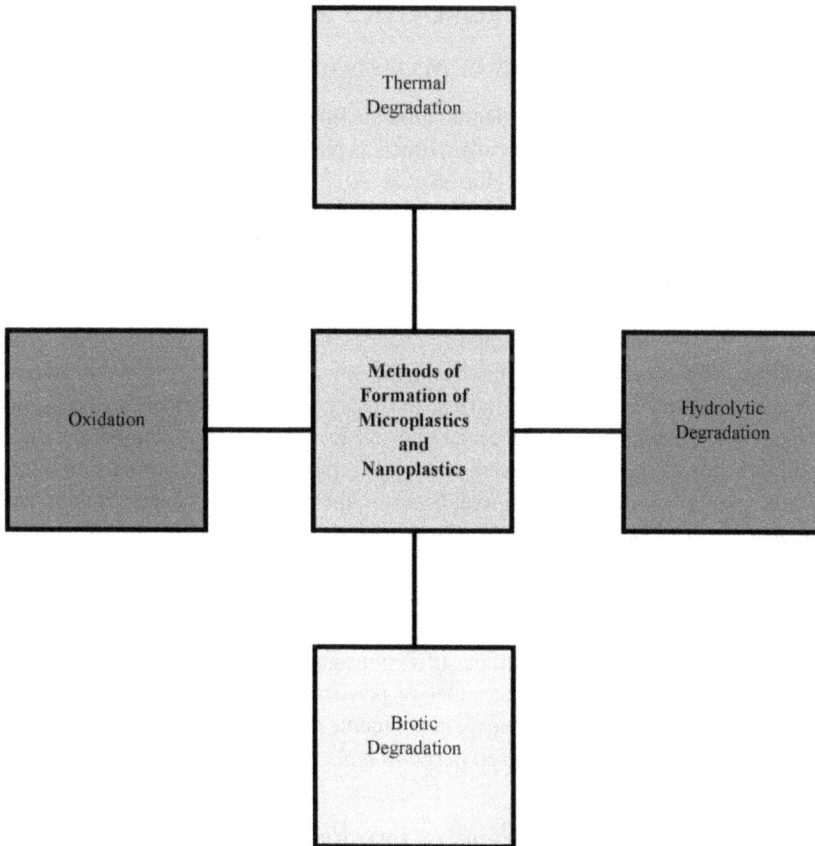

FIGURE 10.2 Schematic diagram of different methods of formation of microplastics and nanoplastics.

breakdown of plastics by breaking bonds of a covalent nature, which in turn leads to chain scission reactions and cross-linking that generate free radicals. The degradation of the plastic will be accelerated once free radicals are produced (Lambert et al. 2014; Boyle and Örmeci 2020).

Hydrolytic degradation is possible only when the polymer contains covalent character groups such as ether and ester (Boyle and Örmeci 2020). Hydrolytic action targets covalent bond groups, changing the molecular weight and the plastic's strength (Lambert et al. 2014).

Biotic degradation is caused by the activity of the organisms. The enzymes that are secreted by organisms hydrolytically cleave the polymer chains of plastic, resulting in a decrease in molecular (GMM) weight (Lambert et al. 2014). The decrease of molar mass causes further biodegradation, which is ramped up by exposure to oxygen and water, resulting in the persistent loss of molecule structure as time passes (Boyle and Örmeci 2020).

10.3   PROPERTIES OF MICROPLASTICS AND NANOPLASTICS

10.3.1   SIZE AND MORPHOLOGY OF MICROPLASTICS AND NANOPLASTICS

Microplastics' bioavailability and degradation are influenced by their shape, color, size, and density. Microplastics are typically defined as plastics 5 mm (Barnes et al. 2009) in size or between 1 mm and 5 mm (Horton et al. 2017). Common forms of microplastics present in the environment are broken edges, pellets, granules, fragments, filaments, fibers, and irregularly shaped microplastics. Microbeads are also found in cleaning products that are washed down drains every day and also in hygiene and beauty products. Once airborne, fibers and microbeads are easily unloaded in terrestrial and freshwater. Three key factors influence microplastic shapes: the indigenous plastic form, plastic breakdown, and their age in the environment. Microplastics are extremely dynamic due to their variety of shapes and sizes, reactivity, and high surface area-to-volume ratio. Because each particle's properties improve its bioavailability, quantification and sampling become tough. When exposed to water, plastics with a specific gravity (SG) of 1 sink and those with an SG >1 float; however, the SG will vary depending on hetero aggregation and establishment of microbes. Owing to the actions of the microbes, biofouling and biofilm formation attract algae and invertebrates, altering the SG even more.

Nanoplastics are similar to microplastics in that they lack a standard size definition, and they are so small that they cannot be seen with the naked human eye. Figure 10.3 (Wang et al. 2021) shows morphologies of various nanoplastics. However, nanoplastics are more commonly referred to particles of plastics of less than 0.1 μm in size (Ng et al. 2018). Nanoplastics are primarily the outcome of micro- and macroplastics being degraded into nano- and micro-sized particles – plastics' simplest particles.

10.3.2   CHEMICAL METHODOLOGIES OF IDENTIFICATION
OF MICROPLASTICS AND NANOPLASTICS

Table 10.1 (Kumar et al. 2021) represents the characteristics of various analytical identification of micro- and nanoplastics.

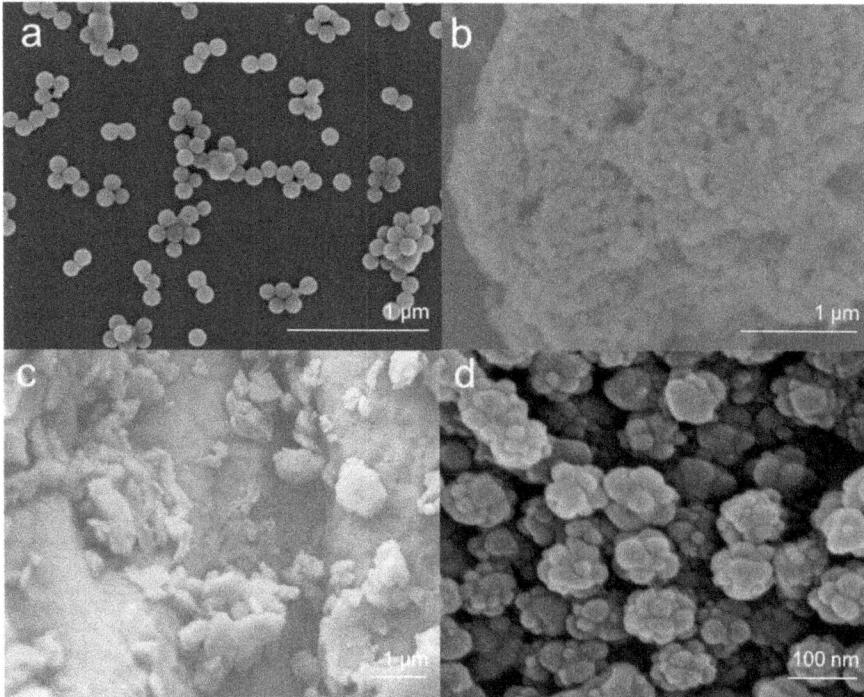

FIGURE 10.3 Morphologies of different nanoplastics: (a) Polystyrene (PS) nano-bead particles that are commercially available, (b) polytetrafluoroethylene (PTFE) nanoparticles with a diameter of 120 nm that are commercially available, (c) nano-sized PS particles were attached to the surface of polystyrene spherules for a month before being fragmented from the expanded polystyrene spherules by accelerated mechanical abrasion, and (d) a raspberry-colored synthetic metal-doped polyacrylonitrile (PAN) nanoparticle (reprinted with permission from Wang et al. 2021).

TABLE 10.1

A Few Methods of Identification and Characterization of Micro/nanoplastics in Environmental Samples (Kumar et al. 2021)

Analytical Method	Micro- and Nanoplastics Size (Diameter)	Elucidation	Advantages	Limitations
Fourier-transform infrared spectroscopy (FT-IR)	Large size	Functional groups	Fast and simple to use	Not being able to classify tiny micro- and nanoplastics
X-ray photoelectron spectroscopy (XPS)	Large size	Chemical characterization of the surface	Simple to operate	Not being able to classify tiny micro- and nanoplastics

(Continued)

TABLE 10.1 *(Continued)*

A Few Methods of Identification and Characterization of Micro/nanoplastics in Environmental Samples (Kumar et al. 2021)

Analytical Method	Micro- and Nanoplastics Size (Diameter)	Elucidation	Advantages	Limitations
Raman microspectroscopy (RM)	>100 nm	Identify the shape and size, chemical characterization	Examine the dimensions, morphology, and chemical compositions simultaneously with spatial clarity than FT-IR	The diffraction limit of the laser spot makes the detection of smaller micro- and nanoplastics qualitative rather than quantitative
Scanning electron microscopy (SEM)	Any range	Assess the geometry, size, and surface structure	High-definition images	Costlier, difficult to quantify
Transmission electron microscopy (TEM)	Any range	Evaluate the micro- and nanoplastics' various components as well as their form and size	High-definition images	Costlier, difficult to quantify
Energy-dispersive X-ray spectroscopy (EDS)	Any range	Elemental constitution	Simple to interconnect to TEM and SEM	Unable to correctly identify some components, such as carbon
Dynamic light scattering (DLS)	1–3000 nm	Utilize Brownian motion fluctuations to determine the size	Quick, simple, and in situ	Does not offer any chemical details
Electrophoretic light scattering (ELS)	1–3000 nm	Calculate the surface charge by observing the change in the electric field brought on by the motion of micro- and nanoplastics	Quick, simple to use, and on-site	Receptive to environmental factors
Multiangle laser scattering (MALS)	10–1000 nm	Measure the dimensions of the particles using scattered laser light from various angles	Simple to combine with separation techniques	Receptive to environmental factors
Laser diffraction (LD)	10 nm–10 mm	Determine the dimension of the particles using the Fraunhofer diffraction theory	A wide size range, simplicity of use, and speed	Micro- and nanoplastics cannot be characterized
Nanoparticle tracking analysis (NTA)	10–1000 nm	Analyze particle size using an image sensor and a microscope to study it in accordance with the Brownian motion theory	Visualizing micro- and nanoplastics in movement	Micro- and nanoplastics cannot be characterized
Pyrolysis gas chromatography mass spectrometry (PyGC-MS)	Not applicable	Chemical identification	Simple coupling to separation techniques	A good standard is necessary

10.3.2.1 Vibration Spectroscopy and Microspectroscopy

10.3.2.1.1 Stereo- (or Dissecting) Microscopy

The stereomicroscope is frequently used to identify MPs whose sizes are in the hundreds of micron range because it enables three-dimensional examination by taking two slightly different images of the sample to create stereoscopic vision. As a result, objects can be primarily observed using low-magnification reflected light. MPs are recognized using a stereomicroscope based on their outward appearance. Using this fast-screening method, instantaneous identification of the particles' form, size, and color is possible before further characterization by other techniques.

10.3.2.1.2 Fluorescence Microscopy

Studies of microplastics have made use of fluorescence microscopy. The samples that are energized by a certain wavelength produce fluorescent emission, which is collected by the fluorescence microscope. Identification of MPs based on their natural propensity to emit fluorescence is a valuable method, especially for white and clear polymers (Mofijur et al. 2021).

10.3.2.1.3 Scanning Electron Microscopy

With the help of high-resolution photographs of the surface state, the surface topography of MPs is characterized using scanning electron microscopy (SEM).

Since SEM has EDS detectors, it can provide information about the materials' chemical composition (energy-dispersive X-ray spectroscopy). Since it may be fitted with a STEM holder to analyze samples on a grid, the SEM is a flexible tool that enables the characterization of NPs as well. SEM evaluation showed that the predominant morphologies were fragments, pellets, and fibers, with sediment particle sizes averaging 11 mm and water suspension particle diameters averaging 130 mm.

10.3.2.1.4 Transmission Electron Microscopy

The most popular method for characterizing nanomaterials is transmission electron microscopy (TEM), which offers chemical data and images of nanomaterials with an atomic-scale spatial resolution. The electron microscope can observe material of tiny sizes due to its extremely high resolution. Because of their amorphous nature, NPs are difficult to visualize with TEM; since NPs and MPs are not electron-dense, heavy-metal stains are necessary. For less amorphous particles, TEM successfully visualizes NPs. Elements that make up polymers typically exhibit poor contrast in TEM because of the weak elastic interactions they have with electrons. Incoming light passes through a thin foil material and is transformed to either elastically or inelastically scattered electrons when an e-beam interacts with it. The technique known as electron energy loss spectroscopy (EELS) allows for the spectroscopic analysis of these inelastically scattered electrons.

In fact, TEM is frequently employed in research on how MPs affect model systems (Kiran et al. 2022).

It is necessary to combine microscopy with additional methods like spectroscopy.

10.3.2.2 Analytical Techniques

10.3.2.2.1 Raman Spectroscopy

One of the techniques most frequently employed in the literature to analyze micro-plastics is Raman spectroscopy, particularly the portion of particles less than 20 μm (María and Rodríguez 2021). This method makes use of laser radiation that interacts with molecules' vibrational movements to reemit light at wavelengths that are unique to that particular atomic group.

Since Raman provides a more precise fingerprint reader than FT-IR and more spectral coverage, it is attracting a lot of attention for its direct target visualization. The inelastic scattering of photons from incident radiation by the molecules in the sample produces the Raman spectrum (Kiran et al. 2022). Amorphous carbon can be recognized using this technology since the entire wavelength region can be examined and used for identification (María and Rodríguez 2021).

Wider spectral coverage, better resolution, and less water interference are the three key advantages of Raman spectroscopy over IR spectroscopy (Boyle and Örmeci 2020).

10.3.2.2.2 Fourier-Transform Infrared Spectroscopy

The permanent dipole moment of a chemical bond changes when functional groups in a molecule absorb IR light, producing the FT-IR signal. As a result, it is simple to detect polar functional groups, such as those found in polymers. With a spatial resolution of only 5 μm, contemporary FT-IR microscopes may combine optical and chemical analysis of microparticles (María and Rodríguez 2021). Carbon-based plastics may be recognized by FT-IR, and their bond compositions can be separated from those of other inorganic and organic moieties. FT-IR is frequently employed, along with Raman spectroscopy, to characterize MPs (Kiran et al. 2022).

10.3.2.2.3 Thermal Analysis

The temperature stability of plastic polymers varies as well. The process is based on identifying the polymer based on the by-products of its decomposition. Techniques used in thermal analysis include thermogravimetry (TGA) and differential scanning calorimetry (DSC).

The foundation of DSC is the measurement of the heat flow difference between the sample being analyzed and a reference sample. In this instance, the sample and reference materials are heated independently so that their temperatures are maintained constant while being linearly increased or lowered. It is mostly used to identify primary MPs with well-known properties, such as PE microbeads. However, due to the fact that these plastics melt at a variety of temperatures, DSC cannot be used to analyze plastics like polycarbonate (PC), PS, polymethyl methacrylate (PMMA), and acrylonitrile butadiene (ABS), among others.

While enthalpy changes usually signal the start of the breakdown of polymeric materials, mass loss is detected in TGA. DSC measurements can be used to obtain enthalpy changes, which cannot be obtained using TGA. In MP analysis, a mix of two approaches is therefore recommended (Kiran et al. 2022).

Microspectroscopy, which combines vibration spectroscopy and optical microscopy, may reveal both the visual characteristics of plastic particles and their chemical makeup. The size of the particles under study affects the signal obtained (María and Rodríguez 2021).

Mass-based techniques should be employed below the 1 μm limit since the single-particle characterization is insensitive below this size.

10.3.2.3 Mass Spectroscopy Methods

The methods based on mass spectroscopy analyze the sample in bulk. Given that there is a sufficient mass of each nanoplastic in the sample, the fundamental benefit of these approaches is their capacity to simultaneously identify nanoparticles of different polymers. These techniques' signals primarily depend on the bulk of plastics. These methods are broadly suitable for qualitatively identifying the types of plastics in a sample (María and Rodríguez 2021). Two types of mass spectroscopy methods are used.

10.3.2.3.1   Thermal Desorption Coupled with Gas
Chromatography-Mass Spectrometry

The sample is heated from ambient temperature to 600°C at a steady rate and under constant circumstances (with air or inert gases (nitrogen or argon) as purge gases), in a thermogravimetric analyzer (TGA), to first achieve thermal extraction and adsorption. The polymer-specific, gaseous breakdown products are captured on a solid-phase adsorber after the sample has been pyrolyzed. Thermal desorption gas chromatography-mass spectrometry (TDS-GC-MS) can then be used to chromatographically separate and evaluate the trapped degradation products.

10.3.2.3.2   Pyrolysis-Gas Chromatography-Mass Spectrometry

Pyrolysis-gas chromatography-mass spectrometry (Py-GC-MS) is used to determine the chemical properties of polymers. By introducing MP/NPs actively into the pyrolyzer in the presence of an inert gas, usually helium, gaseous thermal degradation products are created, and samples are evaluated, preventing contamination throughout sample processing. They are then chromatographically separated and subjected to mass spectrometric analysis (Boyle and Örmeci 2020). Due to their quicker processing, these solutions could be more appealing.

10.3.2.4   Surface Identification Methods

10.3.2.4.1   Time of Flight Secondary Ion Mass Spectroscopy

Given its exceptional mass resolution and high spatial resolution, time of flight secondary ion mass spectroscopy (TOF-SIMS) is a flexible tool for the surface examination of materials. For a variety of materials, it might offer a wealth of information.

10.3.2.4.2   Matrix-Assisted Laser Desorption/Ionization-Time
of Flight Mass Spectroscopy

A stainless-steel sample plate can be used to vaporize and ionize the samples, and mass spectrometry is used to isolate the analytes and detect them using their mass-to-charge ratios. With its high benefits of sensitivity, throughput analysis, and ease of use, matrix-assisted laser desorption/ionization-time of flight mass spectroscopy (MALDI-TOF-MS) has also received a lot of interest in finding and quantifying new environmental pollutants.

10.3.2.4.3   Inductively Coupled Plasma Mass Spectroscopy

This method provides data on particle size distribution, chemical composition, particle number density, and particle mass concentration. With functionalized gold (Au) nanoparticles coupled to NPs, this technique was used to identify and measure the NPs. Here, the individual NPs are counted by the adsorbed Au particle formed in the SP inductively coupled plasma mass spectroscopy (ICP-MS) signal, providing an exact quantification (Kundu et al. 2021).

10.3.2.5   New Identification Strategies

At the microscale, chemical analysis of polymers has frequently been done using an FT-IR or Raman spectroscopy coupled with a microscope. Additionally, they enable the detection of MPs with tens of micron-sized dimensions. Analytical techniques demand expensive equipment and a lengthy analysis period. As a result, it's important to create new, alternative methods that make it quick and easy to identify MPs for field monitoring as well as laboratory research to determine their toxicity, accumulation, aging, etc.

A staining process might provide a unique solution or another approach to handle these issues. "Nile Red" (9-diethylamino-5-benzo[a]phenoxazinone; NR) is a dye, because its absorption, selectivity, and fluorescent qualities enable the quick detection and quantification of MPs by adhering to polymer composites more frequently than organic ones. When exposed to blue light, the dye adheres to plastic surfaces and turns them luminous.

This technique, which is still among the most cutting-edge ones currently accessible, can detect a polymeric particle on its own without the need for additional spectroscopic examinations (Kiran et al. 2022). The diffraction limit of IR spectroscopy has been overcome by a device developed by Photothermal Spectroscopy Corporation. This is accomplished by inducing photothermal activity at the sample surface using a pulsing mid-IR laser. Due to its ability to gather the Raman-shifted light generated by photothermal activity, this microspectroscopy technology can perform IR and Raman concurrently. By doping nanoplastics with metals, researchers have found a solution to the difficulty of recognizing and retrieving nanoplastics. These metals are tracers that make detection easier and even enable measurement (María and Rodríguez 2021).

10.4   SOURCES OF NANOPLASTICS

There are many different ways that plastics end up in the environment. Micro- and nanoplastics can infiltrate the environment both directly and indirectly.

Primary MPs are unintentionally released into the environment by a number of industrial processes from products and applications including cosmetics, paints, etc.

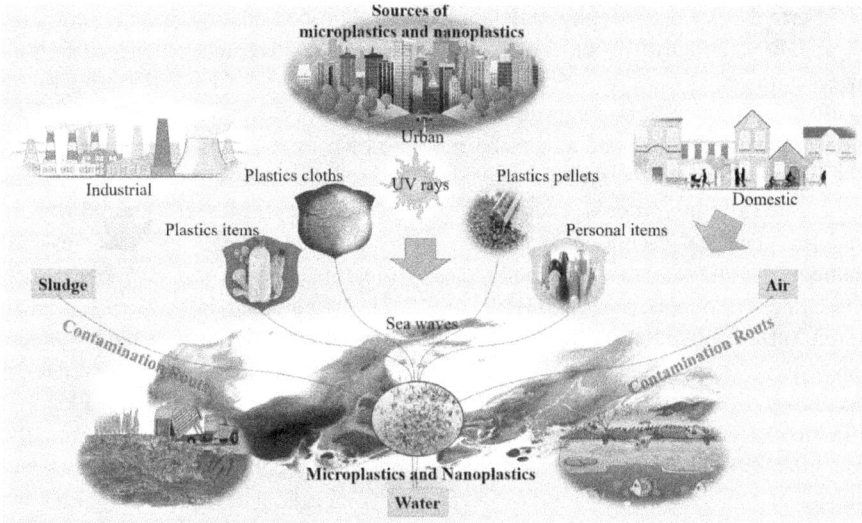

FIGURE 10.4 Representation of various sources of microplastics and nanoplastics in the water system through the use of arrows from the source to the water system (reprinted with permission from Kumar et al. 2021).

Secondary MPs are the outcomes of the MPs' surface deterioration brought on by biotic and abiotic weathering.

Figure 10.4 (Kumar et al. 2021) shows sources of microplastics and nanoplastics.

Plastic bottles are also mentioned as a potential source of microplastics. According to research, samples of bottled water include 91% microplastic contamination. Microplastic was discovered to be two times as prevalent in bottled water than in tap water. The thermal cutting of polystyrene foam, agricultural polyethylene foils, and clothes drying are some NP sources. Textiles, tyres, and city dust are the main sources of microplastics, accounting for more than 80% of all environmental microplastic pollution. There is a definite risk of contamination when plastic is intentionally placed or utilized, such as in plastic waste, greenhouse construction materials, and soil conditioners. Consumer goods may contain nanoplastics as a by-product of polydisperse raw materials or the breakdown of microplastics during processing. More specific sources of micro- and nanoplastics include municipal wastewater, textiles, the fashion industry, metalworking, commercial fishing, and packing and plastic bottles. Coastal tourism, aquatic aquaculture and fishing, and land-based input are the principal sources of microplastics in the marine environment. Due to their small size, traditional wastewater treatment methods are unable to completely remove these plastic particles and cause a substantial influx of microplastic debris into the marine ecosystem (Shen et al. 2019).

In comparison to terrestrial ecosystems, plastic debris is more easily fragmented into microplastic in the marine ecosystem due to the effects of salt concentrations and microbes. Non-biodegradation and biodegradation are the two distinct processes that make up the mechanism of degradation.

Thermal deterioration, physical degradation, photodegradation, thermo-oxidative degradation, and hydrolysis are the principal non-biodegradation processes that affect plastics.

Plastics frequently decompose during biodegradation by the action of microorganisms. The polymer chains can be broken down by extracellular enzymes produced by live microbes. Smaller plastic particles with various structural variations are created by this process, eventually generating nanoplastics. It's crucial to note that the chemical components used to create the nanoplastics included residual monomers, plasticizers (such as phthalates), surfactants, oxidation preventers, colors, and flame retardants (María and Rodríguez 2021). Irrigation and the use of sewage in agricultural applications are two important indirect paths that micro- and nanoplastics follow. An indirect source for the oceans would be human activities that release plastic into the water from the land. Plastics can be remobilized by local weather and climatic patterns once they have been deposited on soil or sediment. Therefore, the accumulation of micro- and nanoplastics is probably what pollutes freshwater and terrestrial ecosystems.

10.5 RESEARCH TRENDS IN THE PAST TEN YEARS

Table 10.2 presents the number of research publications (excluding the reviews) for each environmental compartment (Allen et al. 2022).

10.5.1 MARINE

Spreading across the coast to the depth of the seas, marine ecosystems are considered a vital pollution sink by plastics. Owing to 2010 data, plastic was estimated to have entered the world's oceans in quantities ranging from 4.8 to 12.7 million MT (metric tonnes), with microplastics accounting for more than 10% of the budget

---

TABLE 10.2

Number of Research Publications (Excluding the Reviews) for Each Environmental Compartment (Allen et al. 2022)

All the publications	Marine	Lake/ freshwater	Soils	Air/ atmosphere	Fauna/flora/ biota	Human Health
Macro + micro + nanoplastics	3860	3049	1165	673	2379	1390
Plastics pollution	4939	3152	3021	3148	2764	2130
Only the research articles						
Macro + micro + nanoplastics	3142 (2995)	2391 (2268)	851 (773)	507 (459)	1830 (1730)	889 (820)
Plastics pollution	4065 (3568)	2469 (2127)	2264 (1609)	2264 (1093)	2221 (1883)	1487 (1144)

of marine plastic (Koelmans et al. 2017). It was estimated that 19–23 million MT or 11% of plastic waste generated all around the world in 2016 entered ecosystems, making these estimates now considered conservative. A 2018 study of a Vancouver's wastewater treatment plants stated that the micro- and nanoplastics' loss in the environment is 1.76 trillion particles annually and 0.3 trillion micro- and nanoplastics particles are submitted to the marines. By investigating effluent discharge from Finnish wastewater treatment plants to Finland's Gulf, this study corroborates a previous study that found wastewater treatment plants' sewage discharge to be a micro- and nanoplastics transport pathway to the marines. When microplastic particles are covered by biofilm, they accumulate faster, minimizing buoyancy, accumulation, and settling on the seafloor. Over the last ten years, micro- and nanoplastics have been discovered in the marines' sediment and it has been said that they act like a sink as well as a pathway with changing hotspots (Allen et al. 2022).

10.5.2 FRESHWATER

In 2016, effluent samples from 17 wastewater treatment plants' examination demonstrated that the microbeads' average discharge from US municipal wastewater treatment plants was 13 billion particles/day (Mason et al. 2016). As a result of fishing activities, tourism and recreation may also be the important origin of micro- and nanoplastics. During the dry seasons, elevated temperatures facilitates the breakdown of plastics in freshwater systems. During the wet season, high flows aggravate plastic pollution in these bodies of water and resuspend locked-in sediments in micro- and nanoplastics. Contamination with micro- and nanoplastics may endanger both human health and the environment (Allen et al. 2022).

10.5.3 SOIL

Because of the huge content of organic matter and its complex composition, it is very difficult to evaluate micro- and nanoplastics. Micro- and nanoplastics can be found all over the soil structure. Soils, which is a direct route into the food web of micro- and nanoplastics, are believed to contain 4–23 times the level of micro- and nanoplastics found in the oceans. With a concentration of 7 microplastics/m^3 in Germany, soils have been stated to be a plastics contributor to the atmosphere via wind erosion and to groundwater via infiltration (Mintenig et al. 2019). Due to transportation and ingestion via their activities and excretions, terrestrial worms were the first to be noted as soil micro- and nanoplastics transport vectors, and since then micro- and nanoplastics multicellular soil fauna terrestrial assessments have expanded to include springtails, nematodes, and beetles as well (Allen et al. 2022).

10.5.4 ATMOSPHERE

The air mass concentration studies and deposition were launched in cities throughout Europe, the Middle East, and Asia, starting in Paris, resulting in the identification of atmospheric micro- and nanoplastics concentrations in the range of 1–5700 micro- and nanoplastics/m³ (Zhang et al. 2020) and atmospheric deposition of micro- and nanoplastics in the range of 3100 microplastics/m²/day to over 6000 micro- and nanoplastics/m² (outdoors to indoors). Originally assumed to be localized near cities and industries, efficient evaluation of transportation of pollution through the atmosphere started in 2019 led to the characterization of micro- and nanoplastics in the atmosphere in distant regions quantitatively. Particle history analysis and backward air mass are used to determine the "path of flight" of micro- and nanoplastics found in samples of the atmosphere, indicating potential atmospheric source locations as well as elevation and atmospheric transport duration (Allen et al. 2022). Micro- and nanoplastics can advance huge distances across countries and oceans in the boundary layer of the planet and in the free troposphere in a relatively short time, demonstrating that the atmosphere is an efficient pathway for transportation than fluvial transport and oceanic currents. Micro- and nanoplastics in water from rain drop effect and the bubble burst ejection technique discovered in sea sale aerosol ejection has been recognized in new research. The transport assessments and concentration in the atmosphere bring to light the information gap in the particles and dynamics of transport for micro- and nanoplastics. Micro- and nanoplastics are the type of particles that have lesser density than dust or soil, have a charge, and act as a carrier for viruses, bacterial elements, and environmental contaminants (Allen et al. 2022).

10.5.5 BIOTA

Before 2012, the focus of marine biota research was mainly on ingestion and entanglement; however, current studies have diversified to consider micro- and nanoplastics translocation to circulatory systems and other organs. Micro- and nanoplastics have been obtained in gills, feces, hemolymph, skin, muscle, and the circulatory system, and its permeation effects in endocrine disruption, survival, tissue inflammation, behavioral changes, decreased growth, and reproductive have been assessed. Chronic exposure has been assessed thanks to a toxicological evaluation of micro- and nanoplastics' impact on ecosystems and individual species. Exposure studies done earlier used extremely high concentrations of micro- and nanoplastics that were not thought to be reflective of the prevailing environmental micro- and nanoplastics pollution, culminating in a recent shift in biota impact assessments to ecotoxicological concentrations of micro- and nanoplastics. Neurotoxicity transfer to the offsprings from the parents (*Caenorhabditis elegans*), trans-generational toxicity in the reproductive system, multi-generational increased mortality, and oxidative stress have all been studied. Micro- and nanoplastics biotic impact assessments are being expanded on an individual and group justification. Nanoplastics were noticed to transfer to water fleas from an alga, then to a consumer fish (secondary), and finally to an end consumer fish under controlled conditions, with behavioral and histopathological changes observed in the end and secondary consumer fish. This trophic transfer assessment is now becoming a part of the ecosystem service debate, with preliminary findings indicating that micro- and nanoplastics contamination may have a detrimental effect on biota provisioning, regulation, and maintenance services (Allen et al. 2022).

10.5.6 HUMAN HEALTH

Before 2017, a large percentage of human health studies were based on the micro- and nanoplastics' present in the food chain and, as a result, their role in human ingestion. Despite the fact that the micro- and nanoplastics' physical impact on the human system had been speculated and discussed, in vitro or in vivo research didn't start completely until 2017. Following the acknowledgment with the aid of using state policies and investment in agencies overlooking the problems and dangers associated with human micro- and nanoplastic uptake, the research into human fitness effects has grown in the latest years. The researchers started by figuring out the motion and effect of micro- and nanoplastics on unique cellular lines, then went directly to laboratory experiments performed on mammals, which are often used as signs or proxies in human fitness studies. Semi-quantitative checks of human inhalation and ingestion of micro- and nanoplastics by humans advocate an annual inhalation uptake of approx. 48,000 (5000–109,000) (Cox et al. 2019; Allen et al. 2022). There has been full-size improvement within the examination of micro- and nanoplastics' transportation effect on human cells within a previous couple of years, with research figuring out inflammatory responses, tight dysregulation of the junction, and cytotoxic effects. Micro- and nanoplastics have been discovered in human lung tissue, stool, the circulatory system, and the placenta, as well as a variety of rodent organs. The investigation of micro- and nanoplastics' acute and chronic effects on human health is gaining momentum. Even so, research into the possible impact of micro- and nanoplastics uptake on human health is still in the initial stages (Allen et al. 2022).

10.6 EFFECTS OF MICROPLASTICS AND NANOPLASTICS

10.6.1 TO ORGANISMS IN TERRESTRIAL ECOSYSTEMS

NPs have been shown to have an impact on the soil microbiome. Polystyrene NPs (0.1–1 mg/kg) can suppress the enzymes' activities in the carbon, nitrogen, and phosphorus cycles (Wang et al. 2021). The key biomes' activities that dominate the nitrogen cycle decreased. The abundance of Rhizobiaceae, Xanthobacteraceae, and Isosphaeraceae in the soil gut of the oligochaete *Enchytraeus crypticus* decreased significantly after feeding it (10 wt%) polystyrene nanoplastics-added oatmeal (Kim et al. 2020). Polystyrene nanoparticles (530 nm) were found to be toxic to the nematode (*Caenorhabditis elegans*). When the concentration of nanoplastics increased to 10 mg/kg, the number of offsprings significantly lessened ($p < 0.05$). In plants, nanoplastics can be taken up, sequestered, and deviated to aboveground tissues by roots. In the cytoplasm, bigger sized nanoplastics can accumulate, whereas smaller nanoplastics (~30 nm) can get into the nucleus and disrupt chromatin structure and function, causing genotoxicity. Interestingly, plant growth may be enhanced due to the internalization of nanoplastics. Root elongation of *Triticum aestivum* L. (wheat) significantly increased ($p < 0.01$) by 89–123% on exposure to polystyrene (0.01–10 mg/L) nanoplastics compared to the control (Lian et al. 2020). Further observation saw an increase in plant biomass, nitrogen, and carbon content. Wheat seedlings grew faster after being subjected to nanoplastics without being stressed. This was most likely due to polystyrene nanoplastics increasing the amylase's activity as a nanocatalyst, thereby speeding up the derivation of soluble sugars from starch granules. Nanoplastics, on the other hand, were accumulated in wheat tissues, which indicates a potential threat along the food chain to higher trophic levels (Wang et al. 2021).

10.6.2 To Organisms in Aquatic Ecosystem

The ecotoxicity of microplastics and nanoplastics in the marine environment has been extensively researched and reviewed (Wang et al. 2021). Nanoplastics can harm organisms at all trophic levels, including algae, bacteria, bivalves, echinoderms, rotifers, arthropods, and fish (Wang et al. 2021). Some massive concerns in this field include nanoplastics bioaccumulation in tissues, nanoplastics' effects on growth and reproduction, nanoplastics-induced immune system damage, neurotoxicity, and changes in metabolic pathways. Polystyrene nanoparticles (52 nm, 5 mg/L) were found to cause embryonic development, which was abnormal and inhibited reproduction in *Daphnia galeata*, a crustacean of freshwater. The investigation on the effects of polystyrene nanoparticles on the growth of macrophytes *Myriophyllum spicatum* and *Elodea* sp. was conducted. The biomass of the root ($p < 0.05$) was observed to undergo a significant increase (van Weert et al. 2019). It is presumed that nanoplastics' sorption on the surface of the root hinders nutrient uptake. To combat this stress, the root biomass of the macrophytes increased by enhancing root diameter, root length, and root number, resulting in enhanced nutrient uptake and transport. Polystyrene nanoparticles caused sex differentiation in gametophytes of *Ceratopteris pteridoides*. Environmental stress induced by nanoplastics led to a rise in male gametophytes, which will have dire implications for reproductive success (Wang et al. 2021).

10.6.3 To Human Beings

Nanoplastics can be exposed through inhalation, dermal exposure, or ingestion (Figure 10.5) (Wang et al. 2021). Because of the inhalation of nanoplastic-containing aerosols and nanoplastic infiltration into the capillary bloodstream, this contaminant propagates all through the human body. Nanoplastics can get into contact with human skin via exposure to polluted water or air or via the use of personal care products. Due to their hydrophobic property, nanoplastics have a difficult time penetrating the skin in water. Polystyrene nanoplastics with a range of diameters from 20 to 200 nm were found to only infiltrate the stratum corneum to 2–3 μm (Campbell et al. 2012).

However, ingredients in personal care products might favor nanoplastics' penetration (Wang et al. 2021). The primary route of exposure to nanoplastics is through the consumption of contaminated water and food. The primary site for uptake of nanoplastics is the gastrointestinal tract (GI), which has a surface area of 32 m^2 approximately (Helander and Fändriks 2014). In vitro studies show that nanoplastics

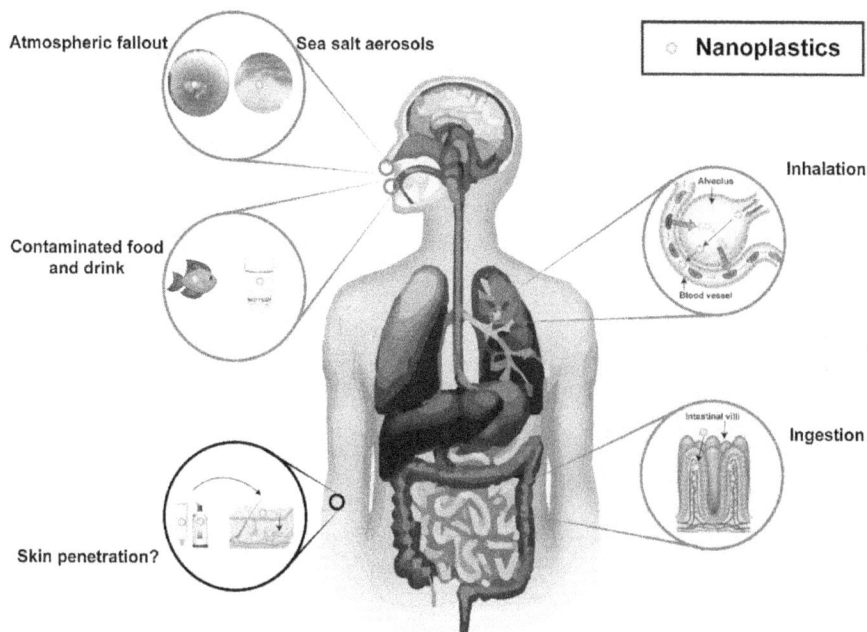

FIGURE 10.5 Nanoplastics exposure pathways in humans through atmospheric fallout, sea-salt aerosols, skin penetration, ingestion, and inhalation. ((reprinted with permission from Wang et al. 2021).

can permeate the intestinal villi, further entering the blood vessels, and this leads to the formation of a protein-plastic complex. Because protein-coated nanoplastics interact with tissues and organs rather than bare nanoplastics, this process is important to the toxicity of nanoplastics in organisms. A study of human blood cells (in vitro) found that protein-coated nanoplastics have a higher genotoxic and cytotoxic effect than virgin nanoplastics. This is owing to the formation of a biomolecular corona on the surface of nanoplastics, which aids in their getaway from the immune system, leading to a longer life span in the circulatory system. The mechanism by which the corona of the protein binds to nanoplastic is unknown, but hydrogen bonding, van der Waals force, and delocalized bonding are assumed to play their roles here. Very few in vivo and in vitro rodent and human studies have shown that nanoplastics can harm the immune system. Toxic effects of nanoplastics on human cells comprise cytokines' induction involved in gastric pathologies, disruption of iron transport, oxidative stress, induction of apoptosis, and endoplasmic reticulum stress. Extrapolations from nanoparticles are one feasible method for improving our understanding of the toxicology of nanoplastics (Wang et al. 2021). However, it was suggested that extrapolating information from nanoparticles to nanoplastics be done with caution because nanoplastics have clearly different surface chemistry (Wang et al. 2021).

10.7 RISK MANAGEMENT STRATEGIES FOR MICROPLASTICS AND NANOPLASTICS

Both land and aquatic creatures may be at risk of nanoplastics being present in the environment. Nanoplastics may potentially endanger human health by entering the food chain or by being directly inhaled or exposed through the skin (impacts). Governments and policymakers are prioritizing micro- and nanoplastics as a result of the growing plastic pollution, Therefore, it is crucial to explore risk reduction strategies with respect to nanoplastics pollution, such as designing new remediation strategies, implementing regulations, and promoting sustainable education (responses) (Wang et al. 2021). Figure 10.6 (Mofijur et al. 2021) shows prospective strategies to control nanoplastics pollution.

Origin management is the most successful tactic that can lessen or perhaps stop the pollution caused by plastic. Governments and decision-makers should enact laws and

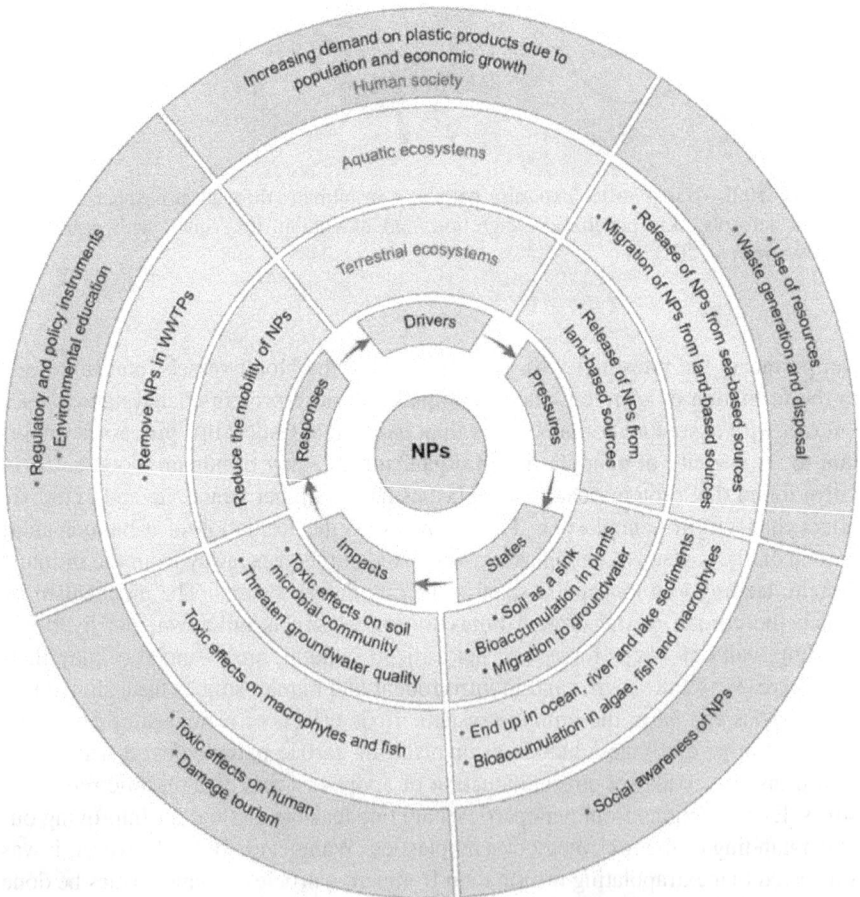

FIGURE 10.6 Prospective strategies to control nanoplastics pollution, depicted in a circular chart (reprinted with permission from Wang et al. 2021).

create policies to prevent the manufacturing of plastic products that may contribute to plastic pollution. The government should also push for a decrease in the use and careless disposal of plastic products, increase public understanding of environmental protection, and draw attention to the dangers of plastic pollution (Kumar et al. 2021).

- Finding a safe alternative to replace plastic is the greatest option to minimize plastic pollution. More money needs to be invested in scientific and technological research as well as plastic substitutes. The increasing substitution of conventional non-biodegradable nanoplastics with ecologically friendly biodegradable nanoplastics is made possible by advances in biotechnology.
- One of the really important strategies is to increase the recycling of plastic to lessen the pollution caused by plastic trash. The most efficient engineering technique to stop nanoplastics from entering aquatic habitats is to remove nanoplastics in a wastewater treatment facility.
- Biodegradation is one possible method to reduce MPs in the marine environment. Microbes have a big part to play in the biodegradation process that breaks down MPs. It provides a practical and eco-friendly implementation strategy that would enable the management of MPs and nanoplastics without causing any harm.
- According to the literature, adding compost (such as biochar) may be able to stop nanoplastics from migrating through the porous material, reducing hazards through retention and settling.

     The avoidance of plastic waste (i.e., improving durability, inventing biodegradable plastics), monitoring (i.e., sensors and instruments), and reprocessing and repurposing (i.e., new technologies, energy recovery) of plastic trash will all depend critically on research and innovation (Boyle and Örmeci 2020).
- Advanced oxidation methods like the Fenton agent, H_2O_2-UV systems, ozone-UV systems, and ultrasound-UV systems have the ability to remove MPs as well since they can break the chains of MPs and cause them to lose their physicochemical properties like molecular weight, density, and size.
- Field flow fractionation (FFF) is another efficient method for separating environmental nanoplastics, and it may be modified to separate nanoplastics of different sizes. By using this technique, nanoplastics in ambient water and aquatic food were also separated.

Researchers have examined the danger evaluations and offered a thorough data foundation to comprehend the nanoplastics pollution methods in relation to

   i. development of cutting-edge cleanup technologies,
   ii. making policy, and
   iii. increasing public awareness and recommending the most effective method to stop nanoplastics from entering marine ecosystems (Kundu et al. 2021).

International agencies, authorities, businesses, and social organizations must collaborate in order to come to an agreement on and implement a protracted and efficient plastic waste management strategy. Adopting a life cycle approach that incorporates recovery, reuse, recycling, and end-of-life disposal will be important. Experts and administrators are urged to collaborate so that pertinent discoveries can be incorporated into regulations as soon as humanly possible.

The "Plastic Waste Management Rules, 2016" were put into effect in India with the goal of eliminating all disposable plastic goods by 2022 and outlawing the production of plastics smaller than 50 μm.

10.8 CHALLENGES IN ASSESSING AND MITIGATING THE PLASTICS

It was estimated that when consumers use 1 g of these commercial products, they release approximately 300 billion plastic particles having a diameter lower than 100 nm (María and Rodríguez 2021). Currently, the lowest detection achieved is approximately 1 μm. There is currently no unified standard technology for micro- and nanoplastics analysis as an emerging pollutant.

Micro- and nanoplastics are extensively spread in many habitats, and micro- and nanoplastics in different environments migrate to one another, and thus micro- and nanoplastics study must be global, and each environment must be examined appropriately in order to evaluate the impact of micro- and nanoplastics pollution more precisely on the environment.

The risk assessment of microplastics and nanoplastics still has numerous problems since there are huge knowledge gaps on the number of MPs and nanoplastics in the surroundings, environmental practices, and potential health risks.

- Micro- and nanoplastics that are introduced to lower trophic levels may go up the food chain and amplify at higher trophic levels, eventually having negative impacts on people. However, only a small number of studies revealed its biomagnification propensity. More research is needed to fully understand the biomagnification efficacy of micro- and nanoplastics.
- Micro- and nanoplastics make it possible to absorb contaminants like heavy metals and organic pollutants, changing their toxicity and bioavailability in the process and imposing detrimental consequences on living things as a whole. More studies are required to explicate the co-pollutant impacts of micro- and nanoplastics and several other contaminants (Kumar et al. 2021).
- However, collection and analytical techniques do have significant limitations, which require additional investigation.

  Future research should concentrate on the following to learn more about the harmful impact of nanoplastics:

  a. Effective ways to identify, measure, and evaluate the presence of plastic particles in the surroundings: The majority of existing sample techniques for microplastics have purposefully ignored the detection of nanoplastics because of their small particle sizes, and these techniques were not appropriate for nanoplastics. As a result, there is a pressing need for the development of novel technologies and methodologies for the identification and measurement of nanoplastics in the environment.

b. The degradation process and rates from big plastic fragments to nano-
    plastics: The degradation processes alter the chemical characteristics
    on the surface of the nanoplastic particles in addition to reducing the
    particle size. However, the methodologies for characterizing nanoplas-
    tics have not yet been published; hence, novel techniques and technolo-
    gies must be created and used (Shen et al. 2019).

    The deterioration of nanoplastics, however, needs to be a key sub-
    ject of investigation. Other issues, such as the impact of oligomers and
    monomers on the environment, will arise as nanoplastics degrade fur-
    ther. Which would be more dangerous, oligomers or nanoplastics?

c. How do harmful compounds desorb from, adsorb to, and/or bind to
    micro- and nanoplastics?

d. The propagation of nanoplastics and their hazardous impacts on sev-
    eral trophic levels in typical aquatic ecosystems: It is best to conduct
    biological experiments along food chains or webs rather than just on a
    single creature or at the individual level. For instance, habitats harboring
    nanoplastics must be exposed to phytoplankton, zooplankton, protozoa,
    metazoans, fish, or higher trophic levels. Although data on the toxicity of
    nanoplastics to species in the food chain can be obtained through labora-
    tory tests, its application to actual environmental conditions can be highly
    challenging. As a result, possible future paths still have to be investigated.

e. Testing nanoplastic amounts in the surroundings: The majority of
    studies on the toxicity and risk evaluations of nanoplastics were fre-
    quently conducted utilizing PS materials. This will result in a mis-
    understanding of the ecotoxicity of nanoplastics. In order to identify
    nanoplastics in the environment, new methods and techniques must
    be developed.

f. What is the ideal level of nanoplastics in seawater, and would this level
    have a substantial impact on marine life and the food chain, ultimately
    affecting human health (Shen et al. 2019)?

To recognize nanoplastic debris in the environment and the body, new technical
techniques are required in light of these unsolved problems. It is important to keep
in mind that there are still no reliable ways to immediately identify microplastics.
It is crucial to create fresh, trustworthy, and useful techniques that can point to the
direction in which techniques and tools for identifying microplastics will be developed in the future.

Without involving the general public, socioeconomic sectors, the tourism industry,
waste management companies, government regulation, and legislation, it is impossible to reduce the negative effects of micro- and nanoplastics. Emerging remediation technology, public policy, education, and attention are the main nanoplastics
mitigation strategies.

The public needs to be informed widely about this issue, which brings us to our
final point. Science dissemination is crucial if we wish to reduce the production of
plastic. It is important for consumers to understand what happens to their products
once they are in the environment. Smaller plastics are formed from larger plastic.

The environment contains and retains these tiny polymers. Even if we were successful in getting rid of macroplastics from the environment, getting rid of micro- and nanoplastics would be quite difficult.

10.9   CONCLUSIONS AND FUTURE PERSPECTIVES

Polymers are among the most important families of the twenty-first century since they can be found just about everywhere and have a significant impact on our everyday lives in a variety of ways. In addition, these substances in general are one of the most significant sources of pollution to which mankind is subjected. Recent research has shown that the procedure of degradation of plastic also leads to the creation of nanoplastics with physicochemical properties distinct from the bulk materials. They are widespread and are deposited in various environmental compartments via lakes, rivers, stormwater runoff, sludge, sewage, and WWTP (wastewater treatment plant). The transport and fate of microplastics/nanoplastics in soil and water are heavily influenced by the physicochemical properties of the plastics (Kiran et al. 2022). Microplastics/nanoplastics accumulation in soil has an impact on microorganism fecundity, soil productivity, and plant growth. The trophic transfer of microplastics/nanoplastics to humans causes cancer, digestive problems, and cardiovascular disease. Plastic-derived contaminants' transmission and buildup in soil microbiome, trophic transfer in the food web, and crop plants can be highlighted for investigation. On the quantitative front of their analysis, researches must continue to focus on advancements in simple and precise techniques, proofs, and repercussions in identification, categorization, and measurements. The indigenous use of FT-IR and Raman is being overshadowed by more efficient techniques such as hyperspectral imaging techniques and GC-MS performed with TDS or pyrolysis. To meet the detection limit of the analytical instruments, quantifying and assessing microplastics/nanoplastics in the soil-microbiome-plant system still requires interventions in pre-treatment, sample preparation, and pre-concentration. Analyte sampling, pre-treatment, and characterization of nanoplastics to this day lack uniformity and standardization. Furthermore, matching libraries must be established in order to identify microplastics/nanoplastics from samples from the environment. Plastic particles from domestic, industrial, and surface runoff enter WWTPs before being released into the environment. In a sludge digestion system, enhancing their loads reduces process efficiency and raises operating costs. Restoration technologies that prioritize sustainable waste management, recycling, awareness, and education incorporating circular models, influxing made from renewable materials, policy, legislation, and a proper plan are some of the critical prerequisites for the development of environmentally friendly practices in the plastic management domain.

REFERENCES

Allen, Steve, Deonie Allen, Samaneh Karbalaei, Vittorio Maselli, and Tony R. Walker. 2022. "Micro(Nano)Plastics Sources, Fate, and Effects: What We Know after Ten Years of Research." *Journal of Hazardous Materials Advances* 6 (February): 100057. DOI: 10.1016/j.hazadv.2022.100057.

Barnes, David K A, Francois Galgani, Richard C Thompson, and Morton Barlaz. 2009. "Accumulation and Fragmentation of Plastic Debris in Global Environments." *Philosophical Transactions of the Royal Society of London. Series B, Biological Sciences* 364 (1526): 1985–98. DOI: 10.1098/rstb.2008.0205.

Boyle, Kellie, and Banu Örmeci. 2020. "Microplastics and Nanoplastics in the Freshwater and Terrestrial Environment: A Review." *Water (Switzerland)* 12 (9). DOI: 10.3390/w12092633.

Campbell, Christopher S J, L Rodrigo Contreras-Rojas, M Begoña Delgado-Charro, and Richard H Guy. 2012. "Objective Assessment of Nanoparticle Disposition in Mammalian Skin after Topical Exposure." *Journal of Controlled Release* 162 (1): 201–7. DOI: https://doi.org/10.1016/j.jconrel.2012.06.024.

Cox, Kieran D, Garth A Covernton, Hailey L Davies, John F Dower, Francis Juanes, and Sarah E Dudas. 2019. "Human Consumption of Microplastics." *Environmental Science & Technology* 53 (12). American Chemical Society: 7068–74. DOI: 10.1021/acs.est.9b01517.

Gangadoo, Sheeana, Stephanie Owen, Piumie Rajapaksha, Katie Plaisted, Samuel Cheeseman, Hajar Haddara, Vi Khanh Truong, et al. 2020. "Nano-Plastics and Their Analytical Characterisation and Fate in the Marine Environment: From Source to Sea." *Science of the Total Environment* 732 (June). DOI: 10.1016/j.scitotenv.2020.138792.

Helander, Herbert F, and Lars Fändriks. 2014. "Surface Area of the Digestive Tract – Revisited." *Scandinavian Journal of Gastroenterology* 49 (6): 681–89. DOI: 10.3109/00365521.2014.898326.

Horton, Alice A, Alexander Walton, David J Spurgeon, Elma Lahive, and Claus Svendsen. 2017. "Microplastics in Freshwater and Terrestrial Environments: Evaluating the Current Understanding to Identify the Knowledge Gaps and Future Research Priorities." *The Science of the Total Environment* 586 (May): 127–41. DOI: 10.1016/j.scitotenv.2017.01.190.

Kim, Shin Woong, Dasom Kim, Seung-Woo Jeong, and Youn-Joo An. 2020. "Size-Dependent Effects of Polystyrene Plastic Particles on the Nematode Caenorhabditis Elegans as Related to Soil Physicochemical Properties." *Environmental Pollution* 258: 113740. DOI: https://doi.org/10.1016/j.envpol.2019.113740.

Kiran, Boda Ravi, Harishankar Kopperi, and S. Venkata Mohan. 2022. *Micro/Nano-Plastics Occurrence, Identification, Risk Analysis and Mitigation: Challenges and Perspectives. Reviews in Environmental Science and Biotechnology.* Vol. 21. Springer Netherlands. DOI: 10.1007/s11157-021-09609-6.

Koelmans, Albert A., Merel Kooi, Kara Lavender Law, and Erik Van Sebille. 2017. "All Is Not Lost: Deriving a Top-Down Mass Budget of Plastic at Sea." *Environmental Research Letters* 12 (11). DOI: 10.1088/1748-9326/aa9500.

Kumar, Manish, Hongyu Chen, Surendra Sarsaiya, Shiyi Qin, Huimin Liu, Mukesh Kumar Awasthi, Sunil Kumar, et al. 2021. "Current Research Trends on Micro- and Nano-Plastics as an Emerging Threat to Global Environment: A Review." *Journal of Hazardous Materials* 409: 124967. DOI: 10.1016/j.jhazmat.2020.124967.

Kundu, Aayushi, Nagaraj P. Shetti, Soumen Basu, Kakarla Raghava Reddy, Mallikarjuna N. Nadagouda, and Tejraj M. Aminabhavi. 2021. "Identification and Removal of Micro- and Nano-Plastics: Efficient and Cost-Effective Methods." *Chemical Engineering Journal* 421 (P1): 129816. DOI: 10.1016/j.cej.2021.129816.

Lambert, Scott, Chris Sinclair, and Alistair Boxall. 2014. "Occurrence, Degradation, and Effect of Polymer-Based Materials in the Environment." *Reviews of Environmental Contamination and Toxicology* 227: 1–53. DOI: 10.1007/978-3-319-01327-5_1.

Lian, Jiapan, Jiani Wu, Hongxia Xiong, Aurang Zeb, Tianzhi Yang, Xiangmiao Su, Lijuan Su, and Weitao Liu. 2020. "Impact of Polystyrene Nanoplastics (PSNPs) on Seed Germination and Seedling Growth of Wheat (Triticum Aestivum L.)." *Journal of Hazardous Materials* 385: 121620. DOI: https://doi.org/10.1016/j.jhazmat.2019.121620.

María, Laura, and Hernández Rodríguez. 2021. "Sources and Fate of Microplastics and Nanoplastics : Detection and Characterization Methods," no. August.

Mason, Sherri A, Danielle Garneau, Rebecca Sutton, Yvonne Chu, Karyn Ehmann, Jason Barnes, Parker Fink, Daniel Papazissimos, and Darrin L Rogers. 2016. "Microplastic Pollution Is Widely Detected in US Municipal Wastewater Treatment Plant Effluent." *Environmental Pollution* 218: 1045–54. DOI: https://doi.org/10.1016/j. envpol.2016.08.056.

Mintenig, S M, M G J Löder, S Primpke, and G Gerdts. 2019. "Low Numbers of Microplastics Detected in Drinking Water from Ground Water Sources." *Science of The Total Environment* 648: 631–35. DOI: https://doi.org/10.1016/j.scitotenv. 2018.08.178.

Mofijur, M., S. F. Ahmed, S. M. Ashrafur Rahman, SK Yasir Arafat Siddiki, A. B.M. Saiful Islam, M. Shahabuddin, Hwai Chyuan Ong, T. M.I. Mahlia, F. Djavanroodi, and Pau Loke Show. 2021. "Source, Distribution and Emerging Threat of Micro- and Nanoplastics to Marine Organism and Human Health: Socio-Economic Impact and Management Strategies." *Environmental Research* 195 (February): 110857. DOI: 10.1016/j.envres.2021.110857.

Ng, Ee-Ling, Esperanza Huerta Lwanga, Simon M Eldridge, Priscilla Johnston, Hang-Wei Hu, Violette Geissen, and Deli Chen. 2018. "An Overview of Microplastic and Nanoplastic Pollution in Agroecosystems." *The Science of the Total Environment* 627 (June): 1377–88. DOI: 10.1016/j.scitotenv.2018.01.341.

Peng, Bo Yu, Zhibin Chen, Jiabin Chen, Huarong Yu, Xuefei Zhou, Craig S. Criddle, Wei Min Wu, and Yalei Zhang. 2020. "Biodegradation of Polyvinyl Chloride (PVC) in Tenebrio Molitor (Coleoptera: Tenebrionidae) Larvae." *Environment International* 145 (August): 106106. DOI: 10.1016/j.envint.2020.106106.

Alice Pradel. 2022. "Environmental Fate and Behavior of Nanoplastics: Implication of Physico-Chemical Processes HAL Id : Tel-03597424," no. November 2021.

Shen, Maocai, Yaxin Zhang, Yuan Zhu, Biao Song, Guangming Zeng, Duofei Hu, Xiaofeng Wen, and Xiaoya Ren. 2019. "Recent Advances in Toxicological Research of Nanoplastics in the Environment: A Review." *Environmental Pollution* 252: 511–21. DOI: 10.1016/ j.envpol.2019.05.102.

Wang, Liuwei, Wei Min Wu, Nanthi S. Bolan, Daniel C.W. Tsang, Yang Li, Muhan Qin, and Deyi Hou. 2021. "Environmental Fate, Toxicity and Risk Management Strategies of Nanoplastics in the Environment: Current Status and Future Perspectives." *Journal of Hazardous Materials* 401 (June 2020): 123415. DOI: 10.1016/ j.jhazmat.2020.123415.

Wayman, Chloe, and Helge Niemann. 2021. "The Fate of Plastic in the Ocean Environment—A Minireview." *Environmental Science: Processes and Impacts* 23 (2): 198–212. DOI: 10.1039/d0em00446d.

Weert, Sander van, Paula E Redondo-Hasselerharm, Noël J Diepens, and Albert A Koelmans. 2019. "Effects of Nanoplastics and Microplastics on the Growth of Sediment-Rooted Macrophytes." *Science of The Total Environment* 654: 1040–47. DOI: https://doi. org/10.1016/j.scitotenv.2018.11.183.

Zhang, Yulan, Shichang Kang, Steve Allen, Deonie Allen, Tanguang Gao, and Mika Sillanpää. 2020. "Atmospheric Microplastics: A Review on Current Status and Perspectives." *Earth-Science Reviews* 203: 103118. DOI: https://doi.org/10.1016/ j.earscirev.2020.103118.

11 Bioremediation and Biodegradation
Importance and Recent Development

Debajyoti Kundu, Palas Samanta,
Sukhendu Dey, Deblina Dutta, Rahul Rautela,
Anjani Devi Chintagunta, N.S. Sampath Kumar,
Rahul Mishra, Knawang Chhunji Sherpa,
Srushti Muneshwar, Ankit Motghare,
and Sunil Kumar

11.1 INTRODUCTION

Anthropogenic activities, such as mining and industrial processes, are key contributors to extensive environmental degradation around the globe (Megharaj and Naidu, 2017). The massive amounts of organic and inorganic pollutants released on the earth's surface, either intentionally through industrial operations or accidentally through leakage, cause their concentration to exceed permissible levels (Verma and Kuila, 2019). As a result, several sites worldwide have been severely contaminated, necessitating remediation. The price of cleaning up contaminated places is too expensive. Incineration, the use of oxidants (potassium permanganate or hydrogen peroxide), soil washing, landfilling, transport, excavation, digging, and dumping are some conventional methods for remediating contaminated soil (Megharaj and Naidu, 2017). Many contaminated commercial sites were left or idled rather than remediated due to the expensive solutions of remediation. As a result, using microorganisms to remediate hazardous environmental contaminants is better than using traditional approaches. To put it another way, bioremediation is a method for restoring the natural ecosystem by eliminating contaminants from the environment and minimising further contamination. Contemporary advancements in bioremediation techniques aim to effectively rejuvenate damaged areas in a sustainable and economically efficient manner. (Azubuike et al., 2016). Bioremediation is both cost-effective and environmentally friendly in comparison to other remediation processes such as physical and chemical. Various bioremediation approaches have been proposed by researchers. Nevertheless, based on the environment and different types of contaminants, no single bioremediation technique functions as a "viable solution" for restoring damaged sites (Verma and Jaiswal, 2016). The method of contaminant removal

DOI: 10.1201/9781003352396-11

173

largely relies on the pollutant types that can include chlorinated compounds, sewage, plastics, radioactive waste, hydrocarbons, heavy metals, dyes, and agrochemicals. When selecting a bioremediation technique, factors such as the severity and extent of the contaminants, the contaminants types, the environment, the cost, and policies are taken into account (Smith et al., 2015). Prior to a bioremediation project, performance criteria such as pH, temperature, oxygen, and nutrient concentrations as well as other abiotic elements that determine the effectiveness of the bioremediation processes are also significantly taken into account.

The contaminants' toxicity can be reduced via bioremediation, which uses the metabolic ability of microorganisms to convert highly harmful chemicals into less harmful compounds. Intensification of heavy metals or chemically polluted soil remnants with organic modification aids in improving the soil's physical characteristics, as well as increasing the bioavailability of nutrients for microbes (Park et al., 2011). In bioremediation, contaminants are used as a source of nutrients by microbial populations. Bioremediation can be implemented either directly at the site of contamination (in situ) or for contaminants that have been moved from their initial location (ex situ).The treatment of toxic substances in the place where they are identified is known as in situ bioremediation. As a result, people are becoming more interested in microbial bioremediation of pollutants as they attempt to find sustainable ways to clean up contaminated sites (Kumar et al., 2016). The breakdown of several xenobiotic chemicals like nitrated aromatic compounds, strongly halogenated compounds, and a few insecticides using microorganisms has not yet been reported (Gangola et al., 2019). Various factors such as availability and physiological characteristics of the environment, contaminants' chemical nature, and concentration affect the microorganism's efficiency. As a result, the factors that influence the ability of microorganism degradation are either environmental parameters or nutritional needs.

This chapter provides an overview of the history of bioremediation technologies, mechanisms involved in the process, techniques employed, role of different living agents and enzymes, and case studies on bioremediation.

11.2   HISTORY OF BIOREMEDIATION AND BIODEGRADATION

"Bioremediation" is a metabolic process that uses biological entities to eliminate or neutralise harmful substances. The phrase "bioremediation" consists of two components: "bios," which denotes "life," and "remediation," which implies "to remedy a situation" (Sardrood et al., 2013). The biological entities, namely, bacteria, algae, fungi, or living plants, are generally employed to remediate the environmental burden. Bioremediation is a notion that has been around for a long time. The Romans and other ancient civilizations began utilizing biological entities, especially microorganisms, as early as 600 B.C. for the purpose of decontaminating wastewater. This method is still utilised to treat industrial effluents, but it's been broadened to handle a variety of additional substances. Although biologists have already been researching the process since the 1940s, the bioremediation technique has been utilised commercially for nearly three decades (Hoff, 1993). A bioremediation device was first utilised commercially in 1972 to wipe up a Sun Oil Pipeline leakage in Ambler, Pennsylvania. The scientific interest in bioremediation techniques was ignited in

1989 during the cleanup of the Exxon Valdez oil spill in Prince William Sound, Alaska. Since 1989, works on remediation have appeared in a variety of publications, including science publications, trade publications for the dangerous waste and environmental industries, and popular press articles. These articles cover its rising popularity as an environmental technology with a variety of uses, including toxic waste degrading, contaminants removal, soil remediation, and oil spill cleaning.

Simply put, remediation is the biological mechanism of decontaminating a damaged area. The habitat can be terrestrial, watery, or a combination of both. However, the following is the more thorough meaning of bioremediation (Baggott, 1993; Mentzer and Ebere, 1996; Sharma, 2020):

> Bioremediation is a means of cleaning up contaminated environments by exploiting the diverse metabolic abilities of microorganisms to convert contaminants to harmless products by mineralization, generation of carbon (IV) oxide and water, or by conversion into microbial biomass.

It's important to note that bioremediation and biodegradation are not interchangeable terms. Biodegradation is a naturally occurring phenomenon in which bacteria or other organisms change and decompose organic compounds into certain compounds (Verma and Jaiswal, 2016). On the other hand, bioremediation is a technology that might incorporate biodegradation as among the underlying mechanisms or be used in the remediation process. For instance, the introduction of nutrients or other elements to polluted settings or modification of polluted media using processes like temperature regulation or aeration are examples of bioremediation (Azubuike et al., 2016). Additionally, for cleaning soils/sediments polluted with environmental xenobiotics, and also coastal ecosystems harmed by oil spillage, bioremediation offers a wide range of uses in both terrestrial and freshwater ecosystems. Further, only a few pollutants are recyclable, and only a small percentage of pollutants can be degraded by microbes (Kundu et al., 2020). As a result, it'd be worthwhile to investigate microorganisms' biodegradation capacity.

A detailed chronological achievement in bioremediation technology is shown in Figure 11.1. In this context, the chronology of bioremediation in remediating environmental xenobiotics can indeed be categorised into the following three developmental phases:

i. **Research period (before 1989)**: During this time period, the bioremediation technique was mostly unknown beyond the microbiologist and toxic waste communities. This stage is referred to as the "courtship" phase. During this phase, the individuals as researchers are engaged to find out the capability or importance of biological entities to degrade environmental contaminants.

ii. **Celebration period (1989–1991)**: During this time period, bioremediation as a method drew widespread interest and higher emphasis. After the celebration period ended, there had been a disappointment period, as the technology's promises had not always been borne out by its application in real-life settings. This stage can also be referred to as the "honeymoon/developmental" phase.

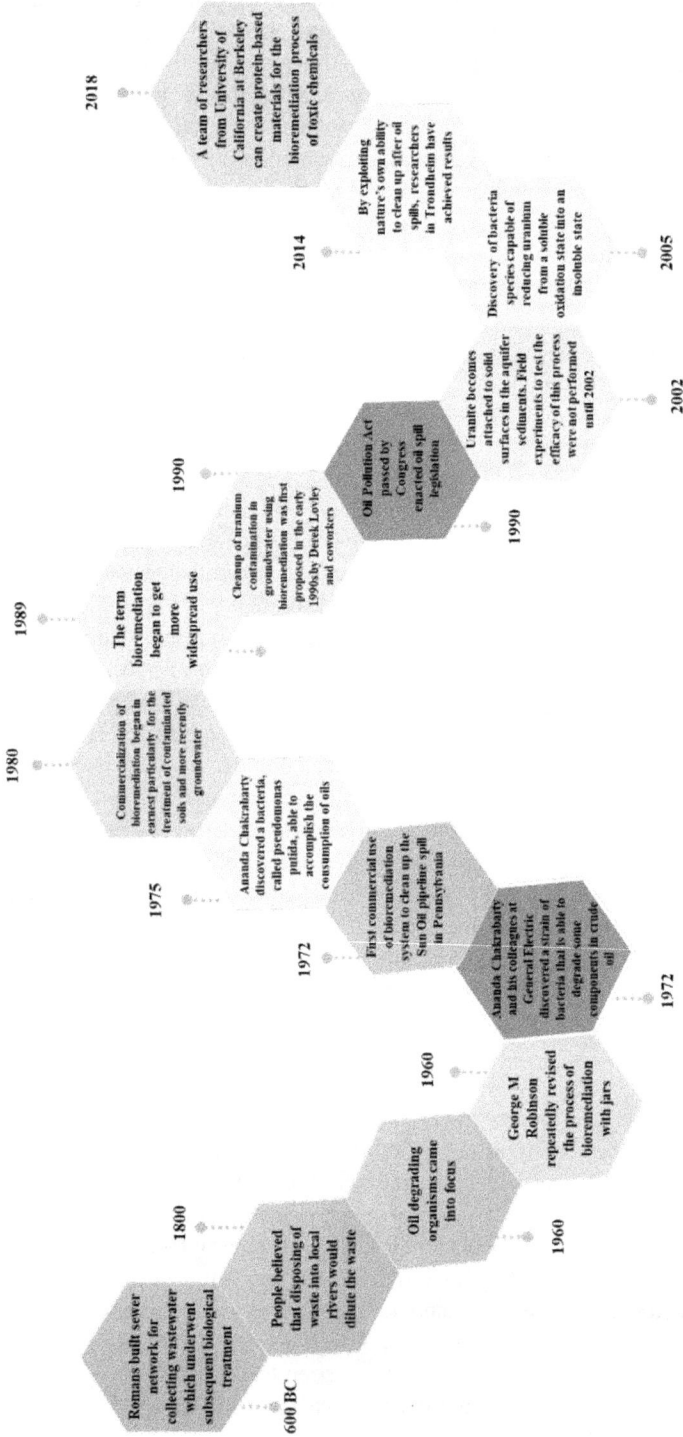

FIGURE 11.1 Chronological achievement in bioremediation.

iii. *Establishment period (from 1992 onwards)*: During this time, bioremediation has gained a higher degree of acceptance, with much more realistic ideas than before, yet curiosity and focus have declined significantly. This stage can also be referred to as the "mature" phase. The government's role was just getting started from this period.

11.3 MECHANISM OF BIOREMEDIATION AND BIODEGRADATION

11.3.1 BIOSORPTION

Biosorption is a potential biotechnological strategy for contaminant removal, and this is referred to simply as the expulsion of compounds from solution using biological weapons including dead mass (Figure 11.2). Essentially, this is a physicochemical technique that reduces pollutant concentrations by utilising either alive or dead biomass. The biosorption process differs from adsorption as in biosorption the dissolution or permeation of solids/liquids (the absorbent) is governed by a solvent (better known as absorbate) (Jin et al., 2018). Adsorption, on the other hand, is a superficial occurrence, whereas absorption affects the entire substance volume. The biosorption process includes a wide range of techniques for removing contaminants such as surface precipitation, chemisorption and ion exchange, the creation of persistent compounds involving organic ligands, and redox phenomena (Jin et al., 2018). Due to its effectiveness, affordability, and trash end products, it's been utilised since too many years and is regarded as advantageous over the current conventional ion-exchange technique (Samanta et al., 2020; Wang and Chen, 2009). Biosorption is a physiologically (metabolic) passive activity since it does not require energy, and the contaminant quantity that may be eliminated or degraded by a sorbate is determined

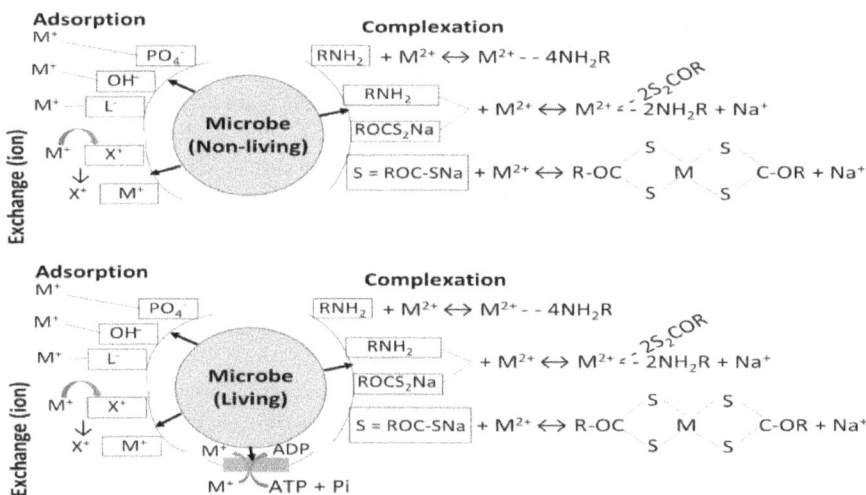

FIGURE 11.2 Schematic demonstration of microbial biosorption mechanism.

by kinetic equilibria and cellular surface characteristics of the sorbent. The biosorption process comprises two distinct mechanisms: the metabolism-dependent process, conducted through the cell membrane, and the metabolically independent process, governed by the cell wall. (Jin et al., 2018; Saraswat et al., 2020). More specifically, in a metabolically independent process the presence of carboxylic, amine, hydroxyl, and phosphonic groups on the outer layer of the microbe's cell wall delivers negative charges at reaction sites which helps to absorb positively charged pollutants through complexion. On the other hand, in the metabolism-dependent process the contaminants form complexes with the cell membrane's functional units, namely, oxygen, sulphur, phosphorus, and nitrogen, which act as coordinating molecules. The process is believed to be reversible in nature and is aided by the existence of functional units such as carboxylic, amine, hydroxyl, and phosphonic. Gram-positive bacteria have a higher sorption capacity owing to the influence of a thick peptidoglycan layer because of processing greater sorption sites. Temperature, redox potential, pH, the existence of metals, and the solute concentration all perform a critical role during the sorption process (Jin et al., 2018; Saraswat et al., 2020).

11.3.2   PRECIPITATION

Precipitation is another widely used technique for remediating the contaminants. Precipitation, also called microbially induced carbonate precipitation (MICP) based on biomineralisation, is a fascinating research area as well as a novel method for remediating a wide range of pollutants. The process uses living bacteria to convert contaminants into minerals from an aqueous solution through crystalline precipitates, eliminating contaminants from the solution and allowing for decontamination and bio-recovery (Saraswat et al., 2020). The process generally employs two techniques, namely, biologically controlled mineralisation (BCM) and biologically induced mineralisation (BIM) during bio-recovery (Jain and Arnepalli, 2019). Microbes' unregulated metabolism in the BIM process governs the geochemical processes in the surrounding environment, contributing to extracellular mineral development. On the other hand, bio-recovery is governed either intracellularly or epicellularly in the BCM process, and the mechanism is entirely controllable (Jain and Arnepalli, 2019). Microbes of various phyla utilising distinct metabolic processes synthesised over 64 different types of carbonates, oxides, silicates, phosphorites, sulphides, sulphates, oxalates, and amorphous silica deposits (Jain and Arnepalli, 2019). MICP is the most studied among all microbial precipitation [direct mineralisation, microbially induced phosphate precipitation (MIPP), sulphide mineralisation, etc.] to date, as carbonate precipitation occurs abundantly in the natural environment and involves a wide range of microbes. More specifically, one aerobic and two anaerobic processes are generally governed by the autotrophic mechanisms of MICP. Calcium carbonate mineral precipitation, on either hand, involves a number of heterotrophic mechanisms, including sulphur or nitrogen cycle, and iron reductions. Three separate mechanisms in the N cycle, viz., nitrate reduction, amino acid ammonification, and uric acid/urea breakdown, all increase pH and generate carbonate ions (Jain and Arnepalli, 2019). Among all the approaches, MICP via the urea hydrolysis process or urease-positive microorganisms (Figure 11.3a) is gaining popularity in ecological

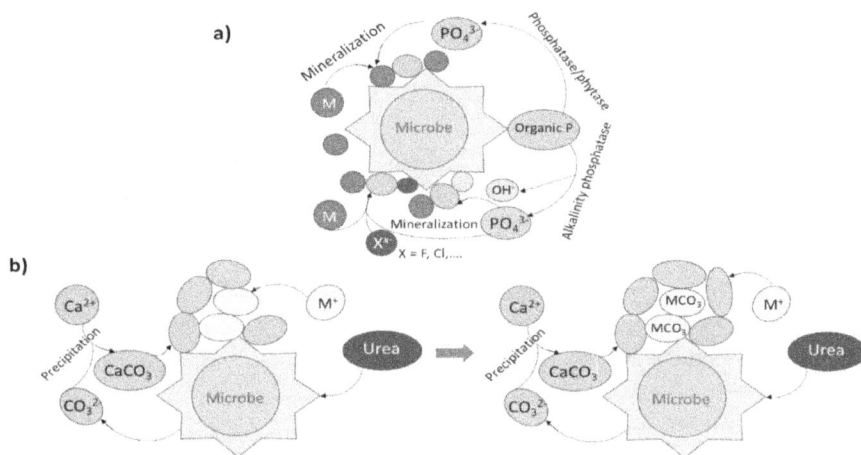

FIGURE 11.3 Schematic representation of (a) MICP and (b) MIPP assisted bioremediation of contaminants (M represents contaminants).

rehabilitation because of its diversity of species, lower energy consumption, and environmentally friendly nature (Fujita et al., 2010). The reactions involving urea hydrolysis process-driven MICP are as follows (Shan et al., 2021):

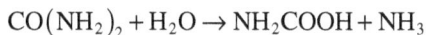

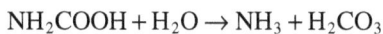

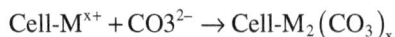

$$CO(NH_2)_2 + H_2O \rightarrow NH_2COOH + NH_3$$

$$NH_2COOH + H_2O \rightarrow NH_3 + H_2CO_3$$

$$H_2CO_3 \leftrightarrow HCO_3^- + H^+$$

$$2NH_3 + 2H_2O \leftrightarrow 2NH_4^+ + 2OH^-$$

$$HCO_3^- + H^+ + 2NH_4^+ + 2OH^- \leftrightarrow CO_3^{2-} + 2NH_4^+ + 2H_2O$$

$$Cell + M^{x+} \rightarrow Cell\text{-}M^{x+}$$

$$Cell\text{-}M^{x+} + CO3^{2-} \rightarrow Cell\text{-}M_2(CO_3)_x$$

MIPP is another widely used bioremediation technique for contaminants (Shan et al., 2021). The MIPP process aids organic phosphate to precipitate as insoluble phosphates on the microbe's cellular surface (Figure 11.3b).

Additionally, the MIPP process also helps to remediate actinides from aqueous solution. The reactions involving MIPP-driven precipitation technique are as follows (Shan et al., 2021):

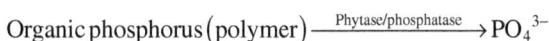

$$Organic\ phosphorus\ (polymer) \xrightarrow{\ Phytase/phosphatase\ } PO_4^{3-}$$

$$M^{x+} + \text{Cell} \rightarrow \text{Cell-}M^{x+}$$

$$\text{Cell-}M^{x+} + PO_4^{3-} + X^{x-} \rightarrow \text{Cell-}M(PO_4)X$$

where X^{x-} indicates hydroxide groups (OH^-) and halogen ions (mainly Cl^-, F^-, and Br^-). Three separate mechanisms in MIPP-driven bioprecipitation process have been demonstrated at present, namely, (i) microbes-derived metabolites aid MIPP driven biomineralisation, (ii) extracellular proteins react with phosphate to accomplish the operation of mineralisation, and (iii) phosphate nutrients biomineralise with toxicants in the cell wall (Roeselers and Van Loosdrecht, 2010; Xu et al., 2020; Zhang et al., 2019). Redox potential, pH, nutrient availability, metallic presence, and solute concentration all perform a critical role during the sorption process (Jin et al., 2018; Saraswat et al., 2020).

11.3.3 DETOXIFICATION

Detoxification is another widely used technique for remediating or breaking down hazardous contaminants into non-toxic or less toxic substances. It is a very promising method for remediating a wide range of pollutants. Detoxification is generally performed via mechanical, chemical, or biological approaches (Wang et al., 2019). However, throughout the removal process, mechanical/chemical decontamination modifies the feed's nutritional structure, and the detoxifying agents might produce secondary environmental pollution (Dixit et al., 2015). Physical processes encompass precipitation techniques like flocculation, coagulation, sedimentation, adsorption, filtration, ultra-filtration, reverse osmosis, and nano-filtration. On the other hand, chemical additions or reactions like oxidation, reduction, ion exchange, and neutralisation are generally employed in chemical techniques. Biological decontamination, on the other hand, has quite an excellent sensitivity, generates innocuous by-products, and, under the right circumstances, could even result in total decontamination (Wang et al., 2019). Biological remedies are implemented under aerobic, anaerobic, or a combination of aerobic–anaerobic conditions in the presence of a biocatalyst, contingent upon the oxygen levels within the system. The biocatalysts could be biological agents like bacteria, fungi or actinomycetes, or plants, or any living biomass, or microbial secretases. Enzymes released by microbes decompose organic molecules in the aerobic process by using processes like reduction, oxidation, methylation, and alkylation. More specifically, extracellular enzymes like lignin peroxidase, copper-containing laccase, manganese peroxidase, etc., play a dominant role during bioremediation of a wide substrate range. Contaminants are used by microbial species to generate energy as electron acceptors or donors. More specifically, toxicants and microbes engage in such a basic redox (reduction and oxidation) reaction in the detoxification process; microbes serve as an oxidant for contaminants, losing electrons, which could then be acquired by alternate acceptors (sulphate, nitrate, and ferric oxides). Contaminants in their oxidation state generally act as terminal electron acceptors during bacteria's anaerobic metabolism (i.e., microorganisms use reducing electron acceptors to reduce organic pollutants). Accordingly, the enzyme-mediated reduction of ionic species potentially leads to the emergence of low hazardous contaminants (Uday et al., 2016). On the other hand, in aerobic settings, oxygen serves as an electron acceptor. Microbial enzyme-assisted methylation is another promising mechanism for contaminant removal as methylation improves permeation capability across microbial cell membranes, resulting in toxicity reductions of contaminants through oxidation-reduction reactions. It has been demonstrated that parameters like pH, temperature, starting pollutant level, etc., have played a crucial role during detoxification of pollutants (Uday et al., 2016).

11.3.4 ENZYMATIC BIOCONVERSION

Enzymes, especially extracellular enzymes derived from microbes, play a pivotal role during bioconversion or biotransformation of different toxic pollutants to benign molecules by oxidative coupling (mainly through polymerisation or copolymerisation) in presence of oxidoreductases (Tegene and Tenkegna, 2020). Microbes extract energy through energy-yielding biochemical reactions facilitated by enzymes. These enzymes play a crucial role in cleaving chemical bonds and aiding the transfer of electrons from a reduced organic substrate (donor) to another chemical compound (acceptor). (Karigar and Rao, 2011). Both oxidising and reducing enzymes play a dominant role during pollutant bioconversion. Oxidising enzymes like LiP, laccase, veratryl alcohol oxidase, and tyrosinase and reductase enzymes like riboflavin, DCIP, azo reductase, and Green HE4B, in this regard, are prominent ones. In addition to this, microbial enzymes are generally classified into two categories, namely, oxidoreductases [oxygenase (monooxygenase, dioxygenase), laccase, peroxidase (lignin peroxidase, manganese peroxidase, versatile peroxidase)] and hydrolase (lipase, cellulase, protease) (Karigar and Rao, 2011). Peroxidases, specifically, catalyse the production of radicals from phenolic compounds by employing hydrogen peroxide as an electron donor. Laccase oxidise a wide variety of organics like polyphenols, phenols, and anilines, as well as recalcitrant environmental pollutants (Tegene and Tenkegna, 2020; Xiao et al., 2010) by utilising only one electron transfer pathway. Not only Mn but also phenolic and nonphenolic aromatic substances, including dyes, can be oxidised by versatile peroxidases (VP) (Tegene and Tenkegna, 2020). The mechanism of some specific enzyme-assisted bioconversion is presented in Figure 11.4.

11.3.5 DEGRADATION/DECOMPOSITION

Degradation/decomposition is the biological process by which organic contaminants are converted or degraded into simpler/low toxic inorganic/organic substances. Both aerobic and anaerobic microbial activities can be used in environmental remediation. In aerobic degradation, oxygen molecules are introduced during substrate reactions through the action of microbial enzyme secretions. These enzymes, such as hydroxylases, monooxygenases, oxidative dehalogenases, dioxygenases, or chemically reactive O_2 molecules produced by enzymes like peroxidases or ligninases, play a key role in the process. On the other hand, initial activation events are accompanied by oxidative catabolism regulated by an anoxic acceptor (electron) during the anaerobic breakdown of pollutants (Tegene and Tenkegna, 2020). In particular, aerobic degradation is preferable to degrade or decompose a wide range of contaminants.

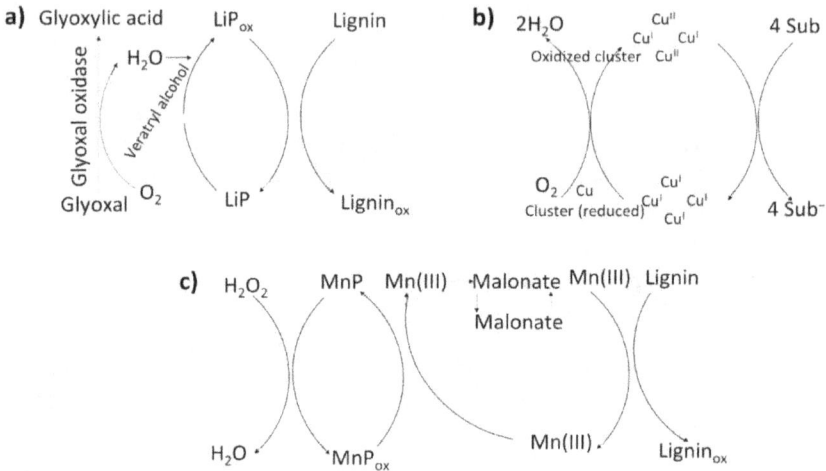

FIGURE 11.4 Mechanism of action for (a) lignin peroxidase, LiP; catalytic action of (b) laccases and (c) manganese peroxidase, *MnP*. ox denotes enzyme's oxidised state.

For instance, during degradation of organic contaminants, the earliest intracellular assault is indeed an oxidative mechanism, with peroxidases and oxygenases catalysing the activation and insertion of O_2 molecules as a primary enzymatic response. The tricarboxylic acid cycle (TCA) degradation route is the prime decomposition method as major precursor molecules, such as acetyl-CoA, pyruvate, and succinate, are used to synthesise cellular mass. Other techniques include (1) microbial cell adhesion to substrates and (2) biosurfactant synthesis (Tegene and Tenkegna, 2020). In particular, the microbial decomposition of chlorophenols follows three distinct routes. Monooxygenase-catalysed hydroxylation to generate hydroquinones is the first mechanism. Hydroxylation of heavier chlorinated phenolics to chlorinated hydroquinones is the second process and reductive dechlorination under anaerobic circumstances is the final process (Mishra and Thakur, 2012). The degradation of chlorinated dioxins attacking the ring close to ether oxygen by angular dioxygenases is the prime pathway. In aerobic conditions, a significant mechanism involves the co-metabolism of chlorinated dioxins with extracellular lignin peroxidases produced by white-rot fungi. (Mishra and Thakur, 2012). For agriculture-derived lignocellulosic waste remediation, composting is the widely accepted bioconversion method that uses thermophilic/mesophilic microbes as biological agents to decompose waste residues (Tegene and Tenkegna, 2020). Figure 11.5 shows the generalised pathway of microbial aerobic biodegradation of hydrocarbon substances.

11.4 TYPES OF BIOREMEDIATIONS

In past decades, bioremediation is the ultimate sustainable pollution removal as sustainable, being able to effectively restore pollutants, and is very cost-effective. Several researchers have developed bioremediation techniques. Two different types of bioremediation techniques (ex situ and in situ) are used based on the aeration of

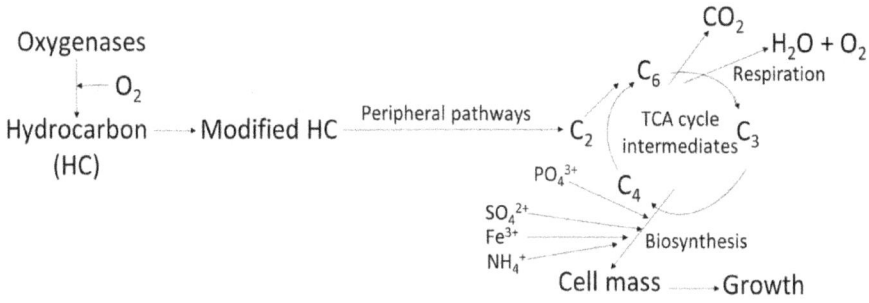

FIGURE 11.5 Schematic mechanism of microbial aerobic biodegradation of hydrocarbon.

an area and also the degree of saturation. Ex situ bioremediation techniques are basically used in the groundwater and soil pollution sites where the pollution load is very high (Xing et al., 2022). But at the same time in situ bioremediation techniques are basically used in the groundwater and soil pollution sites where the pollution disturbance is very minimal. In some cases, microorganisms are used to enhance the bioremediation methods to minimise the pollution load from the polluted sites (Haripriyan et al., 2022). Figure 11.6 depicts the types of bioremediation techniques employed in pollution remediation.

11.4.1 Ex Situ

Ex situ bioremediation techniques are used in pollutants, generated from polluted sites simultaneously transferred into treatment sites for proper treatment. Usually, ex situ bioremediation techniques are used for types of pollutants, depth of the pollutants, costs of the entire treatment procedure, geology of the pollution sites as well as the geographical location of the pollution sites. The ex situ bioremediation process is discussed in detail below (Haripriyan et al., 2022).

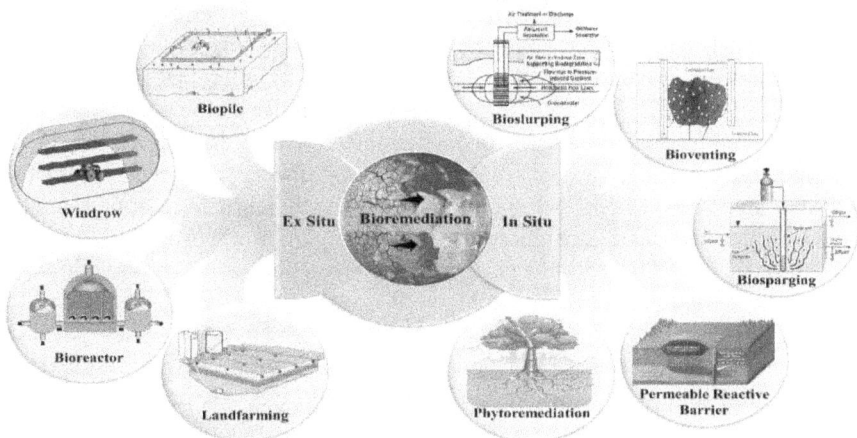

FIGURE 11.6 Types of bioremediation techniques.

11.4.1.1 Biopile

The biopile bioremediation technique acts as an excavated polluted soil above the ground with the help of soil nutrient. In some cases, aeration increases bioremediation by enhancing microbial activities. Some vital components are needed in this technique like nutrients, irrigation, aeration, leachate collection container, and finally treatment system. The main benefit of the said technique is in enhancing the constrictive shape like cost-effective biodegradation depends on the temperature, aeration, and nutrition are sufficiently controlled as established by Whelan et al. (2015). The solicitation of biopile technique in pollution sites which may be helped by the boundary of the volatilisation of very low molecular weight pollutants can also benefit collectively the bioremediation in absolutely polluted environments like cold regions (Dias et al., 2015; Gomez and Sartaj, 2014; Whelan et al., 2015). In connection with this, Gomez and Sartaj (2014) established in their studies that the outcome of the various application rates (3–6 mL/m^3) of microbial association and mature content (5 to 10%) on total petroleum hydrocarbon (TPH) reduction in the field sample biopiles at very low temperature by the help of response surface methodology act as a factorial design of the experiment. Completion of the entire experiment (94 days) TPH 90.7% reduction in the biostimulation or bioaugmentation process were reduced approximately 48% compared to the control condition in TPH removal. A very high average percentage of the TPH reduction is attributed to the interaction between biostimulation and bioaugmentation, thus details showing the natural flexibility of biopiles for bioremediation technique. Dias et al. (2015) reported in their study that about a 71% reduction of the total hydrocarbon content, and then shifted to bacterial culture over 50 days of experiments followed by the pretreatment of contaminated soil samples for prior biopile production, and simultaneously biostimulation by fishmeal.

11.4.1.2 Windrows

Windrows are one of the techniques for ex situ bioremediation, and this can apply to the treatment of pollutants as a reducing mode heaver to lower. Petroleum hydrocarbon impacts soils, which could turn into a treatment area, where the mixing effect on different agents enhances the degradation by microbes (Michel et al., 2022). Windrows rely a periodic turning, the former relies on the periodic rotation of piled contaminated soils to increase bioremediation by increasing the degradation activity of native and/or transient hydrocarbonoclastic bacteria present in contaminated soils, as one of the bioremediation techniques. Windrow treatment shows as biopile treated as higher hydrocarbon pollutant removal, it could influence the type of soil (Creegan et al., 2022). However, periodically using windrows may not be an optimal choice for soil remediation, particularly in association with methane (CH$_4$), which can impact removal efficiency in anaerobic zones. Aeration can be employed to mitigate soil pollution caused by piles. Periodic rotation of contaminated soils, with the addition of water, increases ventilation, pollutants, nutrients, and degradation activity, thus accelerating the rate of bio-reform, which can be accomplished through assimilation, bio-conversion, and mineralisation (Creegan et al., 2022).

11.4.1.3 Bioreactor

Bioreactor is an important bioremediation process, implies to raw material can converted into a specific product, where applied several operation sites such as a batch reactor, fed-batch reactor, sequencing batch reactor, and continuous-multi-stage reactor. Actually, it was applied due to easy operating mode and market economic and capital expenditure (Shannon et al., 2014). In microbial applications, the batch bioreactor has proven effective in controlling pollutant degradation, even with less efficient indigenous microorganisms. The environmental conditions within the bioreactors are sustained by the growth of indigenous microbes or the ionic distribution, influenced by factors such as pH. The bioreactor is fed with pollutant substances, which can be in the form of dry matter or slurry (Carniato et al., 2012). Bioreactors can be controlled with bioprocess parameters such as temperature, pH, agitation and aeration rates, and substrate and inoculum concentrations (Pörtner et al., 2005). This application process in bioremediation has a greater advantage due to ability to mineralise pollutants or change less toxic forms using a bioreactor. For example, microbial bioremediation is used to control food sources, and another as phytoremediation, plants to bind extract, reduced from petroleum hydrocarbons (PAH), metals (Cd, Hg, Pb), pesticides, and chlorinated solvents.

11.4.1.4 Land Farming

Land farming is one of the bioremediation techniques which deals with low cost and minimum applied equipment and no chemical utility. It is applied to either ex situ or in situ bioremediation (Silva-Castro et al., 2012). In land farming, several pollutants on soil can be excavated and/or tilled, if it is on-site to be called the in situ remediation; otherwise, it is chosen in an external site to transfer as ex situ bioremediation. This management is easily separated through plants and microbes, general terms, organic pollutants can be mineralised by microbes such as oily sludge, PAHs, and pesticides. As a rule of thumb, higher molecules come into lower form, which uptakes the plant during farming conditions. As ex situ bioremediation, there are some limitations due to the large operation area and reduced biodiversity owing to unsuitable environmental conditions. Effectively, ex situ bioremediation as pollutant inhomogeneity observed the excavation process by controlled parameters, namely, temperature, pH, and mixing to enhance the bioremediation.

11.4.2 In Situ

11.4.2.1 Bioslurping

This technique is actually in situ bioremediation; indigenous microbes can be involved to control and stimulate airflow by oxygen at the unsaturation zone (vadose). In the bioventing process, amendments could be generated by several nutrients and moisture, thus increasing bioremediation to achieve the ultimate goal especially microbes can transfer into harmless products (Wu et al., 2014). Bioslurping consists of two remedial aspects, one is bioventing and another is vacuum-enhanced to recover free products (Kim et al., 2014). It is found in the aerobic bioremediation process to stimulate hydrocarbon-contaminated soils. Biosparging process, at injecting pressurised air or gas transferred into the contaminated area for increasing in situ microbial activity under aerobic conditions. It is not affected by soil of capillary fringe and the saturated zone. When increased soil air permeability, bioslurping work down gradually, and moisture decreases to inhibit microbial activity. Temperature is the vital factor, slow remediation can be observed at low temperatures.

11.4.2.2 Bioventing

Bioventing, in situ remediation process is applied to treat the polluted groundwater and also soil (Singh and Ram, 2022). This process involves a significant portion of microbes directly degrading the product through direct air injection. This technique is specifically designed to liberate and recover products as light non-aqueous phase liquids (LNAPLs), efficiently remediating both unsaturated and saturated zones (Singh and Ram, 2022). It is controlled by soil moisture and limited air permeability, and microbial activity is reduced when oxygen transfer is decreased. Although the strategy is not suitable for soil treatment with low permeability, it saves costs owing to the low amount of groundwater as a result of the operation, thus reducing the cost of storage, treatment, and disposal. However, field application is a more populated remediation process which takes up to ~90% efficiency (Somu and Paul, 2022).

11.4.2.3 Biosparging

Biosparging can be observed as similar to the bioventing process. Injected air can stimulate the microbial activity for contaminant degradation from polluted areas. The injected air plays a role in the upward movement of the volatile organic compounds (VOCs) from the unsaturated region (Ahmadnezhad et al., 2021). However, two major factors are involved in the biosparging, namely, soil permeability (maintains bioavailability on microbes) and pollutant biodegradability. In principle, soil vapours extraction, bioventing, biosparing are similar to remedial techniques. Biosparging was used in the treatment of aquifer contaminated with petroleum products, such as kerosene. It was previously reported that biosparging applied to clean benzene, toluene, ethylbenzene and xylene resulted in decreasing the contamination level to 70%. For application, combination of zeolite barrier and biosparging techniques for removal of organic pollutant from shallow groundwater (Ahmadnezhad et al., 2021).

11.4.2.4 Phytoremediation

Phytoremediation is a process that uses plants to stabilize, transfer, and degrade pollutants. This process is applied on aqueous and soil contaminants in generally, basically contaminant soil can induce to uptake the metallic substance and transferred into the organic state. In aqueous solution phytoremediation can applied in previously, such as dye pharmaceutical drugs, oil, heavy metal radio elements, and other pollutants. Some important factors are taken role to phytoremediation which are root system, ground biomass, plant survival and its adaptability (Schwitzguébel, 2017). The plants uptake the contaminant through root system by xylem flow and distributed to all parts of plant by translocation. On the other hand, nutrient management through plant growth–promoting bacteria such as rhizobacteria, generally bioagglomerated the pollutants and arrest the pollutant mobility (Wang et al., 2011). It is a very easy method for sustainable bioremediation process. The benefit is that pollutant remediate along with nutrient management to promote growth and stimulate plant health. It was previously reported that transgenic plant can improved the phytoremediation by gene transference. In addition, biosurfactant generated by *Serratia marcescens*, onto gasoline-contaminated soil and plant on *Ludwigia octovalvis*, shows 90% petroleum hydrocarbon remediate by phytoremediation methods, which increases bioavailability in microbial consortia (Al-Mansoory et al., 2017).

11.4.2.5 Permeable Reactive Barrier

It is an important technique to perceive that physical methods can be applied into contaminated groundwater and biological reaction mechanism such as biodegradation, precipitation, and biosorption. In biotechnological context, the biosorption and biodegrading can help the microbes, a passive bioreactive barrier induces the bioaccumulation of the pollutants (Carniato et al., 2012). In principle, the permeability barriers are used to trap the pollutants, potentially preventing contaminants accessible to the surrounding environment. The effectiveness of this process can influence media, type of pollutant, biogeochemical and hydrogeological, environmental factors, and mechanical stability, and sustain cost (Obiri-Nyarko et al., 2014). This concept is applied into the flow path on the contaminated groundwater, which influenced the natural hydraulic gradient using zero valent iron (Obiri-Nyarko et al., 2014).

11.5 LIVING MACHINE IN BIOREMEDIATION

Microorganism is an important bioremediation factor for nutritional chains, which is played in ecological balance in environment. Bacteria, fungi, algae, and yeast can take part in bioremediation (Wang et al., 2011) (Figure 11.7). Microbes can grow at certain temperature in presence of hazardous compounds. This microbial degradation has two characteristic features, that is, adaptability and biological system, which consists carbon requirement as feedstock. Microorganisms act as a significant pollutant removal tool in soil, water, and sediments; mostly due to their advantage over other remediation procedural protocols. Microorganisms are restoring the original natural surroundings and preventing further pollution.

11.5.1 Bacteria

Heavy metals and trace elements are used by microbes as terminal electron acceptors, providing them with the energy they need to detoxify metals through enzymatic and non-enzymatic mechanisms. Additionally, bacteria have the potential to

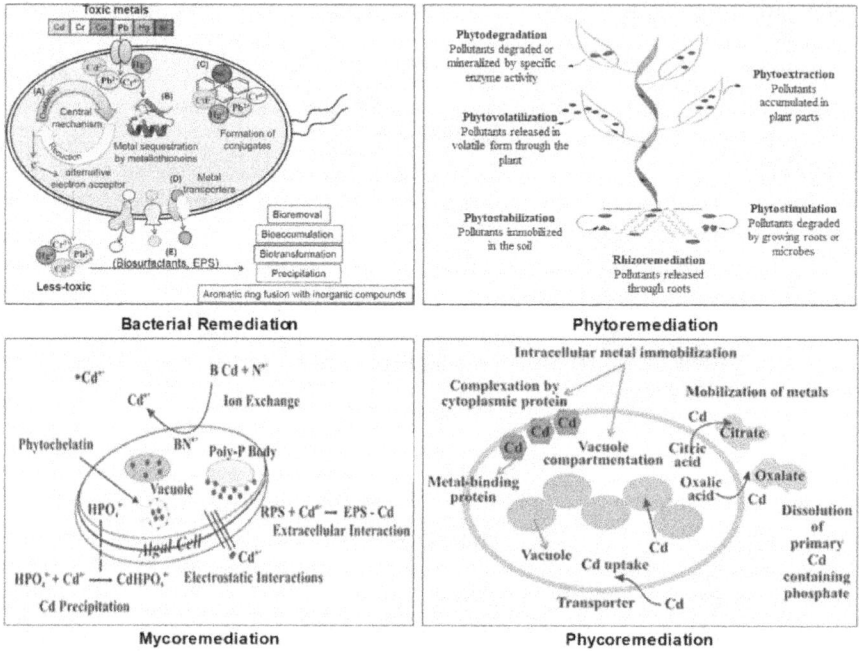

FIGURE 11.7 Bioremediation process in bacterial, phytoremediation (plants), mycoremediation (fungi), and phycoremediation (algae) (modified from Kumar et al., 2021).

bioaccumulate or store heavy metal ions in both particulate and insoluble forms, as well as their by-products. Exopolysaccharide (EPS) is the most crucial component in such bacterial cells with ion sequestration ability. With fewer amounts of protein and uronic acid, EPS is primarily made up of complex, high-molecular-weight organic macromolecules such polysaccharide. EPS defends the bacteria from environmental challenges such as salinity, drought, and heavy metal toxicity. Genera of microbes that produce EPS include *Acetobacter xylinum*, *Pseudomonas* spp., *Leuconostoc*, *Zygomonas mobilis*, *Bacillus* spp., *Xanthomonas campestris*, *Alcaligenes faecalis*, and *Agrobacterium* sp. The utilisation of non-neutral, negatively charged EPS (EPS packed with plentiful anionic functional groups) to be included as a suitable biosorbent must be the main emphasis of heavy metal remediation procedures using bacterial EPS. Commercial bacteria that produce EPS that meet the necessary anionicity requirements include alginate (*Pseudomonas aeruginosa*, *Azotobacter vinelandii*), gellan (*Sphingomonas paucimobilis*), and hyaluronan (*Pseudomonas aeruginosa*, *Pasteurella multocida*, *Streptococci attenuated strains*). Production of EPS is linked to procedures such biofilm formation, which is necessary for the biosorption and biomineralisation of metal ions. Table 11.1 provides the details of bacteria involved in the degradation of different pollutants.

TABLE 11.1
Bioremediation of Different Types of Pollutants by Bacteria (Abatenh et al., 2017)

Species	Compound
Organic Pollutant	
Gloeophyllum striatum	Striatum Pyrene, anthracene, 9-metil anthracene, dibenzothiophene lignin peroxidase
Acinetobacter sp., *Pseudomonas* sp., *Ralstonia* sp., and *Microbacterium* sp.	Aromatic hydrocarbons
Gleophyllum striatum	Striatum Pyrene, anthracene, 9-metil anthracene, dibenzothiophene lignin peroxidase
Acinetobacter sp., *Pseudomonas* sp., *Ralstonia* sp., and *Microbacterium* sp.	Aromatic hydrocarbons
Gleophyllum striatum	Striatum Pyrene, anthracene, 9-metil anthracene, dibenzothiophene lignin peroxidase
Cyanobacteria, green algae and diatoms, and *Bacillus licheniformis*	Naphthalene
Candida viswanathii	Phenanthrene, benzopyrene
Tyromyces palustris, Gloeophyllum trabeum, Trametes versicolor	Hydrocarbons
Alcaligenes odorans, Bacillus subtilis, Corynebacterium propinquum, Pseudomonas aeruginosa	Phenol
Coprinellus radians	PAHs, methylnaphthalenes, and dibenzofurans
A. niger, A. fumigatus, F. solani, and *P. funiculosum*	Hydrocarbon
Phanerochaete chrysosporium	Biphenyl and triphenylmethane
Pseudomonas putida	Monocyclic aromatic hydrocarbons, e.g., benzene and xylene
P. alcaligenes, P. mendocina, P. putida, P. veronii, Achromobacter, Flavobacterium, Acinetobacter	Petrol and diesel polycyclic aromatic hydrocarbons toluene
Penicillium chrysogenum	Monocyclic aromatic hydrocarbons, benzene, toluene, ethyl benzene and xylene, phenol compounds
Oil	
Pseudomonas cepacia, Bacillus cereus, Bacillus coagulans, Citrobacter koseri, and *Serratia ficaria*	Diesel oil, crude oil
Pseudomonas aeruginosa, P. putida, Arthobacter sp., and *Bacillus* sp.	Diesel oil
B. brevis, P. aeruginosa KH6, B. licheniformis, and *B. sphaericus*	Crude oil
Aspergillus niger, Candida glabrata, Candida krusei, and *Saccharomyces cerevisiae*	Crude oil
Bacillus cereus	Diesel oil

(Continued)

TABLE 11.1 *(Continued)*
Bioremediation of Different Types of Pollutants by Bacteria (Abatenh et al., 2017)

Species	Compound
Alcaligenes odorans, *Bacillus subtilis*, *Corynebacterium propinquum*, *Pseudomonas aeruginosa*	Oil
Fusarium sp.	Oil
Dye	
Bacillus firmus, *Bacillus macerans*, *Staphylococcus aureus*, and *Klebsiella oxytoca*	Vat dyes, textile effluents
Exiguobacterium indicum, *Exiguobacterium aurantiacums*, *Bacillus cereus*, and *Acinetobacter baumanii*	Azo dyes effluents
Bacillus spp. ETL-2012, *Pseudomonas aeruginosa*, and *Bacillus pumilus* HKG212	Textile dye (Remazol Black B), sulfonated di-azo dye Reactive Red HE8B, RNB dye
Micrococcus luteus, *Listeria denitrificans*, and *Nocardia atlantica*	Textile azo dyes
Penicillium ochrochloron	Industrial dyes
Pycnoporus sanguineous, *Phanerochaete chrysosporium*, and *Trametes trogii*	Industrial dyes
Myrothecium roridum IM 6482	Industrial dyes
B. subtilis strains NAP1, NAP2, NAP4	Oil-based based paints
Heavy Metals	
Aerococcus sp., *Rhodopseudomonas palustris*	Pb, Cr, Cd
Pseudomonas aeruginosa, *Aeromonas sp.*	U, Cu, Ni, Cr
Bacillus safensis (JX126862) strain (PB-5 and RSA-4)	Cd
Geobacter spp.	Fe(III), U(VI)
Aspergillus versicolor, *A. fumigatus*, *Paecilomyces* sp., *Paecilomyces* sp., *Terichoderma* sp., *Microsporum* sp., *Cladosporium* sp.	Cd
Microbacterium profundi strain Shh49T	Fe
Lysinibacillus sphaericus CBAM5	Co, Cu, Pd, and Cr
Pseudomonas fluorescens and *Pseudomonas aeruginosa*	Fe^{2+}, Zn^{2+}, Pb^{2+}, Mn^{2+}, and Cu^{2+}
Cunninghamella elegans	Heavy metals
Saccharomyces cerevisiae	Heavy metals, lead, mercury, and nickel
Pesticides	
Enterobacter	Chlorpyrifos
Bacillus, *Staphylococcus*	Endosulfan
Pseudomonas putida, *Acinetobacter* sp., *Arthrobacter* sp.	Ridomil MZ 68 MG, Fitoraz WP 76, Decis 2.5 EC, malathion
Enterobacter sp., *Pseudomonas* sp., *Acenetobactor* sp., and *Photobacterium* sp.	Methyl parathion and chlorpyrifos

11.5.2 Fungi

Fungi are promising candidates for bioremediation at various sites due to their wide range of habitats and capacity to secrete a wide variety of enzymes (Deshmukh et al., 2016). Fungi are also beneficial for industrial and other uses such as bioremediation, drug production, and food industry. The majority of research has shown that the white-rot fungi *Pleurotus* sp., *Bjerkandera adjusta*, *Trametes versicolor*, and *Phanerochaete chysosporium* have the ability to bioremediate by virtue of releasing various ligni-nolytic enzymes as such laccases and peroxidases. The ligninolytic enzymes have been used to transform a variety of organic contaminants, including pesticides, from contaminated wastewaters with the use of a biopurification system (BPS) (Rodríguez-Rodríguez et al., 2013). Endophytic fungus can improve the host species capacity for survival, whereas mycorrhizal fungi can improve the intake of nutrients. Moreover, fungi in coastal habitats, especially those capable of breaking down hydrocarbons, play a crucial role. It can be employed in bioremediation to clean up extremely pol-luted regions like oil spills in oceans or river bodies (Yakop et al., 2019). Keratin, the primary constituent of keratinous substrate (nails, hoof, horn, wool, feather, hair, cornified epidermis, etc.), is specifically degraded by keratinolytic fungus. During the breakdown of proteins, these fungi also exhibit lipolytic activity and clear petroleum hydrocarbons from the solution. In the biopile at the refinery, keratinolytic fungi were relatively common, with the geophilic dermatophyte *Trichophyton ajelloi* (teleomorph *Arthroderma uncinatum*) being the main species (Ulfig et al., 2003).

11.5.3 Algae

Pollutants can be removed from the environment using algae or changed into harm-less forms (Table 11.2). Algal bioremediation is highly preferred because the bio-mass it produces is used in biofuel production. Algae increase the biological oxygen

TABLE 11.2
Bioremediation of Organic Pollutants by Various Algae (Chekroun et al., 2014)

Species	Organic Pollutant
Selenastrum capricornutum	Benzene, toluene, naphthalene, phenanthrene, pyrene
Agmenellum quadruplicatum	Bisphenol
Scenedesmus obliquus	GH2 crude-oil degradation
Phytoplankton	Chlorinated hydrocarbons
Scenedesmus quadricauda	Fungicides (dimethomorph and pyrimethanil) and herbicide (isoproturon)
Consortium of *Chlorella sorokiniana* and *Pseudomonas migulae*	Phenanthrene
Chlamydomonas reinhardtii	Herbicide (prometryn)
Pediastrum tetras, Ankistrodesmus fusiformis, Amphora coffeaeformis	Herbicide (mesotrione)
Monoraphidium braunii	Bisphenol
Chlamydomonas reinhardtii	Herbicide (fluroxypyr)

demand (BOD) in contaminated water through photosynthesis and fix carbon dioxide and release oxygen (Singhal et al., 2021). Due to their crucial function in carbon dioxide fixation, microalgae have gained a lot of interest in the bioremediation of coloured wastewater. Additionally, the algal biomass produced has excellent promise as a feedstock for the manufacture of biofuels (Huang et al., 2010).

11.5.4 PLANT

Bioremediation of organic pollutants or heavy metals using plants is known as "Phytoremediation." Phytoremediation is a cheap, environmentally friendly, and effective way to clean up polluted areas, particularly those that include heavy metals. However, at any contaminated site, the amount of metal pollutants in the soil and the capacity of plants to aggressively uptake metals from the soil all impact the efficacy of phytoremediation. It involves several methods like phytoextraction, phytoaccumulation, phytostimulation, phytovolatilisation, and phytostabilisation (Ojuederie and Babalola, 2017). Phytoextraction/phytoaccumulation is based on the mechanism of hyperaccumulation, and phytoextraction entails the uptake and transport of metal contaminants in the soil through plant roots into above-ground components of the plants. Hyperaccumulator plants transport and accumulate metals in organs above ground at concentrations between 100 and 1000 times higher than those found in non-hyperaccumulating species without experiencing any apparent phytotoxic effects. They do this by absorbing large amounts of metals from contaminated soils. Approximately 500 taxa have been recognized as hyperaccumulators of specific metals, meeting certain criteria. Notable members come from various plant families, including *Flacourtiaceae*, *Cunouniaceae*, *Poaceae*, *Cyperaceae*, *Asteraceae*, *Euphorbiaceae*, *Lamiaceae*, and *Fabaceae*, among which *Brassicaceae* is also included. The process of phytostimulation involves raising microbial activity to break down organic pollutants in exudates from plant roots. Low concentrations of ethylene promote root elongation, whereas excessive concentrations prevent DNA synthesis and cell division. This entails using plant roots to absorb contaminants from the soil, hold them in the rhizosphere, and separate and stabilise them, making the pollutants harmless and halting their spread throughout the environment. Phytostabilisation is a superior method of capturing metals in situ because the pollutants are not absorbed into plant tissues and do not spread to the environment. Only the rhizosphere's ability to sequester heavy metals is the main emphasis. Phytovolatilisation is the process of hazardous pollutants being absorbed by plants from the soil, turning into volatile compounds, and then being released into the atmosphere. It is based on the utilisation of particular plants that absorb hazardous pollutants like Hg, Se, and As, convert them into volatile elements with low or no toxicity, and release them into the environment by evapotranspiration through the stomata, leaves, or stems. In contrast to other methods, the key benefit of this one is that it eliminates pollutants from the soil without harvesting the plant (Raklami et al., 2022).

11.5.5 ENGINEERED MICROORGANISMS

Nowadays, genetically engineered organisms are used in the bioremediation process to remove toxins effectively where native bacteria cannot. Genetically modified organisms (GMOs) are useful for cleaning up industrial waste, lowering the toxicity

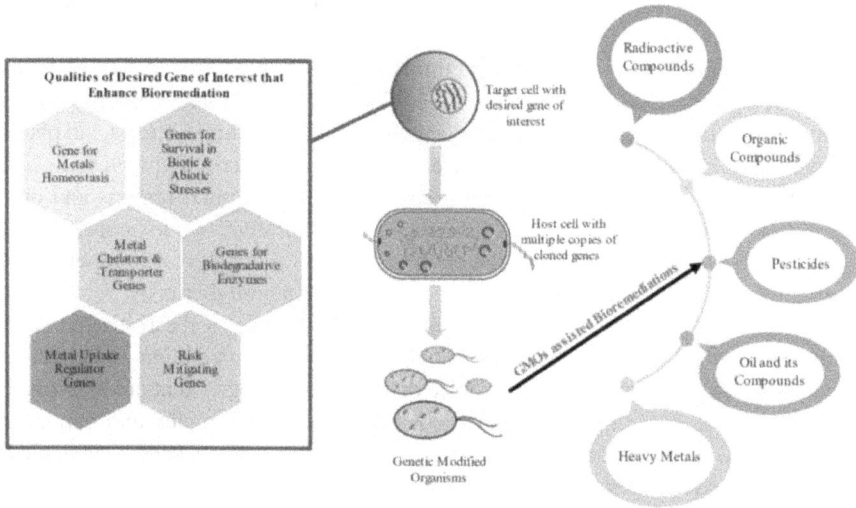

FIGURE 11.8   Role of genetic engineered organisms in bioremediation (modified from Pant et al., 2021; Singh et al., 2011).

of some dangerous substances, and removing pollution from hydrocarbon and gasoline discharges. Role of GMO in pollutant removal is explained in Figure 11.8. For the successful creation of GMOs, a range of molecular methods are available, including molecular cloning, horizontal DNA transfer in bacteria, electroporation, protoplast transformation, biolistic transformation, conjugation, and transformation of competent cells. By introducing a gene into the bacterium, a unique strain is created that can quickly remove hydrocarbon pollutants from the environment (Kumar et al., 2018). In order to create novel strains of microorganisms with desirable characteristics for the bioremediation processes, recent advancements in the field of molecular biology have been used for microorganisms. Despite the fact that GMOs have several benefits for bioremediation, their usage in the environment is still constrained due to the instability of the inserted genetic material. There are two main causes for this: first, the ability of GMOs to transport genetic material in a steady manner is essential to their effectiveness; and second, the transfer of genetic material to native organisms is seen negatively even though it is a common occurrence among native organisms. These elements have encouraged research into the persistence, competition, and survival of GMOs in the environment, as well as the potential dangers associated with their use. This is in addition to the major advancements in development and application that have already been made (Perpetuo et al., 2011).

11.6   ENZYMES FOR THE BIOREMEDIATION OF POLLUTANTS

Environmental pollutants are mostly categorised as inorganic and organic pollutants. Organic pollutants are found mostly in the form of polycyclic aromatic hydrocarbons (PAHs); polychlorinated biphenyls (PCBs); insecticides such as dichlorodiphenyl-trichloroethane (DDT) and aldrin; pesticides like chlordane; and fungicides such as hexachlorobenzene (Elekwachi, 2014). A large number of persistent organic pollutants (POPs) which are extremely harmful are discharged into the milieu as a result of the enhanced urbanisation. Agricultural activities remain one of the primary producers of environmental contaminants, despite the threat posed by POPs. Excessive use of pesticide, herbicide, insecticide, etc. has resulted in environmental pollution. As these contaminants are poorly destroyed by natural bacteria and plants, the pollutants can last for decades in the environment. Furthermore, wind and ocean currents can transport them over vast areas, including the polar regions and open oceans. Because of their long-term residual, bioaccumulative, and toxic nature, pollutants are a source of concern in the environment. As a result, research pertaining to effective pollutant degrading processes has become a global priority.

Traditionally, waste was disposed of in pits but because of the restricted space for dumping, this method is less used. High-temperature incineration and chemical decomposition are considered novel waste disposal technologies that are effective in lowering a wide range of contaminants, but they are both complex and unpopular approaches. Due to the shortcomings of these approaches, efforts have been made to harness the current bioremediation process as a viable alternative. Bioremediation was considered to be the safest and most environmentally friendly method of combating anthropogenic chemicals in ecosystems. In this method, breakdown of pollutants into non-hazardous or less hazardous compounds occurs in presence of microorganisms such as bacteria, fungi, algae, and plants. Remediation of pollutants by the plants is referred to as phytoremediation and genetically modified plants are mostly employed for this purpose (Shankar et al., 2011). For instance, remediation of arsenic from the soil has been reported by the plant *Arabidopsis thaliana* which was incorporated into two bacterial genes. One gene is responsible for modifying arsenate into arsenite, whereas the other binds the arsenite and stores it in the vacuoles (Leung, 2004).

Besides, bioremediation by the microbes mainly depends upon the environmental conditions such as temperature, pH, moisture, oxygen content, presence of toxins, nutrient availability, and soil structure that favour microbial growth. Most of the bacteria belonging to the genus *Achromobacter, Acinetobacter, Bacillus, Pseudomonas, Enterobacter, Burkholderia, Flavobacterium, Alcaligens*, etc. are involved in bioremediation. In particular, bacteria from the genera such as *Alcanivorax, Cycloclasticus, Oleispira, Thallassolituus*, and *Marinobacter* are exclusively known for the degradation of hydrocarbons (Yakimov et al., 2007). Apart from bacteria, fungi belonging to genera *Aspergillus, Rhizopus, Trichoderma, Fusarium, Penicillium, Drechslera, Mucor*, etc. are also well known for degrading aromatic hydrocarbons. Fungi have ability to produce enzymes such as catalases, peroxidases, laccases, etc. that have ability to degrade organic contaminants. Fungi metabolises pollutants by the following three strategies:

Strategy 1: Fungi utilise the target pollutants as carbon and energy source (Mougin et. al. 2012.

Strategy 2: Fungi completely mineralises recalcitrant compounds into CO_2 and other simple compounds. Fungi enzymatically attacks the target compound which leads to minor structural changes that reduce the bioavailability and toxicity of the pollutant (Baker et al., 2019).

Strategy 3: The toxic pollutant is taken up by the fungi and gets concentrated within the organism instead of being metabolised.

11.6.1 Organic Substrates Biodegradation

Enzymes belonging to the classes oxidoreductase and hydrolase, namely, monooxygenase, dioxygenase, laccase, lignin peroxidase, manganese peroxidase, versatile peroxidase, lipase, cellulase, and protease, are mostly involved in bioremediation. Table 11.3 provides a list of enzymes with a focus on their source, specific substrate, the reactions catalysed by them, and their applications (Bhandari et al., 2021; Karigar and Rao, 2011; Mousavi et al., 2021).

TABLE 11.3
Types of Microbial Enzymes Employed in Bioremediation Process and Their Source Functions and Applications (Bhandari et al., 2021; Karigar et al., 2011; Mousavi et al., 2021)

Enzyme	Source	Substrate	Reaction	Application
Oxidoreductase				
Monooxygenase	• *Stenotrophomonas maltophilia* PM102 • *Pseudomonas cepacia* G4	Alkane, aromatic compounds, fatty acid, and steroids	The enzyme is involved in the amalgamation of O_2 into the substrate which will be further used as the reducing agent. Besides, the enzyme is involved in denitrification, ammonification, desulfurization, dehalogenation, and hydroxylation of substrate	• Bioremediation • Protein engineering • Synthetic chemistry, etc.
Dioxygenase	*Pseudomonas* sp.	Aromatic compounds	This enzyme incorporates two O_2 into the substrate resulting in extradiol and intradiol cleavage along with formation of aliphatic product	• Bioremediation • Pharmaceutical industry • Synthetic chemistry, etc.,
Laccase	• *Agaricus bisporus* • *Cerrena unicolor* • *Pyricularia oryzae* • *Pycnoporus cinnabarinus* • *Trametes hispida* • Plants	Ortho and para di phenols, aminophenols, polyphenols, polyamines, lignins, and aryldiamines	• Catechol oxidation • Phenol detoxification • Degradation of azo dyes 2,4-dichlorophenol, benzopyrenes, textile effluent, chlorophenols, urea derivatives, etc. • Decolouration of dyes	• Bioremediation • Synthetic chemistry • Food industry • Paper and pulp industry • Textile and cosmetics industry

(Continued)

TABLE 11.3 *(Continued)*
Types of Microbial Enzymes Employed in Bioremediation Process and Their Source Functions and Applications (Bhandari et al., 2021; Karigar et al., 2011; Mousavi et al., 2021)

Enzyme	Source	Substrate	Reaction	Application
Peroxidase				
Lignin peroxidase	• *Phanerochaete chrysosporium* • *Chrysonilia sitophila*	Halogenated phenolic compounds, polycyclic aromatic compounds, and other aromatic compounds	Oxidation of substrate in the presence of H_2O_2 and mediator such as veratryl alcohol	• Bioremediation • Pharmaceutical industry • Synthetic chemistry, etc. • Food industry • Paper and pulp industry • Textile industry
Manganese peroxidase	*Phanerochaete chrysosporium*	Lignin and other phenolic compounds	In the presence of Mn^{2+} and H_2O_2, the co-substrate catalyses oxidation of Mn^{2+} to Mn^{3+} which results in an Mn^{3+} chelate oxalate, which in turn oxidizes the phenolic substrates	• Bioremediation • Pharmaceutical industry • Synthetic chemistry, etc. • Food industry • Paper and pulp industry • Textile industry
Versatile peroxidase	*Pleurotus eryngii Pleurotus* and *Bjerkandera* sp.	Methoxybenzenes and phenolic aromatic	The enzyme catalyzes the electron transfer from an oxidizable substrate	Bioremediation
Hydrolase				
Cellulase	*Cellulomonas, Cellvibrio, Micrococcus* sp. *Pseudomonas fluorescens, Bacillus subtilus, E. coli*, and *Serratia marscens*	Cellulose	Hydrolyses the substrate to simple carbohydrates	• Paper and pulp industry • Textile industry • Detergent production • Bioremediation
Protease	*Arthrobacter, Streptomyces, Flavobacterium* sp., *Bacillus* sp., *S. cerevisiae, Conidiobolus* spp., *Aspergillus*, and *Neurospora* spp.	Protein	Enzymes that hydrolyse peptide bonds in aqueous environment	• Bioremediation • Leather and laundry
Lipase	*Pseudomonas* sp.	Lipid	The hydrolysis of triacylglycerols to glycerols and free-fatty acids	• Bioremediation • Control of oil spills • Detergent production • Baking industry • Paper and pulp industry • Personal care products

Apart from the above-mentioned enzymes, esterase catalyses the cleavage of the ester bonds in various chemicals such as pesticides, herbicides, polyurethanes, and aliphatic and aromatic polyesters. Microbes such as *Escherichia coli* and *Pichia pastoris* produce thermostable esterase enzyme that acts on phthalate esters (Ufarte et al., 2015). Another group of enzymes that involve in the hydrolyses of herbicides, plastics, and polymers is nitrilases. This enzyme is mostly produced by *Fusarium solani*, *Aspergillus* sp., *Streptomyces* sp., *Pseudomonas fluorescens*, *Bacillus pallidus*, and *Rhodococcus* sp. Cyanide dihydratase, a nitrilase enzyme, degrades cyanide into formate and ammonia whereas cyanide hydratase metabolises cyanide to formamide. These enzymes are also important in the treatment of waste streams from coal kilns and metal plating baths. Another hydrolase enzyme that is produced by *Pseudomonas diminuta* and effective in remediating organophosphorus compounds is organophosphorus hydrolase. This enzyme acts on P-S, P-O, and P-F bonds and is considered as the enzyme with the fastest catalytic rate.

11.6.2 INORGANIC SUBSTRATES BIODEGRADATION

Many microorganisms have developed definite pathways led by various enzymes for resistance against heavy metals. For instance, the inorganic forms of arsenic are arsenite and arsenate which are toxic and capable of causing carcinoma, keratosis, hemolysis, gangrene, neurodegenerative, and heart diseases. Arsenate and arsenite transform into one another in the presence of arsenate reductase and arsenite oxidase through oxidation-reduction reaction (Kapahi and Sachdeva, 2019). As a result, the solubility of the arsenic is enhanced and it easily leaches out from the soil. Various microbial species that are capable of producing the mentioned enzymes include *Acinetobacter* sp., and *Penicillium* sp., *Pseudomonas* sp., *Bacillus* sp., *Rhizobium* sp., *Trichoderma* sp., *Rhizopus* sp., and *Aspergillus* sp. (Sher and Rehman, 2019).

Another metal which is in general present in minute quantity but is enhanced due to industrialisation is lead. The lead toxicity causes anaemia, appetite loss, and neurological, gastrointestinal, and reproductive disorders. A few microorganisms such as *Cupriavidus metallidurans*, *Staphylococcus epidermidis*, *Aspergillus niger*, *Agaricus bisporus*, *Penicillium canescens*, *Rhizopus nigricans*, and *Saccharomyces cerevisiae* are capable of producing enzymes which are capable of degrading and remediating lead (Rahman and Singh, 2020). The heavy metal that is lethal in both forms (inorganic and organic) is mercury (Hg). Because of various humic activities, this metal spreads rapidly in the soil and water. Organic mercury is toxic and causes neurotoxicity, nephrotoxicity, allergies, and inability to speak, whereas the inorganic mercury is less toxic. Organic mercury has affinity for sulfhydryl groups of proteins and accumulates in living organisms. Mercury-resistant bacteria, namely, *Aeromonas*, *Pseudomonas*, *Bacillus*, *Rhodococcus*, *Staphylococcus*, and *Citrobacter* are capable of producing the enzyme mercuric reductase that can convert into less toxic state from highly toxic organic forms of Hg (Pratush et al., 2018). Additionally, the enzyme that plays a crucial role in the remediation of Hg is organomercurial lyase. This enzyme generally cleaves the bonds between carbon, mercury, and various organo-Hg compounds.

In addition to the above-mentioned metals, chromium (Cr) exhibits high toxicity due to its high oxidative potential. Cr is responsible for cell damage and carcinogenic and mutagenic effects in the cell. Bioremediation is possible by reducing hexavalent Cr to trivalent Cr through aerobic or anaerobic pathways. Chromate reductase, Ni-Fe dehydrogenase, cytochrome c3, nitroreductase, iron reductase, flavin reductases, and quinone reductases produced by *Escherichia* sp., *Bacillus* sp., *Pseudomonas* sp., *Enterobacter* sp., *Shewanella* sp., and *Thermus* sp. are involved in remediation of chromium (Chardin et al., 2003). Hence, pollutant bioremediation by using enzymes seems to be a practically feasible, cost-effective, and efficient approach.

11.7   ADVANTAGES AND DISADVANTAGES OF BIOREMEDIATION

In the bioremediation process, the use of nutrients such as fertilizers to remediate pollutants in contaminated matrix makes the process not only eco-friendly but also economical. It is efficient and economical to use, while destroying harmful contaminants and converting them into harmless ones. Bioremediation technology's major drawback is that it can only be applied to biodegradable compounds. Furthermore, researchers have discovered that in some cases the new compound formed after biodegradation is more toxic than the original substance. Finally, for the ex situ technology, the bioremediation process requires excavation and pumping, which can be time-consuming. Pros and cons of the bioremediation process are as follows (Yap and Peng, 2019; Iosob et al., 2016; Kumar et al., 2018):

Advantages
- It provides a natural process where microbes involved remediate pollution from different domains using the energy from sunlight.
- Energy requirement is lesser than any other technologies.
- There is a possibility of full degradation of organic pollutants into different non-toxic substances.
- It requires minimum equipment in comparison to other technologies.
- In comparison to other methods, bioremediation provides low-cost treatment facility per unit volume of soils.
- On-site remediation can be done.
- Public acceptance is more due to natural attenuation of the process.
- Bioremediation process improves soil stability, diminish water percolation, and movement of organic pollutants.
- It can easily be combined with other treatment technologies.

Disadvantages
- In case of uncontrolled conditions, the breakdown of organic matters can be incomplete.
- This process is sensitive in high level of toxicity and condition of the soils.
- In the ex situ method, control of VOCs is not easy.
- Treatment time is higher in comparison to any other remediation process.
- This technology is limited to biodegradable compounds.
- It is not able to remove complete pollutants from the contaminated matrix.
- Sometimes microbial metabolism of pollutants resulted into more toxic chemical formation.

11.8 BIOREMEDIATION APPLICATIONS/CASE STUDIES

Case Study 1

Bacterial strains labelled as H1, H8, and H10 with excellent Hg tolerance and reduction capacity were identified from an industrial drain connected to the Yellow River in Gansu Province. *Pseudomonas cremoricolorata* strain 1541T, *P. caricapapayae* strain ATCC 33615T, and *Brevundimonas intermedia* strain ATCC 15262T were found to be quite related to the isolates. In general, identified bacterial species largely reduced Hg^{2+} via the mer operon, and high efficiency was noticed by the consortium at 1:1:1 ratio. The levels of Hg^{2+} in the industrial sludge returned to the national norm (Hg^{2+} 0.05 mg/L) after 48 h of treatment. Bacteria have demonstrated tolerance to and transformative abilities for heavy metals such as Pb^{2+}, Cr^{6+}, As^{5+} and Cd^{2+}, suggesting their potential application in more complex environments contaminated with these heavy metals. (Zhao et al., 2021).

Case Study 2

Khazaal and Ismail (2021) reported potential performance of *Pseudomonas aeruginosa* and *Staphylococcus* spp. in removing of chemical oxygen demand (COD) and TPH during anaerobic biotreatment of real field petroleum refinery oily sludge. Within 14 days, COD and TPH had extreme elimination efficiencies of 96.7% and 90%, respectively, which is considered the most effective procedure within a limited period.

Case Study 3

Pycnoporus sanguineus, a white-rot fungus, was found to biodegrade 5 mg/L triphenyl phosphate (TPhP) with 62.84% efficacy at pH 6, resulting in a significant reduction in toxicity. TPhP is one of the aryl organophosphorus flame retardants, widely employed in a wide range of industries like plastics, electrical products, construction materials, textiles, paints, and so on. Because TPhP is physically introduced to the end product but not strongly bonded to the polymer matrix, it can readily escape into the environment. This is especially concerning for the general people because TPhP has been linked to toxicity against the nervous system, cardiac system, and endocrine disruption in animals. Interaction amid *P. sanguineus* and various potential indigenous degraders, such as *Sphingobium*, *Burkholderia*, *Mycobacterium*, and *Methylobacterium*, might effectively accelerate TPhP biodegradation in the sediment water bodies which can be considered as a potential bioremediation agent (Feng et al., 2021).

Case Study 4

To remediate chromium (Cr(VI)) and cadmium (Cd(II)), *Aspergillus fumigatus*, *A. flavus*, and *A. fumigatus* with high metal tolerance capability were tested separately and as a consortium. These fungi were isolated from polluted sites, and subsequent screening revealed that they removed more than 70% of Cr(VI) from the liquid medium with the fungi *A. flavus* and *A. fumigatus*. *A. fumigatus*, on the other hand, eliminated 74% of the Cd(II). The reduction of heavy metals in effluents discharged from diverse manufacturing units was enhanced through the use of a microbial consortium. This consortium consisted of 1 mL, containing 1.2×10^6 spores per mL of each member.(Talukdar et al., 2020).

Case Study 5

Yang et al. (2021) found that activated algal-bacterial aerobic granular sludge (AGS) was capable of bioremediation of 85.1% of Cr(VI) within 6 h. With an external electron donor (glucose) supply at pH 6 and 37°C, its efficacy increased to 93.8%. However, despite the availability of external electron donors, the inactivation of AGS decreased Cr removal efficiency to 29.6%, which can be considered a major pitfall in this model. Chemical fractionation revealed that 90.5% of the loaded Cr on algal-bacterial AGS was immobile, indicating that Cr-loaded algal-bacterial AGS has a low environmental risk after biosorption of toxic metals.

Case Study 6

Subashchandrabose et al. (2019) reported that *Chlorella* sp. MM3 and *Rhodococcus wratislaviensis* strain 9 are accomplished by debasing high-molecular-weight PAHs. A study conducted by his team revealed that algal-bacterial system degraded these PAHs in a mud spiked with 50 mg/L phenanthrene, 10 mg/L of pyrene, and 10 mg/L of benzo[*a*]pyrene within 30 days. Furthermore, a residual toxicity test using *E. coli* DH5 (which expresses green fluorescent protein) revealed that PAH-contaminated soil was successfully bioremediated in the slurry phase.

Case Study 7

A system with two symbiotic organisms rhizospheric *Azotobacter* and *Lepidium sativum* (herb) was developed by Sobariu and his team (2017) for phytoremediation of Cr(VI) and Cd(II). Even though the study was confined to laboratory-level experimentation, *Azotobacter* sp. not only stimulated the sprouting efficiency of *L. sativum* by almost 7% in Cd(II) solutions but also reduced tolerance index (TI %) indicating the tolerance of the plant towards cadmium and chromium toxicity.

Case Study 8

Dashti et al. (2019) used agri-based media to cultivate hydrocarbonoclastic, diazotrophic, and heavy metal-resistant bacteria for bioremediation of heavy metal. Following an 8-month inoculation with microorganism-loaded sorbents in nitrogen-free media, approximately 35.3% of the oil was lost. After an 8-month incubation period, the wheat straw, corncobs, and sugarcane bagasse samples absorbed 1.9, 1.1, and 2.5 g oil samples, respectively, and lost 37.8%. Various nitrogenase-coding nifH genes were discovered in total genomic DNA from culture and sorbents. Hg^{2+}, Cd^{2+}, Pb^{2+}, AsO^{43}, and AsO^{33} were all tolerated by pure hydrocarbonoclastic microbial isolates. Even in the presence of oil, several of those isolates thrived well with up to 1000 ppm of Pb^{2+} and 36,000 ppm of AsO_4^3. The examined strains lowered the toxicity of the heavy metals Hg^{2+}, Cd^{2+}, and Pb^{2+} against the hydrocarbon degraders by removing them from the medium. Plant-based sorbents were found to physically remove oil and bioremediate spilled oils with the support of host microbial communities.

Case Study 9

Lindane is a hazardous and persistent organochlorine insecticide that has accumulated in the nature. Raimondo et al. (2020) attempted to remove it using an actinobacteria consortium after biostimulating with sugarcane filter cake. After 14 days of the assay, bioaugmentation of biostimulated soils increased pesticide removal in silty loam (61.4%), clayey soils (70.8%), and sandy soils (86.3%). Hence, in conclusion, the simultaneous application of bioaugmentation with the actinobacteria consortium is a promising technique for regenerating soils polluted with lindane.

Case Study 10

Biosurfactants fits in as a perfect candidate for cleaning of oil spills with a low environmental impact. Rita de Cássia et al. (2021) developed an efficient biosurfactant from *Pseudomonas cepacia* CCT6659 using industrial waste, which can be further extended to remove hydrophobic contaminants from soil and water. The developed tensoactive removed 76.55% of petrol derivatives from soil and 84.5% from sea stones. As a result, the synthesised biosurfactant offers ideal conditions for use as a dispersion agent in the decontamination of oil-saturated terrestrial and aquatic ecosystems in place of chemical and poisonous chemicals.

11.9 CONCLUSION AND FUTURE SCOPE

Bioremediation is a natural phenomenon employed to remediate environmental pollutants without any adverse effects. Different biological agent such as bacteria, fungi, algae, plant, and animals actively participate in the bioremediation process, thus making it eco-friendly, low cost, and safer than any other process. Through different mechanisms such as biosorption, precipitation, detoxification, enzymatic bioconversion, and decomposition, these agents efficiently remove pollutants from the contaminated on-sites or off-sites. Since conventional bioremediation technologies have noticeable limitations, bioremediation is currently regarded as a sustainable, economical, and reliable pollution control method. Further scientific experiments and advancements will enable it to be implemented sustainably and perfectly. An overview of different enzyme-mediated bioremediation approaches for soil and groundwater has been provided. Synthetic biology, rational enzyme design, directed enzyme evolution, and artificial intelligence/machine learning can constitute future directions for enzyme-mediated bioremediation.

Recent breakthroughs in bioremediation technology have brought a dramatic change to current decades. Recent developments in bioremediation have allowed the solution of a wide range of environmental pollution problems. However, even though bioremediation has several advantages, it is not without its disadvantages as well; long time taking, limited degradation of contaminants, difficulty for bench- and pilot-scale studies to be extrapolated as well as absence of potential biological agents, and a lack of adequate knowledge are the main setbacks. Improvements in microbial genomics studies will pave the way for a range of powerful bioremediation strategies for contaminated soils and waters. Although the concept of engineered ecosystems based on biological tools is in its infancy, it has substantial potential for a sustainable world free from pollutants.

Using bioremediation, it is possible to restore contaminated environments in an inexpensive and efficient manner. Various environmental factors play a role in affecting the rate and extent of biodegradation, but they are not fully understood. In many field tests, the design, control, or analysis has been inadequate, leading to uncertainty when choosing response options. Thus, there is a need for future field studies to adopt legitimate scientific approaches and to acquire high-quality data.

There are also a variety of microbes with detoxification capabilities that remain largely undiscovered. Microbes' natural role in the environment and a lack of knowledge regarding their impact on the environment could make their use unacceptably risky. To gain a better understanding of potential oil degraders and to understand their genetics and biochemistry, it is essential to understand the diversity of microbial communities living in petroleum-contaminated environments. This will result in the development of bioremediation strategies that will preserve the long-term sustainability of natural terrestrial and marine ecosystems.

There may be ways to design bioreactors or products that are more efficient and feasible. Moreover, these systems might be able to remove pollutants completely from the environment. In addition, useful compounds are produced as by-products.

REFERENCES

Abatenh, E., Gizaw, B., Tsegaye, Z., & Wassie, M. (2017). The role of microorganisms in bioremediation-a review. *Open Journal of Environmental Biology*, 2(1), 038–046.

Ahmadnezhad, Z., Vaezihir, A., Schüth, C., & Zarrini, G. (2021). Combination of zeolite barrier and bio sparging techniques to enhance efficiency of organic hydrocarbon remediation in a model of shallow groundwater. *Chemosphere*, 273, 128555. https://doi.org/10.1016/j.chemosphere.2020.128555

Al-Mansoory, A. F., Idris, M., Abdullah, S. R. S., & Anuar, N. (2017). Phytoremediation of contaminated soils containing gasoline using Ludwigia octovalvis (Jacq.) in greenhouse pots. *Environmental Science and Pollution Research International*, 24(13), 11998–12008. https://doi.org/10.1007/s11356-015-5261-5

Azubuike, C. C., Chikere, C. B., & Okpokwasili, G. C. (2016). Bioremediation techniques–classification based on site of application: Principles, advantages, limitations and prospects. *World Journal of Microbiology and Biotechnology*, 32(11), 1–18.

Baker, P., Tiroumalechetty, A., & Mohan, R. (2019). Fungal enzymes for bioremediation of xenobiotic compounds. In A. Yadav, S. Singh, S. Mishra, & A. Gupta (Eds), *Recent Advancement in White Biotechnology Through Fungi: Volume 3: Perspective for Sustainable Environments* (pp. 463–489). Springer.

Baggott, J. (1993). Biodegradable lubricants. A paper presented at the Institute of Petroleum Symposium "Life cycle analysis and eco-assessment in the oil industry," November 1992, Shell, England.

Bhandari, S., Poudel, D.es K., Marahatha, R., Dawadi, S., Khadayat, K., Phuyal, S., Shrestha, S. et al. (2021). Microbial enzymes used in bioremediation. *Journal of Chemistry*. https://doi.org/10.1155/2021/8849512

Carniato, L., Schoups, G., Seuntjens, P., Van Nooten, T., Simons, Q., & Bastiaens, L. (2012). Predicting longevity of iron permeable reactive barriers using multiple iron deactivation models. *Journal of Contaminant Hydrology*, 142–143, 93–108. https://doi.org/10.1016/j.jconhyd.2012.08.012

Chardin, B., Giudici-Orticoni, M.-T., De Luca, G., Guigliarelli, B., & Bruschi, M. (2003). Hydrogenases in sulfate reducing bacteria function as chromium reductase. *Applied Microbiology and Biotechnology*, 63(3), 315–321.

Chekroun, K. B., Sánchez, E., & Baghour, M. (2014). The role of algae in bioremediation of organic pollutants. *International Research Journal of Public and Environmental Health*, 1(2), 19–32.

Creegan, E. F., Flynn, R., Torell, G., Brewer, C. E., VanLeeuwen, D., Acharya, R. N., Heerema, R. J., & Darapuneni, M. (2022). Pecan (Carya illinoinensis) and dairy waste stream utilization: Properties and economics of on-farm windrow systems. *Sustainability*, 14(5), 2550. https://doi.org/10.3390/su14052550

Dashti, N., Ali, N., Khanafer, M., & Radwan, S. S. (2019). Plant-based oil-sorbents harbor native microbial communities effective in spilled oil-bioremediation under nitrogen starvation and heavy metal-stresses. *Ecotoxicology and Environmental Safety*, 181, 78–88.

Deshmukh, R., Khardenavis, A. A., & Purohit, H. J. (2016). Diverse metabolic capacities of fungi for bioremediation. *Indian Journal of Microbiology*, 56(3), 247–264.

Dias, R. L., Ruberto, L., Calabró, A., Balbo, A. L., Del Panno, M. T., & Mac Cormack, W. P. (2015). Hydrocarbon removal and bacterial community structure in on-site biostimulated biopile systems designed for bioremediation of diesel-contaminated Antarctic soil. *Polar Biology*, 38, 677–687.

Dixit, R., Malaviya, D., Pandiyan, K., Singh, U. B., Sahu, A., Shukla, R., Singh, B. P., Rai, J. P., Sharma, P. K., & Lade, H. (2015). Bioremediation of heavy metals from soil and aquatic environment: An overview of principles and criteria of fundamental processes. *Sustainability*, 7, 2189–2212.

Elekwachi, O. (2014). Global use of bioremediation technologies for decontamination of ecosystems. *Journal of Bioremediation & Biodegradation*, 5(4), 1–9.

Feng, M., Zhou, J., Yu, X., Wang, H., Guo, Y., & Mao, W. (2021). Bioremediation of triphenyl phosphate by *Pycnoporus sanguineus*: Metabolic pathway, proteomic mechanism and biotoxicity assessment. *Journal of Hazardous Materials*, 417, 125983.

Fujita, Y., Taylor, J. L., Wendt, L. M., Reed, D. W., & Smith, R. W. (2010). Evaluating the potential of native ureolytic microbes to remediate a 90Sr contaminated environment. *Environmental Science & Technology*, 44(19), 7652–7658.

Gangola, S., Joshi, S., Kumar, S., & Pandey, S. C. (2019). Comparative Analysis of Fungal and Bacterial Enzymes in Biodegradation of Xenobiotic Compounds. In *Smart Bioremediation Technologies* (pp. 169–189). Academic Press.

Gomez, F., & Sartaj, M. (2014). Optimization of field scale biopiles for bioremediation of petroleum hydrocarbon contaminated soil at low temperature conditions by response surface methodology (RSM). *International Biodeterioration & Biodegradation*, 89, 103–109.

Haripriyan, U., Gopinath, K. P., Arun, J., & Govarthanan, M. (2022). Bioremediation of organic pollutants: A mini review on current and critical strategies for wastewater treatment. *Archives of Microbiology*, 204(5), 286. https://doi.org/10.1007/s00203-022-02907-9

Hoff, R. Z. (1993). Bioremediation: An overview of its development and use for oil spill cleanup. *Marine Pollution Bulletin*, 26(9), 476–481.

Huang, G., Chen, F., Wei, D., Zhang, X., & Chen, G. (2010). Biodiesel production by microalgal biotechnology. *Applied Energy*, 87(1), 38–46.

Iosob, G. A., Prisecaru, M., Stoica, I., Călin, M., & Cristea, T. O. (2016). Biological remediation of soil polluted with oil products: An overview of available technologies. *Universitatea" Vasile Alecsandri" din Bacău*, 25(2), 89–101.

Jain, S., & Arnepalli, D. N. (2019). Biominerlisation as a Remediation Technique: A Critical Review. In V. K. Stalin, & M. Muttharam (Eds.), *Geotechnical Characterisation and Geoenvironmental Engineering* (pp. 155–162), Lecture Notes in Civil Engineering 16. Springer Nature Singapore Pte Ltd. https://doi.org/10.1007/978-981-13-0899-4_19

Jin, Y., Luan, Y., Ning, Y., & Wang, L. (2018). Effects and mechanisms of microbial remediation of heavy metals in soil: A critical review. *Applied Sciences*, 8, 1336. https://doi.org/10.3390/app8081336

Kapahi, M., & Sachdeva, S. (2019). Bioremediation options for heavy metal pollution. *Journal of Health and Pollution*, 9(24), 191203.

Karigar, C. S., & Rao, S. S. (2011). Role of microbial enzymes in the bioremediation of pollutants: A review. *Enzyme Research*, 2011, 805187. https://doi.org/10.4061/2011/805187

Khazaal, R. M., & Ismail, Z. Z. (2021). Bioremediation and detoxification of real refinery oily sludge using mixed bacterial cells. *Petroleum Research*, 6(3), 303–308.

Kim, S., Krajmalnik-Brown, R., Kim, J.-O., & Chung, J. (2014). Remediation of petroleum hydrocarbon-contaminated sites by DNA diagnosis-based bioslurping technology. *Science of the Total Environment*, 497–498, 250–259. https://doi.org/10.1016/j.scitotenv.2014.08.002

Kumar, A., Chanderman, A., Makolomakwa, M., Perumal, K., & Singh, S. (2016). Microbial production of phytases for combating environmental phosphate pollution and other diverse applications. *Critical Reviews in Environmental Science and Technology*, 46(6), 556–591.

Kumar, N. M., Muthukumaran, C., Sharmila, G., & Gurunathan, B. (2018). Genetically Modified Organisms and Its Impact on the Enhancement of Bioremediation. In *Bioremediation: Applications for Environmental Protection and Management* (pp. 53–76). Springer.

Kumar, V., Shahi, S. K., & Singh, S. (2018). Bioremediation: An Eco-Sustainable Approach for Restoration of Contaminated Sites. In *Microbial Bioprospecting for Sustainable Development* (pp. 115–136). Springer.

Kumar, A., Subrahmanyam, G., Mondal, R., Cabral-Pinto, M. M. S., Shabnam, A. A., Jigyasu, D. K., & Yu, Z. G. (2021). Bio-remediation approaches for alleviation of cadmium contamination in natural resources. *Chemosphere*, 268, 128855.

Kundu, D., Dutta, D., Mondal, S., Haque, S., Bhakta, J. N., & Jana, B. B. (2020). Application of Potential Biological Agents in Green Bioremediation Technology: Case Studies. In *Waste Management: Concepts, Methodologies, Tools, and Applications* (pp. 1192–1216). IGI Global.

Leung, M. (2004). Bioremediation: Techniques for cleaning up a mess. *Journal of Biotechnology*, 2, 18–22.

Megharaj, M., & Naidu, R. (2017). Soil and brownfield bioremediation. *Microbial Biotechnology*, 10(5), 1244–1249.

Mentzer, E., & Ebere, D. (1996). Remediation of hydrocarbon contaminated sites. A paper presented at 8th Biennial International Seminar on the Petroleum Industry and the Nigerian Environment, November, Port Harcourt.

Michel, F., O'Neill, T., Rynk, R., Gilbert, J., Wisbaum, S., & Halbach, T. (2022). Chapter 5—Passively Aerated Composting Methods, Including Turned Windrows. In R. Rynk (Ed.), *The Composting Handbook* (pp. 159–196). Academic Press. https://doi.org/10.1016/B978-0-323-85602-7.00002-9

Mishra, M., & Thakur, I. S. (2012). Bioremediation, Bioconversion and Detoxification of Organic Compounds in Pulp and Paper Mill Effluent for Environmental Waste Management. In T. Satyanarayana et al. (Eds.), *Microorganisms in Environmental Management: Microbes and Environment* (pp. 263–287). Springer Science+Business Media B.V. https://doi.org/10.1007/978-94-007-2229-3_13

Mougin, C., Cheviron, N., Pinheiro, M., Lebrun, J. D., & Boukcim, H. (2012). New insights into the use of filamentous fungi and their degradative enzymes as tools for assessing the ecotoxicity of contaminated soils during bioremediation processes. *Fungi as Bioremediators*, 32, 419–432. https://doi.org/10.1007/978-3-642-33811-3_18

Mousavi, S. M., Hashemi, S. A., Iman Moezzi, S. M., Ravan, N., Gholami, A., Lai, C. W., Chiang, W. H., Omidifar, N., Yousefi, K., & Behbudi, G. (2021). Recent advances in enzymes for the bioremediation of pollutants. *Biochemistry Research International*, 2021. https://doi.org/10.1155/2021/5599204

Obiri-Nyarko, F., Grajales-Mesa, S. J., & Malina, G. (2014). An overview of permeable reactive barriers for in situ sustainable groundwater remediation. *Chemosphere*, 111, 243–259. https://doi.org/10.1016/j.chemosphere.2014.03.112

Ojuederie, O. B., & Babalola, O. O. (2017). Microbial and plant-assisted bioremediation of heavy metal polluted environments: A review. *International Journal of Environmental Research and Public Health*, 14(12), 1504.

Pant, G., Garlapati, D., Agrawal, U., Prasuna, R. G., Mathimani, T., & Pugazhendhi, A. (2021). Biological approaches practised using genetically engineered microbes for a sustainable environment: A review. *Journal of Hazardous Materials*, 405, 124631.

Park, J. H., Lamb, D., Paneerselvam, P., Choppala, G., Bolan, N., & Chung, J. W. (2011). Role of organic amendments on enhanced bioremediation of heavy metal(loid) contaminated soils. *Journal of Hazardous Materials*, 185(2–3), 549–574.

Perpetuo, E. A., Souza, C. B., & Nascimento, C. A. O. (2011). *Engineering Bacteria for Bioremediation*. IntechOpen.

Pörtner, R., Nagel-Heyer, S., Goepfert, C., Adamietz, P., & Meenen, N. M. (2005). Bioreactor design for tissue engineering. *Journal of Bioscience and Bioengineering*, 100(3), 235–245. https://doi.org/10.1263/jbb.100.235

Pratush, A., Kumar, A., & Hu, Z. (2018). Adverse effect of heavy metals (As, Pb, Hg, and Cr) on health and their bioremediation strategies: A review. *International Microbiology*, 21(3), 97–106.

Rahman, Z., & Singh, V. P. (2020). Bioremediation of toxic heavy metals (THMs) contaminated sites: Concepts, applications and challenges. *Environmental Science and Pollution Research*, 27(22), 27563–27581.

Raimondo, E. E., Aparicio, J. D., Bigliardo, A. L., Fuentes, M. S., & Benimeli, C. S. (2020). Enhanced bioremediation of lindane-contaminated soils through microbial bioaugmentation assisted by biostimulation with sugarcane filter cake. *Ecotoxicology and Environmental Safety*, 190, 110143.

Raklami, A., Meddich, A., Oufdou, K., & Baslam, M. (2022). Plants—Microorganisms-based bioremediation for heavy metal cleanup: Recent developments, phytoremediation techniques, regulation mechanisms, and molecular responses. *International Journal of Molecular Sciences*, 23(9), 5031.

Rita de Cássia, F., Luna, J. M., Rufino, R. D., & Sarubbo, L. A. (2021). Ecotoxicity of the formulated biosurfactant from Pseudomonas cepacia CCT 6659 and application in the bioremediation of terrestrial and aquatic environments impacted by oil spills. *Process Safety and Environmental Protection*, 154, 338–347.

Rodríguez-Rodríguez, C. E., Castro-Gutiérrez, V., Chin-Pampillo, J. S., & Ruiz-Hidalgo, K. (2013). On-farm biopurification systems: Role of white rot fungi in depuration of pesticide-containing wastewaters. *FEMS Microbiology Letters*, 345(1), 1–12.

Roeselers, G., & Van Loosdrecht, M. (2010). Microbial phytase-induced calcium-phosphate precipitation—a potential soil stabilization method. *Folia Microbiologica*, 55, 621–624.

Samanta, P., Ghosh, A. R., & Jung, J. (2020). Microbe-Mediated Remediation of Heavy Metal Contamination. In T. Saha, & B. K. Tiwary (Eds.), *Microbes, Environment and Human Welfare* (pp. 27–64). NOVA Science Publishers, Inc.

Saraswat, R., Saraswat, D., & Yadav, M. (2020). A review on bioremediation of heavy metals by microbes. *International Journal of Advanced Research*, 8(07), 200–210.

Sardrood, B. P., Goltapeh, E. M., & Varma, A. (2013). An Introduction to Bioremediation. In E. M. Goltapeh et al. (Eds.), *Fungi as Bioremediators* (pp. 1–27). Springer-Verlag. https://doi.org/10.1007/978-3-642-33811-3_1

Schwitzguébel, J.-P. (2017). Phytoremediation of soils contaminated by organic compounds: Hype, hope and facts. *Journal of Soils and Sediments*, 17(5), 1492–1502. https://doi.org/10.1007/s11368-015-1253-9

Shan, B., Hao, R., Xu, H., Li, J., Li, Y., Xu, X., & Zhang, J. (2021). A review on mechanism of biomineralization using microbial-induced precipitation for immobilizing lead ions. *Environmental Science and Pollution Research*. https://doi.org/10.1007/s11356-021-14045-8

Shankar, K. S., Devi, C. A., & Rao, C. S. V. (2011). Phytoremediation of heavy metals with Amaranthus dubius in semi-arid soils of Patancheru, Andhra Pradesh. *Indian Journal of Dryland Agricultural Research and Development*, 26(2), 71–76.

Shannon, J. M., Hauser, L. W., Liu, X., Parkin, G. F., Mattes, T. E., & Just, C. L. (2014). Partial nitritation ANAMMOX in submerged attached growth bioreactors with smart aeration at 20°C. *Environmental Science: Processes & Impacts*, 17(1), 81–89. https://doi.org/10.1039/C4EM00481G

Sharma, I. (2020). Bioremediation Techniques for Polluted Environment: Concept, Advantages, Limitations, and Prospects. In *Trace Metals in the Environment – New Approaches and Recent Advances*. IntechOpen. https://doi.org/10.5772/intechopen.90453

Sher, S., & Rehman, A. (2019). Use of heavy metals resistant bacteria—a strategy for arsenic bioremediation. *Applied Microbiology and Biotechnology*, 103(15), 6007–6021.

Silva-Castro, G. A., Uad, I., Gónzalez-López, J., Fandiño, C. G., Toledo, F. L., & Calvo, C. (2012). Application of selected microbial consortia combined with inorganic and oleophilic fertilizers to recuperate oil-polluted soil using land farming technology. *Clean Technologies and Environmental Policy*, 14(4), 719–726. https://doi.org/10.1007/s10098-011-0439-0

Singh, J. S., Abhilash, P. C., Singh, H. B., Singh, R. P., & Singh, D. P. (2011). Genetically engineered bacteria: An emerging tool for environmental remediation and future research perspectives. *Gene*, 480(1–2), 1–9.

Singh, J. S., Abhilash, P. C., Singh, H. B., Singh, R. P., & Singh, D. P. (2011). Genetically engineered bacteria: an emerging tool for environmental remediation and future research perspectives. *Gene*, 480(1-2), 1–9.

Singh, R., & Ram, K. (2022). Environmental Remediation Technologies. In V. P. Singh, S. Yadav, K. K. Yadav, & R. N. Yadava (Eds.), *Environmental Degradation: Challenges and Strategies for Mitigation* (pp. 211–225). Springer International Publishing. https://doi.org/10.1007/978-3-030-95542-7_10

Singhal, M., Jadhav, S., Sonone, S. S., Sankhla, M. S., & Kumar, R. (2021). Microalgae based sustainable bioremediation of water contaminated by pesticides. *Biointerface Research in Applied Chemistry*, 12, 149–169.

Smith, E., Thavamani, P., Ramadass, K., Naidu, R., Srivastava, P., & Megharaj, M. (2015). Remediation trials for hydrocarbon-contaminated soils in arid environments: Evaluation of bioslurry and biopiling techniques. *International Biodeterioration & Biodegradation*, 101, 56–65.

Sobariu, D. L., Fertu, D. I. T., Diaconu, M., Pavel, L. V., Hlihor, R. M., Drăgoi, E. N., Curteanu, S., Lenz, M., Corvini, P. F. X., & Gavrilescu, M. (2017). Rhizobacteria and plant symbiosis in heavy metal uptake and its implications for soil bioremediation. *New Biotechnology*, 39, 125–134.

Somu, P., & Paul, S. (2022). Chapter 11—Bioaugmentation as a Strategy for the Removal of Emerging Pollutants from Wastewater. In I. Haq, A. S. Kalamdhad, & M. P. Shah (Eds.), *Biodegradation and Detoxification of Micropollutants in Industrial Wastewater* (pp. 211–230). Elsevier. https://doi.org/10.1016/B978-0-323-88507-2.00007-5

Subashchandrabose, S. R., Venkateswarlu, K., Venkidusamy, K., Palanisami, T., Naidu, R., & Megharaj, M. (2019). Bioremediation of soil long-term contaminated with PAHs by algal–bacterial synergy of *Chlorella* sp. MM3 and *Rhodococcus wratislaviensis* strain 9 in slurry phase. *Science of the Total Environment*, 659, 724–731.

Talukdar, D., Jasrotia, T., Sharma, R., Jaglan, S., Kumar, R., Vats, R., Kumar, R., Mahnashi, M. H., & Umar, A. (2020). Evaluation of novel indigenous fungal consortium for enhanced bioremediation of heavy metals from contaminated sites. *Environmental Technology & Innovation*, 20, 101050.

Tegene, B. G., & Tenkegna, T. A. (2020) Mode of action, mechanism and role of microbes in bioremediation service for environmental pollution management. *Journal of Biotechnology & Bioinformatics Research*. SRC/JBBR-112. https://doi.org/10.47363/JBBR/2020(2)118

Uday, U. S. P., Bandyopadhyay, T. K., & Bhunia, B. (2016). Bioremediation and Detoxification Technology for Treatment of Dye(s) from Textile Effluent. In: *Textile Wastewater Treatment* (pp. 75–92). IntechOpen. https://doi.org/10.5772/62309

Ufarte, L., Laville, E., Duquesne, S., & Potocki-Veronese, G. (2015). Metagenomics for the discovery of pollutant degrading enzymes. *Biotechnology Advances*, 33(8), 1845–1854.

Ulfig, K., Plaza, G., Worsztynowicz, A., Manko, T., Tien, A. J., & Brigmon, R. L. (2003). Keratinolytic fungi as indicators of hydrocarbon contamination and bioremediation progress in a petroleum refinery. *Polish Journal of Environmental Studies*, 12(2). 245–250.

Verma, J. P., & Jaiswal, D. K. (2016). Book review: Advances in biodegradation and bioremediation of industrial waste. *Frontiers in Microbiology*, 6, 1555.

Verma, J. P., & Jaiswal, D. K. (2016). Book review: Advances in biodegradation and bioremediation of industrial waste. *Frontiers in Microbiology*, 6, 1–2. https://doi.org/10.3389/fmicb.2015.01555

Verma, S., & Kuila, A. (2019). Bioremediation of heavy metals by microbial process. *Environmental Technology & Innovation*, 14, 100369.

Wang, J., & Chen, C. (2009). Biosorbents for heavy metals removal and their future. *Biotechnological Advances*, 27, 195–226.

Wang, N., Wu, W., Pan, J., & Long, M. (2019). Detoxification strategies for zearalenone using microorganisms: A review. *Microorganisms*, 7(7), 208. https://doi.org/10.3390/microorganisms7070208

Wang, Z., Xu, Y., Zhao, J., Li, F., Gao, D., & Xing, B. (2011). Remediation of petroleum contaminated soils through composting and rhizosphere degradation. *Journal of Hazardous Materials*, 190(1), 677–685. https://doi.org/10.1016/j.jhazmat.2011.03.103

Whelan, M. J., Coulon, F., Hince, G., Rayner, J., McWatters, R., Spedding, T., & Snape, I. (2015). Fate and transport of petroleum hydrocarbons in engineered biopiles in polar regions. *Chemosphere*, 131, 232–240.

Wu, T.-W., Lee, J.-W., Liu, H.-Y., Lin, W.-H., Chu, C.-Y., Lin, S.-L., Chang-Chien, G. P., & Yu, C. (2014). Accumulation and elimination of polychlorinated dibenzo-p-dioxins and dibenzofurans in mule ducks. *Science of the Total Environment*, 497–498, 260–266. https://doi.org/10.1016/j.scitotenv.2014.07.106

Xiao, P., Mori, T., Kamei, I., & Kondo, R. (2010). Metabolism of organochlorine pesticide heptachlor and its metabolite heptachlor epoxide by white-rot fungi, belonging to genus Phlebia. *FEMS Microbiology Letters*, 314, 140–146.

Xing, Z., Su, X., Zhang, X., Zhang, L., & Zhao, T. (2022). Direct aerobic oxidation (DAO) of chlorinated aliphatic hydrocarbons: A review of key DAO bacteria, biometabolic pathways and in-situ bioremediation potential. *Environment International*, 162, 107165. https://doi.org/10.1016/j.envint.2022.107165

Xu, X. Y., Hao, R. X., Xu, H., & Lu, A. H. (2020) Removal mechanism of Pb(II) by Penicillium polonicum: Immobilization, adsorption, and bioaccumulation. *Scientific Reports*, 10(1), 9079. https://doi.org/10.1038/s41598-020-66025-6

Yakimov, M. M., Timmis, K. N., & Golyshin, P. N. (2007). Obligate oil-degrading marine bacteria. *Current Opinion in Biotechnology*, 18, 257–266.

Yakop, F., Taha, H., & Shivanand, P. (2019). Isolation of fungi from various habitats and their possible bioremediation. *Current Science*, 116(5), 733.

Yang, X., Zhao, Z., Van Nguyen, B., Hirayama, S., Tian, C., Lei, Z., Shimizu, K., & Zhang, Z. (2021). Cr(VI) bioremediation by active algal-bacterial aerobic granular sludge: Importance of microbial viability, contribution of microalgae and fractionation of loaded Cr. *Journal of Hazardous Materials*, 418, 126342.

Yap, C. K., & Peng, S. H. T. (2019). Cleaning contaminated soils by using microbial remediation: A review and challenges to the weaknesses. *American Journal of Biomedical Science and Research*, 2(3), 126–128.

Zhang, K. J., Xue, Y. W., Xu, H. H., & Yao, Y. N. (2019). Lead removal by phosphate solubilizing bacteria isolated from soil through biomineralization. *Chemosphere*, 224, 272–279.

Zhao, M. M., Kou, J. B., Chen, Y. P., Xue, L. G., Fan, T. T., & Wang, S. M. (2021). Bioremediation of wastewater containing mercury using three newly isolated bacterial strains. *Journal of Cleaner Production*, 299, 126869.

12 Bioremediation of Phthalate Esters Contaminated Soil with Augmentation Technique in Bioslurry Reactor

*Maddala Rama Krishna, Shailaja Srinivasan,
and Gangagni Rao Anupoju*

12.1 INTRODUCTION

The planet earth with its beautiful environment is probably the only place of dwelling for the living world. Environment in simple terms consists of hydrosphere, lithosphere, stratosphere, and atmosphere. In the prehistorical periods, the earth and its surroundings were very clean and beautiful, but as the civilization started developing the environmental conditions started deteriorating and several contaminants accumulated (More, 1986). Pollution has spread in the atmosphere, namely, air, water, and soil, and the changes may be temporary or permanent based on the extent to which pollution has taken place. A contaminant is anything added to the environment and a pollutant is a contaminant responsible for causing adverse effects on the environment.

India is a developing country. Population growth, increasing urbanization, and industrialization are common phenomena and have a direct impression on the quality of water bodies, air, social and economic aspects of the citizens, and the health of the living beings. The levels of stack gases increased due to rapid industrialization, and particulate matter has been increasing in the atmosphere. The wastewaters emanating from industries have been affecting the liner water bodies. The solid garbage along with the municipal waste has been forming heaps of unwanted waste on the soil. Garbage, waste paper and packing materials, plastic waste, rejects from households, solid waste from industries, and chemicals like pesticides are the main land pollutants. Materials like plastics, which are difficult to degrade, are posing challenges to scientists and environmentalists in devising suitable and safe disposable methods (Wikipedia, 2006).

12.2 PLASTICWASTE

Plastics have grown into major industries that influence all of our lives. The first man-made plastic developed in the year 1860 by Alexander Parkes, and he had no idea what this material would play in global life. The uses of plastic material

DOI: 10.1201/9781003352396-12

have grown rapidly in space programs, bulletproof vests, and prosthetic limbs, as well as in everyday products like beverage bottles, medical devices, and automobiles (Flory, 1953). The word plastic has come from the word plasticus, a Latin word.

These plastics are long-standing in the environment, and waste plastic material disposed of is a potential source of environmental pollution, having harmful effects on living beings. The fertility condition of soil will be reduced due to the low percolation of water into the earth as plastic sheets or bags make a nonporous layer. This prevents the biodegradation of waste and causes depletion of underground water sources as well (Parvathi, 2006). Polymer or resin and additive materials are the two main components of plastics in which polymer makes up the bulk of the plastic material. Plasticizers are one of the vital additives in the plastic manufacturing process (Ritchie, 1972), which are added to increase the utility and processing of plastics by diminishing second-order transition temperature and structural flexibility (Rosen, 1993) and resins with low molecular weight forms secondary bonds to polymer chains and thereby reduces polymer-polymer chain secondary bonding and gives added flexibility for macromolecules ensuring a moldable material. For the first time in 1920, phthalic acid esters were used as plasticizers and persist to be the widest class of plasticizers. Commercially available phthalate esters which are widely used are given in Table 12.1. Three important esters belonging to phthalates, namely, DEP (diethyl phthalate), DnBP (di-n-butyl phthalate), and DEHP

TABLE 12.1
Commercially Available Phthalates

Phthalate Ester	Abbreviation	CAS No.
Dimethyl phthalate	DMP	131-11-3
Diethyl phthalate	DEP	84-66-2
Diallyl phthalate	DAP	131-17-8
Dipropyl phthalate	DPP	131-16-8
Di-n-butyl phthalate	DnBP	84-74-2
Di-iso-butyl phthalate	DIBP	84-69-5
Butyl benzyl phthalate	BBP	85-68-7
Dihexyl phthalate	DHP	84-75-3; 68515-50-4
Di-n-octyl phthalate	DnOP	117-84-0
Butyl-2-ethylhexyl phthalate	BOP	85-69-8
Di(n-hexyl, n-octyl, n-decyl phthalate)	610P	25724-58-7; 68515-51-5
Di(2-ethylhexyl phthalte)	DEHP	117-81-7
Diisononyl phthalate	DINP	28558-12-0; 68515-49-1
Diisodecyl phthalate	DIDP	26761-40-0; 68515-49-1
Di(heptyl, nonyl, undecyl)phthalate	D711P	3648-20-2; 68515-44-6
Diisooctyl phthalate	DIOP	27554-26-3
Diundecyl phthalate	DUP	3648-20-2
Ditridecyl phthalate	DTDP	119-06-2; 68515-47-9

(diethyl hexyl phthalate), gained a dominant position among plasticizers and were chosen based on their length of the alkyl chain and octanol-water partition coefficient. These phthalates are declared as major pollutants by the US EPA and China National Environmental Monitoring Centre (Wang et al., 1997). The assessment of phthalates to the environmental exposure depends on the release of substances during production of plastics, manufacturing and processing of printing inks, glass fibers, adhesives, etc.

Phthalates are carcinogenic in nature causing liver damages and functional defects in the fetus when exposed. Because of their carcinogenic nature and long persistent nature in the environment, it is essential to remove phthalates. Governments, regulatory agencies, scientists, and social organizations have been striving hard to keep the pollutants in the environment under control. Some of the important treatment methods that are in practice for the degradation of pollutants are physical, chemical, and biological treatments. Again, in biological treatment, aerobic and anaerobic treatments are available.

Adsorption is one of the important physical treatment methods for the elimination of certain pollutants from the environment. A number of materials like activated charcoal, alumina, molecular sieves, bone char, silica gel, bauxite, magnesia, etc., possess adsorption properties. Due to a large number of vast pores and its surface area, activated carbon is considered to be an effective adsorbent.

Ozonation is one of the important chemical methods in which ozone is used as a powerful oxidizing agent for the treatment of wastewaters. It is a pretreatment technique to biological treatment in which the toxic organic compounds are converted into much fewer toxic substances. Biological treatment is to remove the organic substances using biodegradation techniques by aerobic or anaerobic bacteria. As reported by Jung-Shan et at., biodegradable organic substances are converted into gases which are released into the atmosphere (Jung-Shan et al., 2022). The aerobic environment in the reactor is achieved by diffusion or mechanical aeration and the end products are environmentally friendly. Anaerobic degradation of organic wastes involves metabolic processes that are less efficient than aerobic metabolism. Anaerobic bacteria release energy rich in methane at the end of the reaction (Davis and Cornwell, 1991; Venkata Mohan et al., 2006). The end products of the biological treatment can be used as fertilizer based on the quality.

These various biological treatment technologies depend on the biodegradation functions of microorganisms and focus on enhancing biodegradation but are slow in nature. Some technologies like bioremediation that bring chemicals into contact with microorganisms allow rapid degradation. The term bioremediation is applied to the existing sites of the pollution and to clean them (Kincannon and Lin, 1985). According to Alexander (1990), bioremediation means cleaning of contaminated soil or water in an environment using microorganisms. Microorganisms are constantly working in the natural environment by breaking down organic matter while some of the microorganisms in a polluted environment are capable of eating organic pollutants for their survival. In the bioremediation process, the microorganisms are supplemented with nutrients and oxygen which makes the microorganisms work efficiently and encourages the organisms to break down the organic pollutants at a faster rate.

Bioremediation is a technique to remove toxic waste by improving the biodegradation processes that happen in the environment. Bioremediation is dependent on the site and pollutant present and is reported safe and inexpensive when compared with incineration or landfilling methods. The aim of bioremediation is to bring down the levels of the pollutants below the limits prescribed by the regulatory agencies to the detectable or undetectable levels.

Bioremediation is broadly categorized into two methods, namely, in-situ and ex-situ. In the in-situ process, on-site existing microflora will treat groundwater or contaminated soil, whereas in the ex-situ process pumping of groundwater or excavation of contaminated soil will be made subject to the treatment (Overcash et al., 1986). It takes years together in the in-situ remediation technique to achieve the goals and mainly depends on the substrate/contaminant present, the nature of the soil, and the microorganism in the site.

A wide array of contaminants and soil types can be cleaned in ex-situ techniques than in-situ method because of rapid degradation and ease of control. The various steps involved in ex situ are as follows (Lynch and Genes, 1989):

- In bio-pile construction, the steps involved are addition of nutrients and moisture to the excavated soil along with aeration through pipes for enhancing degradation.
- Excavation and spreading of contaminated soils are involved in land farming and aeration is improved by tilling.
- Windrows are constructed by excavation of the contaminated soils using agricultural machinery.
- Soils that are contaminated are mixed with water to make slurry along with nutrients followed by dewatering.

Solid-phase systems requires large amount of space and time but are easy to maintain when compared to slurry phase processes (Saylor et al. 1991). Desorption of contaminants, water, air, and nutrients is the limiting step in solid-phase bioremediation and hence necessary to improve mass transfer by means of slurry bioreactors. The idea of a bioslurry reactor is a considerably modern technique in the area of soil bioremediation to environmental scientists. The reactor uses existing bacteria or a definite organism which is added to change harmful biodegradable materials to carbon dioxide and water under controlled specific conditions (Venkata Mohan et al., 2004; Venkata Mohan et al., 2006). Water and other additives are added to the contaminated soil in a big tank called a "bioreactor "to make it into slurry to keep the microorganisms active and degrade contaminants in the soil.

In designing a bioreactor, the principles to improve bacterial growth in the soil environment by adding nutrients and supplying additional oxygen. To support proper environmental conditions for microbial growth, other factors such as pH and temperature should be maintained. The advantage of the slurry phase reactor is to curtail mass transfer limitations and to inculcate desorption of organic matter from the soil, and thus it used in the cleaning of contaminated soil, specifically in areas where there are low atmospheric temperatures which adversely affect rates of biodegradation (Battersby and Wilson, 1988).

Bioremediation in the bioslurry phase necessitates the contaminated soil treatment (soil, sediment, and sludge) in a closed system which is tri-phasic relating three major components: water, air, and suspended particulate matter. It was considered to be a high-rate biological system economically used for highly concentrated compounds in less contaminated areas(Rama Krishna et al., 2006). Water in a slurry phase reactor helps as a suspending medium where nutrients and trace elements are made available easily for microorganisms. Type of soil, reactor design, pollutant concentration, solid loading rate, temperature, aeration, oxygen demand, surfactant, nutrients, etc. are the parameters relied on in slurry phase bioremediation. Bioslurry systems could be operated under aerobic or anaerobic conditions and inoculums (microflora) can be added if the treated soils have low potent microorganisms. This technique is known as bioaugmentation in which microorganisms are added to enhance biological activity and non-indigenous microorganisms are used in natural or engineered environments to treat contaminated water or soil. The existence and persistence of added inoculums are important aspects for the bioaugmentation system to work well. Biostimulation is the addition of nutritional supplements to promote the metabolism of bacteria by improving in the fields of microbial ecology and bioreactor design.

Many efforts were made to degrade phthalates in polluted environmental matrices. A few studies were reported on the adsorption process for removal of phthalates (Herbert and Zeng, 2004). Phthalates were degraded by a wide range of aerobic and anaerobic bacteria in several samples, namely, wastewater, fresh water, marine water, soils, and sediments (Staples et al., 1997). Studies were carried out for the degradation of phthalates with respect to aerobic, anaerobic, nitrate sinking (Shailaja et al., 2008), and methanogenic conditions (Cheng et al., 2000) with diverse cultures or pure cultures of bacteria. The biodegradation rate of phthalates was related to the alkyl chain length. The degradation characteristics of three phthalates, i.e., DEP, DnBP, and DEHP, in soil augmented with acclimatized sludge were investigated (Jianlong et al., 2004). Investigations were carried out by anaerobic biodegradation and reported to be non-biodegradable under these parameters (Hariklia et al., 2003). It was reported in the literature that DEHP is non-degradable under methanogenic conditions (Sherlton et al., 1984), whereas O'Connor et al., (1989) reported minute mineralization of DEHP under these conditions. Very few studies were reported in the literature for the removal of phthalates from wastewater and there was no information regarding the effect of intractable phthalates on the digester's performance during the treatment of sludge in anaerobic conditions (Chou et al., 2016). Hariklia et al., (2003) mentioned that DEHP accumulation toxicity in the anaerobic digester had a harmful effect on removal rates and the biogas production during anaerobic biodegradation of DEP, DnBP, and DEHP. An extensive literature review on the biodegradation of the three phthalates is presented in Tables 12.2–12.4.

From the above facts, the importance of remediating phthalates from contaminated soil was necessitated for alternative techniques which are faster, cleaner, and economical ways of degradation. During recent years, there has been massive growth in the controlled and practical use of bacteria for the destruction of polluting chemicals. The schematic diagram of the phthalate's degradation is shown in Figure 12.1.

TABLE 12.2
Reported Literature on the Biodegradation of DEP

Test Matrix	Temperature(°C)	Initial Concentration (ppm)	Test Duration (Days)	Degradation (%)	Reference
DEP: Aerobic Synthetic aqueous medium	25	20	7	36	Urushigawa and Yonezawa, 1979
Synthetic aqueous medium; wastewater	–	5	7	>90	Patterson and Kodukula, 1981
Soil	15–32	0.02	51–129	19–99	Overcash et al., 1986
Sludge; soil	–	1000	287	95	Kinconnin and Lin, 1985
Fresh water	20	0.00086	5	95	Furtmann, 1993
Wastewater	25	0.00086	17	0	Furtmann, 1993
Soil; synthetic aqueous medium	25	1	41	86–100	Russell et al., 1985
Anaerobic Synthetic aqueous medium; sludge	37	20–200	17	33–56	O'Connor et al., 1989
Synthetic compost	37	50–250	100	5–90	Eljertsson et al., 1996

TABLE 12.3
Reported Literature on the Biodegradation of DnBP

Test Matrix	Temperature (°C)	Initial Concentration (ppm)	Test Duration (Days)	Degradation (%)	Reference
DnBP: aerobic activated sludge	25	20	7	60	Urushigawa and Yonezawa, 1979
Activated sludge; synthetic aqueous medium; soil	22	20	28	90	Suggat et al., 1984
Soil	23	1000	80	84–99	Jhonson and Lulves, 1975
Soil	30	1000	80	83	Inman et al., 1984
Soil; wastewater	4	1000	53	2	Inman et al., 1984
Activated sludge; synthetic aqueous medium; soil	22	20	28	57	Suggat et al., 1984
Anaerobic sludge	35	20	28	100	Shelton and Tiedje, 1984

(Continued)

TABLE 12.3 *(Continued)*
Reported Literature on the Biodegradation of DnBP

Test Matrix	Temperature (°C)	Initial Concentration (ppm)	Test Duration (Days)	Degradation (%)	Reference
Sludge	35	20	17–70	80–100	Shelton and Tiedje, 1984
Sludge; synthetic aqueous medium	35	72	60	24–59	Battersby and Wilson, 1988
Sludge; synthetic aqueous medium	35	50	56	32–85	Shelton and Tiedje, 1981

TABLE 12.4
Reported Literature on the Biodegradation of DEHP

Test Matrix	Temperature(°C)	Initial Concentration (ppm)	Test Duration (Days)	Degradation (%)	Reference
DEHP: aerobic Synthetic aqueous medium; activated sludge; soil	22	20	28	>99	Suggat et al., 1984
Fresh water	25	1	20	>99	API, 1994
Soil	20	1	100	64–97	Rudel et al., 2003
Sludge; synthetic aqueous mixture	35	20	70	9	Shelton and Tiedje, 1984
Sediment	22	1	30	0	Jhonson and Lulves, 1975

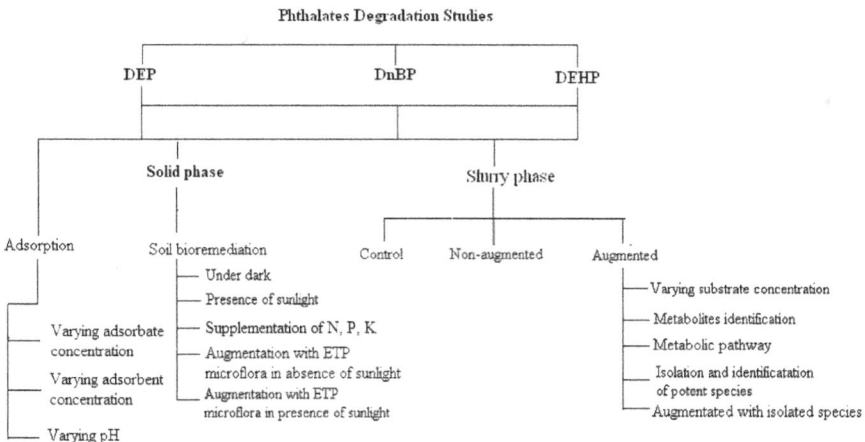

FIGURE 12.1   Schematic diagram of phthalates degradation methodology.

12.3 ADSORPTION

DEP adsorption experiments for the removal of DEP from water using activated charcoal were performed with different substrate concentrations. DEP stock solution was prepared by dissolving 50 mg in 1000 ml of double distilled water and sorption experiments were performed. The pH effect on sorption was measured by performing equilibrium tests at different pH values. Batch isothermal studies were experimented by varying the dose of adsorbent and by maintaining the concentration of the adsorbate (DEP) constant. It is evident from the results that the removal of DEP increased as the time of contact increased for all the experiments studied. After the contact time of 120 min, the removal efficiency remained constant. The percentage removal is observed to be 70.9% for 1.0 mg of DEP, 58.5% for 2.0 mg of DEP, and 53.2% for 3.0 mg of DEP at pH 7.0, and the same trend was observed for all the substrate concentrations studied. In the case of optimization of pH, it was observed that the adsorption process is dependent on the pH and adsorption decreased clearly due to the increased pH of the test solution. The data obtained was found to fit in Langmuir's adsorption isotherm model. However, the adsorption studies could not be performed for the other two selected phthalates (DnBP and DEHP) because of their low aqueous solubility.

12.3.1 SOLID-PHASE BIOREMEDIATION (INSITU)

The experiments are carried out in glass dishes after the soil was air-dried and sieved. To maintain the moisture content of the soil at around 3%, water and effluent treatment plant (ETP) microflora were sprinkled. Subsequently, phthalates (DEP, DnBP, and DEHP) were sprayed onto the soil surfaces accounting for 5.0 mg/g of soil. The experiments were conducted in five different conditions, namely, under dark conditions (which was treated as control), presence of sunlight, addition of urea which is the source of N, P, K, augmented with ETP microflora under dark, and augmented with ETP microflora kept under sunlight. The sunlight experiments were carried out only for 8 h every day. The pH of the soil was maintained at 7.0 ± 0.1. The degradation pattern was analyzed by determining the number of phthalates present in the soils over a period of 45 days. The samples were analyzed by HPLC for phthalates concentration regularly.

In the solid-phase bioremediation studies, the data obtained for all the three phthalates was compared with the control experimental data. The experiments conducted with ETP microflora under the sunlight showed better performance of 51.1% for DEP, 45.3% for DnBP, and 25.4% for DEHP for a period of 45 days than the other treatment studies. In view of the poor performance of solid-phase bioremediation, further experiments were conducted in a soil slurry sequential batch reactor (SSB-SBR) using the bioaugmentation technique.

12.4 REACTORCONFIGURATION

The concept of a bioslurry phase reactor is introduced to remediate phthalates in contaminated soil. The slurry was thoroughly mixed so that the suspended solids and microorganisms were in contact with the soil contaminants. The excavated soil was

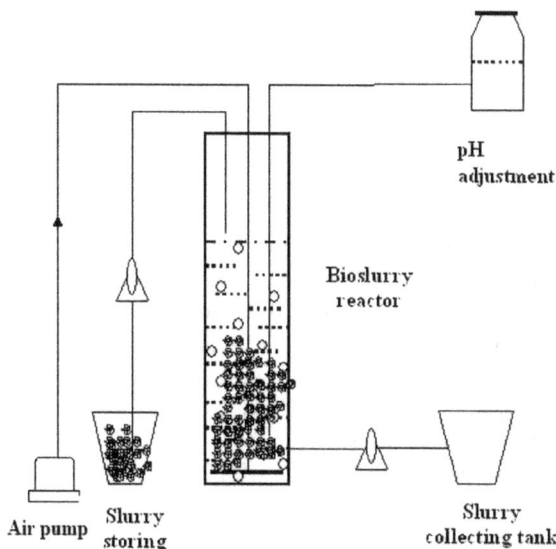

FIGURE 12.2   Schematic representation of soil slurry bioreactor (SSB-SBR).

first refined physically to remove stones and rubble and mixed uniformly with water to make a slurry. All the systems designed were fabricated using Pyrex glass material as their principal component, leakproof scaling with inlet and outlet arrangements, and the reactors were operated in aerobic conditions in which air was supplied to each reactor via diffuser (fabricated) setup at the bottom of each reactor. Feed regulation, recirculation, aeration, and discharge operations were carried out through peristaltic pumps and timers. Reactors were cylindrical in shape with 12 cm of diameter and 30 cm length. Underneath the reactor, a well-distributed sparger network was connected by silicon tubing for supplying through an air diffuser. The slurry phase suspension was kept agitated by a sparger system consisting of four arms, which were equally distant from the center, inculcating regular dispersion of air from underneath the reactor in upward direction. The schematic diagram of the slurry reactor is shown in Figure 12.2.

12.5   SOIL SLURRY PREPARATION

Dried sieved soil was sprinkled with a measured amount of phthalates dissolved in a solvent for uniform sorption onto the soil particles and the solvent was subjected to evaporation for a period of 10h in a fume hood at ambient temperature. The contaminant sprayed soil was used to make a slurry in the ratio of 1:15 (W/V) for the preparation of the slurry in batch mode. In the control reactor, sterile soil and sterile water were used for comparison. ETP microflora was used instead of water as inoculums for the preparation of slurry as an augmentation technique. pH was adjusted to 7.0 and operated at ambient temperature 27 ± 2. The influence of substrate loading rate on the degradation activity of the contaminant with respect to microorganisms was evaluated. Very high amounts of the substrate can be toxic to the microorganisms and low concentration may fail to induce degradation activity. The substrate loading rate for DEP was 20–500 mg/g of soil, for DnBP was 1–60 mg/g of soil, and for DEHP was 1–10 mg/g.

12.6 BIOPROCESS MONITORING

Biochemical processing of the reactor was monitored during the operation of the cycle period by determining pH, oxidation reduction potential (ORP), dissolved oxygen (DO), oxygen uptake rate (OUR), and colony forming units (CFU). The extraction procedure of the slurry for the analysis of aqueous and soil layers is presented in Figure 12.3. The concentration of the DEP, DnBP, and DEHP after the extraction from the slurry reactor was determined periodically by HPLC using a reverse phase C18 column. The injection volume was 20μL and the samples were analyzed at a wavelength of 225nm (UV detector).

Reverse phase semi-preparative column with 250 × 7.7 mm dimensions was used for the separation and collection of metabolites. The aqueous layer was injected using a 100 μl injector loop for the collection of metabolites at the detector outlet and repeated till an adequate amount of the sample was collected. The eluent was concentrated, and the components were further characterized using NMR and MS. Considering the reaction mechanism, metabolites were identified and the metabolic pathway was established.

The better contact established between contaminants and microorganisms in the bioslurry system helped accelerate the degradation of phthalates. Control reactor SSB-SBR1 operated in absence of native soil bacteria to understand the evaporation loss during the operation of the reactor. The DEP concentration during the slurry

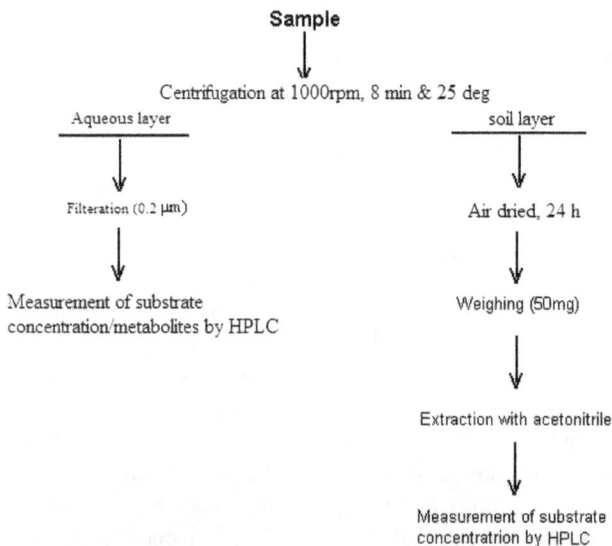

Sample

Centrifugation at 1000rpm, 8 min & 25 deg

Aqueous layer   soil layer

Filteration (0.2 μm)   Air dried, 24 h

Measurement of substrate   Weighing (50mg)
concentration/metabolites by HPLC

Extraction with acetonitrile

Measurement of substrate
concentratrion by HPLC

FIGURE 12.3 Flow chart of slurry extraction procedure.

reactor operation was relatively same up to 48 h as in the control reactor. TheSSB-SBR2 reactor performed with native soil bacteria (7.5×10^3 CFU/g) showed 47.9% DEP degradation. In the first 8 h cycle operation, 9.85% of DEP degradation was observed and reached a maximum of 47.9% in 40h. In augmented reactor SSB-SBR3 with ETP bacteria (2.5×10^7 CFU/ml), the degradation of DEP was observed to be faster when compared to SSB-SBR2, which was a control reactor. 22.45% of DEP degradation was noticed within 8 h of reactor operation and came to a peak 75.4% till the reactor stopped.

The soil bacteria (7.5×10^3 CFU/g of soil) operated reactor SSB-SBR4 was augmented with 2.5×10^7 CFU/ml ETP bacteria and exhibited faster assimilation and comprehensive degradation of the DEP without any delay phase. In the reactor SSB-SBR3, 75.4% of the substrate degradation was attained during 48 h of the cycle period while in the bioaugmented reactor (SSB-SBR4) degradation of DEP was more or less complete. From the experimental data, it was confirmed that the augmented ETP microflora reactor with soil native bacteria showed an improved DEP degradation. Besides rapid uptake of the DEP and negligible lag phase with total dissipation within 48 h of the reactor operation clearly showed the achievement of the slurry reactor with a bioaugmentation technique.

During the operation of slurry phase reactor the partition of DEP in aqueous and soil layers was monitored in the control reactor SSB-SBRI showed the method of partition between two phases because of a non-biodegradable process. During reactor start-up, the DEP desorption was faster from the soil to the aqueous layer, but 24h equalization in substrate partition was observed. However, after 8h of cycle period, less DEP desorption was observed. Owing to the high-water solubility of DEP (1100 mg/l), it desorbed into an aqueous phase in a short period. A highest of 18.24 mg/ml desorbed substrate concentration was observed in the case of the aqueous phase within the 8h cycle period and decreased to 14.49 mg/ml in the 40 h cycle period. This reduction might be attributed to the volatility of substrate and hydrolysis. The DEP partition pattern was totally different in other reactors when compared with SSB-SBR1. The substrate concentration in the SSB-SBR2 reactor was reduced to 0.56 mg/g in the soil layer within 8 h and almost to zero after 32h of subsequent cycle operation. In the aqueous layer, the desorbed DEP after 8h operation showed a constant degradation of 17.47 mg/ml and reached a maximum of 10.43 mg/ml in 32h of the cycle period. The desorption pattern from soil to aqueous phase was similar to that of the SSB-SBR1 reactor up to a cycle period of 8h. In the aqueous phase, the desorbed DEP was further subjected to biodegradation after a cycle period of 8h and this could be possible with soil native microflora.

Other than the degradation rate, the reactors operated for degradation of DEP with ETP microflora in SSB-SBR3 and SSB-SBR2 reactors, partition was more or less the same. In the soil layer, DEP concentration was decreased to 0.32 mg/g within 8h and further equalized till 32h of cycle operation. The substrate desorbed was subjected to degradation from 15.19 mg/ml in 98h to 4.92 mg/ml in 32 h and remained constant till the end of reactor operation. DEP degradation was more effective comparatively with ETP microflora than with soil native microflora. In the reactor SSB-SBR4, i.e., augmented reactor, the partition of DEP was almost the same as the reactors SSB-SBR2 and SSB-SBR3. In the aqueous phase, substrate concentration was decreased from 14.59mg/ml to 0.58 mg/ml and remained stable till the completion of the cycle period. The desorbed substrate in the aqueous layer in the reactor SSB-SBR4 was put through to biological degradation by native soil bacteria, ETP, or both. In addition to the partition effect of DEP, desorption was observed from soil to water.

Some interesting observations were made by evaluating data with different kinetic equations. First, desorption of phthalate from soil surface was confirmed by a better fit zero-order equation (R^2– 0.923) than first- or second-order equations (R^2–0.883). Second, the reactor SSB-SBR4 showed a half-life of DEP degradation of2.52 days when compared with 154.2 days in SSB-SBR1, a control reactor. Similarly, in SSB-SBR4 the half-life period of DEP degradation in the soil phase was found to be 1.19 days when compared with 28.37 days for the control reactor SSB-SBR1. Jianlong et al., (2004) reported 90% degradation in soil matrix in 25 days, while Chang et al., (2004) reported 0.045 l/day and 15.4 days half-life period under the best possible conditions of pH 7.0 and 30°C temperature in anaerobic conditions for DEP in river sediments. The biochemical process was monitored further to understand variations in pH, ORP, DO, and OUR continuously at regular intervals. Due to formation of metabolites and liberation of carbon dioxide, ORP and pH variations were at their peak in SSB-SBR3 and SSB-SBR4. OUR corresponds to organic matter degradation and growth in microorganisms for the aerobic system and hence electron transport system was directly correlated with the metabolic activity of the bacteria. The OUR pattern of the SSB-SBR1 reactor was more or less uniform showing no microbial activity, whereas in all other three reactors pattern was non-linear indicating a positive biological activity. In fact, OUR was high, attributed to multiplication of microorganisms and high degradation rates.

During the analysis of substrate concentration by HPLC, two components other than DEP were observed. The two extra peaks were separated and collected by preparative HPLC for further characterization by MS and NMR. Since the bioslurry reactor was operated in an aerobic environment, initially dihydrolysis of ester groups in DEP was observed. The inherent advantages of bioslurry system in manipulating the operating conditions facilitated rapid degradation compared to in-situ processes where dihydrolysis might have taken place instead of monohydyrolysis. Phthalic acid was formed by the dihydrolysis of phthalate ester groups then to dihydroxy benzoic acid by monocarboxylation. The proposed pathway for the degradation of DEP is shown in Figure 12.4.

To understand the effect of substrate concentration for optimization in a slurry reactor, different reactors with different DEP concentrations (Table 12.5) were carried out and they revealed that the slurry system would sustain its performance up to 300mg/g.

Degradation experiments for DnBP conducted in the slurry reactor (Shailaja et al., 2007) where ETP bacteria was used for microbial augmentation showed enhanced performance by improving the degradation rate when compared with reported work. Wang et al. (1997) and Zhou et al. (2005) reported DnBP degradation in soil to be 92% (100 μg/g soil) and 96% (500 μg of DnBP per gram of soil), respectively, in 30 days. The authors could successfully achieve 99% degradation of DnBP within 72 h at 1 mg/g of soil concentration. Slurry phase bioreactor showed an improved

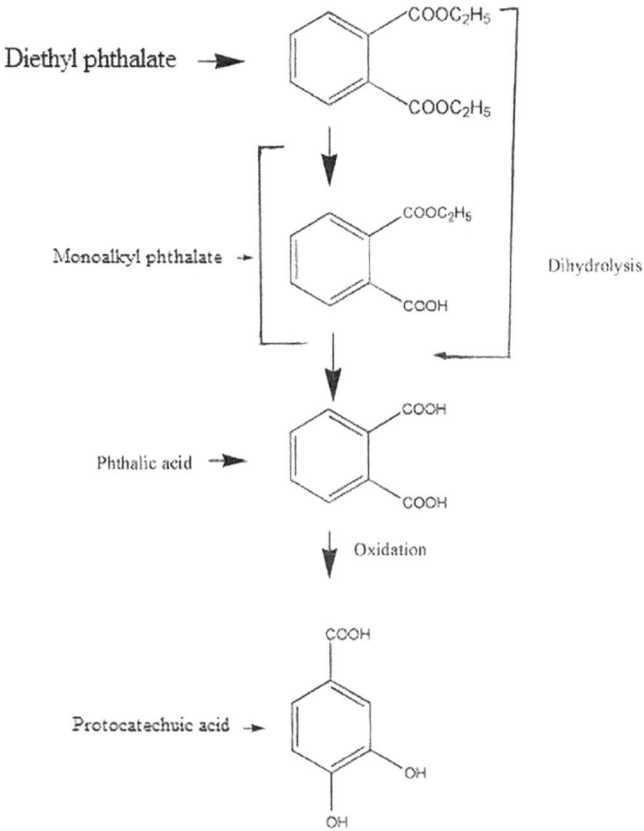

FIGURE 12.4 Degradation pathway of DEP in the bioslurry reactors.

rate of degradation compared to other processes where ETP was used as microbial augmentation. Phthalic acid and dihydroxy benzoic acid were the two metabolites identified during DnBP degradation in a slurry reactor using analytical techniques. The degradation pathway was also established and was similar to DEP. The metabolites obtained were found to be the same for both DEP and DnBP. The potent bacterial strain identified during the degradation of DnBP was isolated and identified as *Pseudomonasaeruginosa*. Initial substrate concentration optimization experiments were conducted to study the efficiency of the slurry phase reactor for the degradation of DnBP. DnBP removal efficiencies with different concentrations (Table 12.5) were studied for a period of 144 h. As the concentration of DnBP increased in the reactor, the rate of degradation decreased. Loading of the highest substrate concentration (60 mg/g of soil) showed the lowest substrate removal (34.92%). The degradation of DnBP in the bioslurry phase reactor apt well in the zero-order model and the kinetic equations gave rate constants from 10.60 to 6.89, which were inversely proportional to the concentration of the substrate. The half-life of DnBP was calculated to be 0.75 days for1 mg/g of soil.

TABLE 12.5
Performance of Reactors (SSB-SBR1to SSB SBR11) with Varying Initial Substrate Concentration

Reactors DEP	DEP Concentration (mg/g)	Degradation (%)	Reactors DnBP	DnBP Concentration (mg/g)	Degradation (%)	Reactors DEHP	DEHP Concentration (mg/g)	Degradation (%)
SSB-SBR3	20	75.4	SSB-SBRB3	1	99.31	SSB-SBRH5	1.0	97.2
SSB-SBR4	20	99.7	SSB-SBRB4	5	96.07	SSB-SBRH6	2.0	96.3
SSB-SBR5	50	99.6	SSB-SBRB5	10	94.37	SSB-SBRH7	5.0	90.5
SSB-SBR6	75	99.4	SSB-SBRB6	30	92.80	SSB-SBRH8	10.0	80.7
SSB-SBR7	100	99.3	SSB-SBRB7	40	57.11	–	–	–
SSB-SBR8	200	90.6	SSB-SBRB8	50	42.31	–	–	–
SSB-SBR9	300	69.2	SSB-SBRB9	60	34.92	–	–	–
SSB-SBR10	400	28.9	–	–	–	–	–	–
SSB-SBR11	500	11.2	–	–	–	–	–	–

Five different bioslurry reactors were operated to study the DEHP degradation (1mg/g soil) in soil and a greater 90% DEHP degradation was achieved in 12 days of reactor operation. pH, DO, OUR, and CFU were used as monitoring parameters for smooth functioning of the bioslurry reactor. CFU is the parameter by which the growth of biomass was continuously monitored. The biomass growth is directly proportional to the rate of substrate degradation. Metabolites were isolated and iden- tified using the analytical techniques during the degradation process. Diethyl hexa- noic acid and diethyl hexanol were the two metabolites identified and a degradation pathway was established (Figure 12.5). Degradation kinetics are also studied and apt well in the zero-order kinetics.

To evaluate the effect of DEHP concentration in the slurry reactor, experiments were carried out by varying initial DEHP concentration in the range of 1.0–10.0 mg/g of soil (Table 12.5). The degradation of DEHP in the bioslurry phase relied on the DEHP concentration. It is clear from Table 12.5 that with an increase in DEHP con- centration, the removal rate was slow. The degradation was found to decrease with an increase in substrate concentration and it was found to be in inverse relationship with the biodegradation rate to substrate concentration. In the reactor SSB-SBRH5,

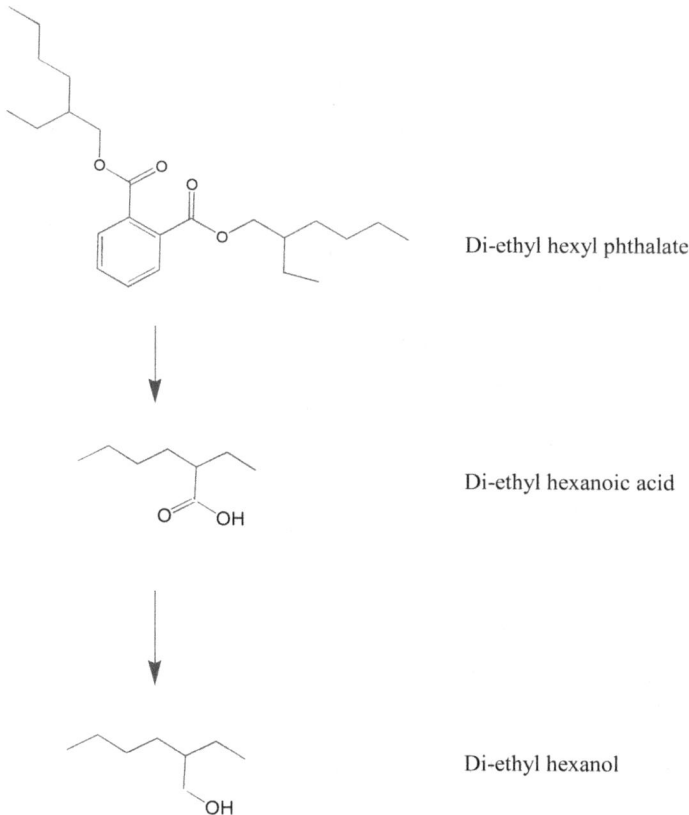

Di-ethyl hexyl phthalate

Di-ethyl hexanoic acid

Di-ethyl hexanol

FIGURE 12.5   Degradation pathway for DEHP in the slurry reactor.

97.2% degradation was observed after 12 days of reactor operation. In the reactors SSB-SBRH6 to SSB-SBRH8, the degradation rate was above 80% (96.3%, 90.5%, and 80.7%, respectively).DEHP removal was fairly slow in the start-up phase of the reactor and was established to be reliant on the DEHP concentration in the slurry phase. With the increase in the reactor operation period, the available microflora might have acclimatized to the new substrate conditions facilitating rapid removal of the organic substrate through mineralization. In the case of reactors operated with high concentrations of the substrate (SSB-SBRH6 to SSB-SBRH8), the removal was low and might lead to inhibition of the system. The experimental studies showed that the slurry system would retain its performance till 5 mg/g of initial DEHP concentration in the given conditions.

12.7 CONCLUSION

For the sustenance of life on earth, water waste and soil components of the environment need to be managed. Phthalate eaters, which are one of the main soil pollutants, have been addressed and tried to decrease their effect on the environment. Phthalic acid molecules joined to variable length alkyl side chains via an ester bond are popularly known are phthalate esters. DMP, DEP, DBP, and BBP are regarded as low molecular weight phthalate esters while DEHP, DINP, and DIDP are high molecular weight. The low molecular weight phthalates are partially water soluble, but the high molecular weight phthalates are moderately soluble. The affinity of phthalates with high molecular weight to the soil particles is much greater than low molecular weight phthalates.

The environmental behavior of phthalate esters is broadly contingent on the alkyl chain of the individual phthalate. Generally, the larger the alkyl side chain length or extent of diversification, the more continual is the compound in the environmental matrices. Phthalate esters have low vapor pressures, in the range of 0.02–1.9 kPa, which reduces as the alcohol sidechain of the ester increases. Phthalate esters tend to volatilize from water very slowly due to the low Henry's law constants. DEHP is easily adsorbed by the soil particles as it has a high octanol-water partition coefficient by which equilibrium is established between water and soil particles in the slurry system. Even though the DEHP solubility is less in water, the quantity present in the surface water is high owing to adsorption onto the organic particles and interlinkage with dissolved organic substance. The bipolar nature of DEHP influenced the high potential for bioaccumulation. The adsorption is high in salt water, particularly on small particles. DnBP, which is one of the important phthalates, slowly volatilizes from surface waters and remains in the water phase at equilibrium. The octanol/water partition coefficient (K_{ow}) of DnBP is high and thereby the balance between water and organic carbon on soil or sediment will be very much in favor of both soils. The high K_{ow} of DnBP shows that the substance has a probability for bioaccumulation. The slurry systems were observed to sustain the performance at higher phthalate concentrations. Moreover, the subunit operations such as settling, aeration, etc., were all integrated into the single reactor making it user-friendly with less reactor volume. The flexibility of sequence phase microenvironment variation and win-through substrate gradient during sequence phase operation with respect to microflora leading to effective and stable bioaugmentation process performance. Furthermore, other phthalate esters may also be experimented in the same way in controlling soil pollution with respect to the environment.

REFERENCES

Alexander, M. Biodegradation and bioremediation, 2nd edition, 1990. Academic press. New York

API American Petroleum Institute, Inter laboratory study of three methods for analyzing petroleum hydrocarbons in soils, Diesel Range Organics (DRO), Gasolene Range Organics (GRO) and Petroleum Hydrocarbons (PHC), 1994.

Battersby, N.S.; Wilson, V. Evaluation of a serum bottle technique for assessing the anaerobic biodegradability of organic chemicals under methanogenic conditions. *Chemosphere*, 1988, 17, 2441–2460.

Chang, B.V.; Liao, C.S.; Yuan, S.Y. Anaerobic degradation of diethyl phthalate, di-n-butyl phthalate and di-(2-ethyl hexyl) phthalate from river sediment in Taiwan. *Chemosphere*, 2004, 55(4), 533–538.

Cheng, H.F.; Chen, S.Y.; Lin, J.G. Biodegradation of di-(2-ethylhexyl) phthalate in sewage sludge. *Water Science Technology*, 2000, 41(12), 1–6.

Chou, H.L.; Hwa, M.Y.; Lee, Y.C.; Chang, Y.J.; Chang, Y.T. Microbial degradation of deca-bromodiphenyl ether (DBDE) in soil slurry microcosms. *Environmental Science Pollution Research Int*, 2016, 23(6), 5255–67.

Davis, M.L.; Cornwell, D.A. Introduction to Environmental Engineering, 2nd edition, 1991. McGraw-Hill, New York.

Eljertsson, G.; Enkaten I praktiken En handbook I enkatmetodik. 1996. Student litteratur, Lund ISBN 91-44-00052-9.

Flory, P.J. Principles of polymer chemistry, 1953. Cornell University Press, New York.

Furtmann, K. Phthalate in der aquatischen unwelt-analytik, Dissertation Universssitat, Gesamthochschule Duisburg, 1993, Verreitung Verbleib, Bewertung.

Hariklia, N.G.; Felipe, A.M.; Iranpour, R.; Birgitte, K.A. Biodegradation of phthalate esters during the mesophilic anaerobic digestion of sludge. *Chemosphere*, 2003, 52(4), 673–682.

Herbert, H.P.; Zeng, F.H. Adsorption of phthalates by activated sludge and its biopolymers. *Environmental Technology*, 2004, 25(7), 757–761.

Inman, I.C.; Strachan, S.D.; Sommers, L.E.; Nelson, D.W. The decomposition of phthalate esters in soil. 1984, *Journal of Environmental Science & Health Part B*, 19, 245–257.

Jianlong, W.; Xuana, Z.; Weizhong, W. Biodegradation of phthalic acid esters (PAEs) in soil bioaugmented with acclimated activated sludge. *Process Biochemistry*, 2004, 39, 1837–1841.

Jung-Shan, H.; Yu, T.-Y.; Wei, D.-J.; Jane, W.-N.; Chang, Y.-T. Degradation of decabromodi-phenylether in an aerobicclayslurrymicrocosmusing a novelimmobilizationtechnique. *Microorganisms*, 2022, 20, 402.

Kincannon, D.E.; Lin, Y.S. Degradation of Hazardous wastes by land treatment, proceedings of the industrial waste conference. 1985, 40, 607–619.

Lynch, J.; Genes, B.R. Petroleum contaminated soils, 1989, 1, 163–174. Lewis Publishers, Chelsea, MI.

More, W. The changing environment, 1stedition, 1986. Springer Verlag, London.

O'Connor, O.Q.; Rivera, D.; Young, L.Y.Toxicity and biodegradation of phthalic acid esters under methanogenic conditions. *Environmental Toxicological Chemistry*, 1989, 8(7), 569–576.

Overcash, M.R.; Weber, J.B.; Tucker, W. Toxic and priority organics in municipal sludge land treatment systems. Water engineering research laboratory, office of research and development. U.S. Environmental Protection Agency, Cincinnati, OH. 1986, EPA/600/2-86/010, 135.

Parvathi, K. Biodegradable plastic and micro organism, The Hindu, Education plus, 2006.

Patterson, J.W.; Kodukula, P.S.; Emission of effluent control: biodegradation of hazardous organic pollutants. Cheical Engineering prog., 1981, 77, 48–53.

Rama Krishna, M.; Shailaja, S.; Siresha, K.; Venkata Mohan, S.; Sarma, P.N. Bioremediation of pendimethalin contaminated soil by bioslurry phase reactor: Bioaugmeting with ETP micro-flora. *International Journal of Environment and Pollution*, 2006, 4, 373–387.

Ritchie, P.D. Plasticizers, Stabilizers and Fillers. The Plastic Institute, 1972. Iliffe Books Limited, London.

Rosen, S.L. Fundamental principles of polymeric materials, 2nd edition, 1993. Wiely Publishers, New York.

Rudel, R.; Camann, D.E.; Spengier, J.D.; Corn, L.R.; Brody, J.G. Phthalates, alkylphenols, polybrominated diphenyl ethers and other endocrine disrupting compounds in indoor air and dust. *Environmental science & technology*, 2003, 37, 4543–4553.

Russell, G.L.; Miller, J.R.; Tsang, L.C.; Seasonal oceanic heat transports computed from an atmospheric model. *Dynamics of atmospheres and Oceans*, 1985, 9, 253–271.

Saylor, G.S.; Fox, R.; Blackburn, J.W. Environmental biotechnology for waste treatment.*Environmental Science Research*, 1991, 41.

Shailaja, S.; Venkata Mohan, S.; Ramakrishna, M.; Sarma, P.N.Biodegradation of di-n-butylphthalate (DnBP) in bioaugmented slurry phase reactor. *Bioresource Technology*, 2007, 98, 1561–1566.

Shailaja, S.; Venkata Mohan, S.; Ramakrishna, M.; Sarma, P.N. Degradation of di-ethylhexyl phthalate (DEHP) in bioslurry phase reactor and identification of metabolites by HPLC and MS. *International Biodeterioration and Biodegradation*, 2008, 62(2), 143–152.

Shelton, D.R.; Tiedje, J.M. General method for determining anaerobic biodegradation potential. *Appl. Environ. Microbial*, 1984, 47(4), 850–857.

Shelton, D.R.; Tiedje, J.M.; Development of a test for determining anaerobic biodegradation potential. EPA report, 1981, 560/5-81-031.

Staples, C.A.; Peterson, D.R.; Parkerton, T.F.; Adams, W.J. The environmental fate of phthalate esters: A literature review. *Chemosphere*, 1997, 35, 667–749.

Venkata Mohan, S.; Shailaja, S.; Ramakrishna, M.; Reddy, K.B.; Sarma, P.N.Bio slurry phase degradation of DEP contaminated soil in periodic discontinuous mode operation: Influence of augmentation and substrate partition. *Process Biochemistry*, 2006, 41, 644–652.

Venkata Mohan, S.; Sirisha, K.; Chandra Sekhar Rao, N.; Sarma, P.N.; Reddy, S.J. Degradation of choloropyrifos contaminated soil by bioslurry reactor operated in sequencing batch mode: Bioprocess monitoring. *Journal of Hazardous Materials Part B*, 2004, 116, 39–48.

Wang, J.L.; Liu, P.; Shi, H.C.; Qian, Y. Biodegradation of phthalic acid ester in soil by indigenous and introduced microorganisms. *Chemosphere*, 1997, 35, 1747–1754.

Wikipedia. 2006, http://en.wikipedia.org/wiki/Canadian_federal_election,_2006

Yoshikuni, U.; Yoshitaka, Y.; *Chemosphere*, 1979, 8(5), 317–320.

Zhou, Q.H.; Wu, Z.B.; Cheng, S.P.; He, F.; Fu, G.P. Enzymatic activities in constructed wetlands and di-n-butyl phthalate (DBP) biodegradation. *Soil Biology and Biochemistry*, 2005, 37, 1454–1459.

Index

Note: Locators in *italics* represent figures and **bold** indicate tables in the text.

A

Abiotic changes, 151
ABS, *see* Acrylonitrile butadiene
Accentuated total reflectance (ATR) mode, 135
Acetobacter xylinum, 188
Acetogenesis, 81
Achromobacter, 194
Acidogenesis, 81
Acinetobacter sp., 194, 197
Acrylonitrile butadiene (ABS), 156
AD process, *see* Anaerobic digestion process
Adsorption, 211, 216
    solid-phase bioremediation (in-situ), 216
Advanced oxidation processes (AOPs), 121
Advanced thermal treatment (ATT), 25–26
Aerobic treatment, 50
Aeromonas, 197
Aesthetic degradation, 92
AFM, *see* Atomic force microscopy
Agaricus bisporus, 197
Agriculture, 57–59
Agriculture-derived lignocellulosic waste
    remediation, 182
Agrobacterium sp., 188
AI-based sorting of waste, *see* Artificial
    intelligence-based sorting
    of waste
Alcaligenes faecalis, 188
Alcaligens, 194
Alcanivorax, 194
Aldrin, 193
Algae, 191–192, **191**
Alginate, 188
Anaerobic beta-oxidation, 54–55
Anaerobic biodegradation, pathway of, *54*
Anaerobic digestion (AD) process, 53, 81
    anaerobic beta-oxidation, 54–55
    fermentation, 53
Anaerobic treatment, 50
Analytical techniques
    Fourier-transform infrared spectroscopy, 156
    Raman spectroscopy, 156
    thermal analysis, 156–157
Animal waste, **7**
Anthropogenic activities, 173
AOPs, *see* Advanced oxidation processes
Aporrectodea rosea, 112

Aquatic ecosystem, effects of microplastics and
    nanoplastics to organisms in, 164
Arabidopsis thaliana, 194
Arthroderma uncinatum, 191
Artificial intelligence (AI)-based sorting of
    waste, 90
Aspergillus sp., 194, 197
    A. nidulans, 114–115
    A. niger, 197
    A. oryzae, 114
ASSOCHAM, *see* Associated Chambers of
    Commerce and Industry of India
Associated Chambers of Commerce and Industry
    of India (ASSOCHAM), 16
Atmosphere, 161–162
Atomic force microscopy (AFM), 119
ATR, *see* Attenuated total reflectance
ATR mode, *see* Accentuated total reflectance
    mode
ATT, *see* Advanced thermal treatment
Attenuated total reflectance (ATR), 119
Azotobacter vinelandii, 188

B

Bacillus sp., 188, 194, 197, 198
    B. pallidus, 197
Bacteria, 187–188, **189–190**
Batteries waste, 12–14
BCM, *see* Biologically controlled mineralisation
Beta-oxidation, 54
BIM, *see* Biologically induced mineralisation
Bio-aerosols, 95
Bioaugmentation, 213
Biochemical pathway, 49
    future prospects, 61–62
    sewage sludge, applications of
        agriculture, 57–59
        biofuels production, 59–60
    sewage sludge, chemical composition of, 51
        anaerobic digestion process, fundamentals
            of, 53–55
        hydrolysis, 52–53
        methane formation, 55
    sewage sludge legislations, 55–56
    sewage sludge production, 49–51
Biodegradable municipal waste (BMW), 25
Biodegradable solid waste, 6

227

For Product Safety Concerns and Information please contact our EU
representative GPSR@taylorandfrancis.com
Taylor & Francis Verlag GmbH, Kaufingerstraße 24, 80331 München, Germany

9 781032 403021